Optical Communications
Second Edition

WILEY SERIES IN TELECOMMUNICATIONS AND SIGNAL PROCESSING

John G. Proakis, Editor
Northeastern University

Optical Communications

Second Edition

Robert M. Gagliardi
Sherman Karp

A Wiley-Interscience Publication
JOHN WILEY & SONS, INC.
New York · Chichester · Brisbane · Toronto · Singapore

Copyright © 1995 by John Wiley & Sons, Inc.

Library of Congress Cataloging in Publication Data:

Gagliardi, Robert M., 1934–
    Optical communications / R.M. Gagliardi, S. Karp. -- 2nd ed.
       p.   cm. -- (Wiley series in telecommunications and signal
processing)
  "A Wiley-Interscience publication."
  Includes bibliographical references and index.
  ISBN 0-471-54287-3
  1. Optical communications. I. Karp, Sherman. II. Title.
III. Series.
TK5103.59.G33 1995
621.382'7--dc20   94-28420

Printed in the United States of America

10 9 8 7 6 5 4 3 2 1

To our parents, wives, and children

CONTENTS

PREFACE

This second edition is an updated version of our earlier text published in 1976. Much of the mathematical modeling and statistical analysis that was detailed in that first edition has been significantly reduced, so as to be more compatible with modern analysis. We have focused on those specific optical technologies that have emerged since that first book. In addition we have inserted new material in the key areas of digital communications, fiberoptics, lightwave networks, atmospheric channels, and space links, all of which are now critical topics in optical applications. We have also inserted brief sections on optical hardware and device descriptions to make the text somewhat self-contained, and to bridge the gap between the theory emphasized in the first edition and the applied analysis emphasized here. We have again chosen to integrate both fiber and space optics in our presentation which, we feel, separates the book from the many recent optics books that concentrate solely on fiber components and fiber communications.

The objective of the new edition remains the same as that of the first — to emphasize the system aspects of optical communications, as opposed to detailed hardware and device description. A reader familiar with our first edition will find that although much of the earlier analytical procedures are still applicable, the present edition is more streamlined and oriented toward modern analysis and design. The material is again presented in a textbook format with a completely new set of homework problems and references to aid the instructor or self-learning reader. The book is aimed at several specific types of readers — the student in electrical engineering or electrophysics, the communication engineer who may wish to become familiar with the potential of optics, and the optical engineer who perhaps has not considered all the theoretical implications of optical information transmission. We consider the text appropriate for a one or two semester course in optical communications at a senior or graduate level. At USC the text is used in an optical communication course complementing undergraduate and graduate companion courses on optics and devices.

Chapter 1 introduces the optical system, defining applications and terminology, and setting the framework for the remaining chapters. A review of optical fields, sources, channels, and signal descriptions are introduced to prepare the reader. Chapter 2 explores optical field reception, field focusing, and optical filtering. The objective is to lead the reader through the necessary analysis to determine power levels in both fiber and space systems.

Chapter 3 is perhaps the most important chapter, since it covers the conversion of optical fields to electronic current flow via photodetection. To the communication engineer, it is here that the important statistical models of the receiver are generated for evaluating system performance in later chapters. Although the topic extended over three chapters in the first edition (photon counting, shot noise theory, and photodetection) it has been consolidated into a single chapter here, de-emphasizing the counting statistics while establishing useable photodetection models. New material on photomultiplication has been inserted.

The next chapters begin the application to system design, and divide naturally into two separate chapters. Direct detection (noncoherent) systems are covered in Chapter 4, and heterodyne (coherent) systems in Chapter 5. The material is presented in the language of the communication engineer, emphasizing demodulation, signal to noise ratios, and performance evaluation. Much of this involves new and updated material of the same topics covered in the first edition.

Chapter 6 specializes to digital communications and data bit transmission, an important area in modern systems. We focus on the important digital formats that have evolved over the last decades, with newer material inserted in the areas of bit error probabilities, coding, and digital clocking. Chapter 7 confines analysis to the fiber optic channel, in which earlier fiber power flow analysis is combined with modulation and signalling to define the overall communication link.

Chapter 8 extends the individual fiber link into combined links forming lightwave networks. This area, which was not covered in our first edition, has progressed rapidly, especially with today's emphasis on lightwave information highways and cable distribution systems. The basics of light distribution, switching, and multiple accessing are presented.

Chapter 9 is devoted entirely to atmospheric optical propagation, and its effect on communication performance in space links. The objective is to guide the communication engineer through available data and graphs to assess the channel effects on link performance. Chapter 10 applies to satellite and space vehicle communications using laser beams to establish crosslinks. Beam pointing, beam acquisition, and beam tracking are discussed and related to the overall link performance. Since beam tracking is generally integrated directly into the communication link, the performance of each is directly interrelated through their individual parameters. This interrelationship is developed in this chapter.

We wish to thank Ms. Milly Montenegro, Ms. Rohini Montenegro, and the staff of the Communication Science Institute at USC for their help in preparing the new edition. In addition, we would like to thank the various practicing engineers and scientists, classroom students, and university instructors for their comments and suggestions on the first edition that aided us in upgrading to this edition.

Robert M. Gagliardi
Sherman Karp

Conversion Formulas

PHYSICAL CONSTANTS

Speed of light, c	$=$	$2.998 \times 10^8 \, \text{m/sec}$
Electron charge, e	$=$	$1.601 \times 10^{-19} \, \text{C}$
Planck's constant, h	$=$	$6.624 \times 10^{-34} \, \text{W-sec/Hz} = (-335.4 \, \text{dBW/Hz}^2)$
Boltzman's constant, k	$=$	$1.379 \times 10^{-23} \, \text{W/}^\circ\text{K-Hz}$

CONVERSION FACTORS

1 micron $= 10^{-6}$ meters $= 10^{-4}$ cm
$1 \, \text{Å} = 10^{-4}$ microns $= 10^{-10}$ meters
1 arc sec $= 2.78 \times 10^{-4}$ degrees $= 4.89 \times 10^{-6}$ radians
Frequency in Hz $= 3 \times 10^{14}$/wavelength in microns
Bandwidth in Hz at center wavelength $\lambda = (c/\lambda^2)$ [bandwidth in wavelength]

OPTICAL FREQUENCIES AND WAVELENGTHS

Violet	$\approx$	$7 \times 10^{14} \, \text{Hz}$	0.38–0.48 microns
Blue	$\approx$	$6 \times 10^{14} \, \text{Hz}$	0.48–0.52 microns
Green	$\approx$	$5.6 \times 10^{14} \, \text{Hz}$	0.52–0.56 microns
Yellow	$\approx$	$5.1 \times 10^{14} \, \text{Hz}$	0.56–0.62 microns
Orange	$\approx$	$4.8 \times 10^{14} \, \text{Hz}$	0.62–0.64 microns
Red	$\approx$	$4.4 \times 10^{14} \, \text{Hz}$	0.64–0.72 microns
Infrared	$\approx$	$3 \times 10^{14} \, \text{Hz}$	0.7–100 microns

Optical Communications
Second Edition

THE OPTICAL COMMUNICATION SYSTEM

The objective of any communication system is the transfer of information from one point to another. This information transfer is accomplished most often by superimposing (modulating) the information onto an electromagnetic wave (carrier). The modulated carrier is then transmitted (propagated) to the destination, where the electromagnetic wave is received and the information recovered (demodulated). Such systems are often designated by the location of the carrier frequency in the electromagnetic spectrum (Fig. 1.1). In radio systems, the electromagnetic carrier wave is selected with a frequency from the radio frequency (RF) portion of the spectrum. Microwave or millimeter systems have carrier frequencies from those portions of the spectrum. In an optical communication system, the carrier is selected from the optical region, which includes the infrared, visible, and ultraviolet frequencies.

The principal advantages in communicating at optical frequencies are (1) the potential increase in modulation bandwidth, (2) the ability to concentrate power in extremely narrow beams, and (3) the significant reduction in component sizes. In any communication system, the amount of information transmitted is directly related to the bandwidth (frequency extent) of the modulated carrier, which is generally limited to a fixed portion of the carrier frequency itself. Thus, increasing the carrier frequency theoretically increases the available transmission bandwidth, and therefore the information capacity of the overall system. This means frequencies in the optical range will have a usable bandwidth approximately 10^5 times that of a carrier in the RF range. This available improvement is extremely inviting to a communication engineer vitally concerned with transmitting large amounts of information. In addition, the ability to concentrate available transmitter power within the transmitted electromagnetic wave also increases with carrier frequency. Thus, using higher carrier frequencies increases the capability of the system to achieve higher power densities, which generally leads to improved performance. Lastly, operation at the extremely small wavelengths of optics produces system devices

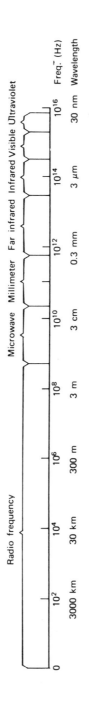

Figure 1.1. The electromagnetic spectrum.

and components that are much smaller than their equivalent electronic counterparts. For these reasons, optical communication has emerged as a field of special technological interest.

Communicating at optical frequencies has several major differences from RF communications. Because optical frequencies are accompanied by extremely small wavelengths, optical component design requires essentially its own technology, completely different from design techniques associated with RF, microwave, and millimeter devices. As a result, optical devices, although emulating equivalent electronic devices, may have performance characteristics significantly different from their electronic counterparts.

Another drawback to optical communications is the detrimental effect of the propagation path on the optical carrier wave. This is because optical wavelengths are commensurate with molecule and particle sizes, and propagation effects are generated that are uncommon to radio and microwave frequencies. Furthermore, these effects tend to be stochastic and time varying in nature, which hinders accurate propagation modeling. A vast amount of experimental data has been collected to aid in understanding this optical propagation phenomenon and, although certain models have been established, continued exploration is required for refinement and further justification.

The development of optical components and the derivation of propagation models, however, are only part of the overall system design. A communication engineer must also be concerned with the choice of components, the selection of system operations, and finally the interfacing or interconnecting of these operations in the best possible manner. These interfacing decisions require reasonably accurate mathematical models, which indicate component performance, anomalies, and degradations, knowledge of which can be used to advantage in system design. It is this aspect of optical communications that this book attempts to elucidate. Our objective is to understand system capability and to formulate system-design procedures and performance characteristics for the implementation of an overall optical communication system.

1.1 OPTICAL SYSTEMS

The block diagram of a generic optical communication system is shown in Figure 1.2. The diagram is composed of standard communication blocks, which are endemic to any communication system. A source producing some type of information (waveforms in time, digital systems, etc.) is to be transmitted to some remote destination. This source has its output modulated onto an optical carrier (a carrier frequency in the optical portion of the electromagnetic spectrum). This carrier is then transmitted as an optical light field, or beam, through the optical channel (free space, turbulent atmosphere, fiberoptic waveguide, etc.). At the receiver, the field is optically collected and processed (photodetected), generally in the presence of noise interference, signal distortion, and inherent background radiation (undesired light fields or other

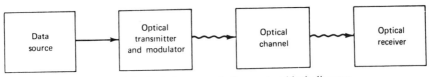

Figure 1.2. Optical communication system block diagram.

electromagnetic radiation). Of course, except for the fact that the transmission is accomplished in the optical range of carrier frequencies, the operations just mentioned describe any communication system using modulated carriers. Nevertheless, the optical system employs devices somewhat uncommon to the standard components of the RF system. These devices have significant differences in their operation and associated characteristics, often requiring variations in design procedures.

The modulation of the source information onto the optical carrier can be in the form of frequency modulation (FM), phase modulation (PM), or possibly amplitude modulation (AM), each of which can be theoretically implemented at any carrier frequency in the electromagnetic range [1]. In addition, however, several other less conventional modulation schemes are also often utilized with optical sources. These include intensity modulation (IM), in which information is used to modulate the intensity (to be defined subsequently) of the optical carrier, and polarization modulation (PLM), in which spatial characteristics of the optical field are modulated.

The optical receiver in Figure 1.2 collects the incident optical field and processes it to recover the transmitted information. A typical optical receiver can be represented by the three basic blocks shown in Figure 1.3, consisting of an optical receiving front end (usually containing some form of lens or focusing hardware), an optical photodetector, and a postdetection processor. The lens system filters and focuses the received field onto the photodetector, where the optical signal is converted to an electronic signal. The processor performs the necessary amplification, signal processing, and filtering operations to recover the desired information from the detector output.

Optical receivers can be divided into two basic types: power detecting receivers and heterodyning receivers. Power detecting receivers (often called direct detection, or noncoherent, receivers) have the front end system shown in Figure 1.4a. The lens system and photodetector operate to detect the instantaneous power in the collected field as it arrives at the receiver. Such receivers represent the simplest type for implementation and can be used whenever the transmitted information occurs in the power variation of the received field.

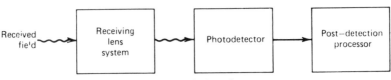

Figure 1.3. The optical receiver.

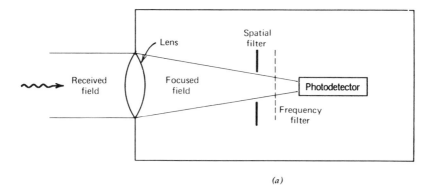

(a)

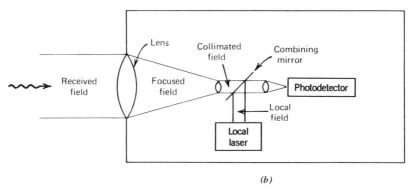

(b)

Figure 1.4. (a) Direct detection receiver. (b) Heterodyne detection receiver.

Heterodyning receivers have the front end system shown in Figure 1.4b. A locally generated lightwave field is optically mixed with the received field through a front end mirror, and the combined wave is photodetected. Such receivers are used whenever information is amplitude modulated, frequency modulated, or phase modulated onto the optical carrier. Heterodyning receivers are more difficult to implement and require close tolerances on the spatial coherence of the two optical fields being mixed. For this reason, heterodyned receivers are often called (spatially) coherent receivers. For either type of receiver, the front end lens system has the role of focusing the received or mixed field onto the photodetector surface. This focusing allows the photodetector area to be much smaller than that of the receiving lens.

The receiver front end, in addition to focusing the optical field onto the photodetector, also provides some degree of filtering, as shown in Figure 1.4. These filters are employed prior to photodetection to reduce the amount of undesired background radiation. Optical filters may operate on the spatial properties of the focused fields (polarization filters, field stops, etc.) or may filter in the frequency domain; that is, they pass certain bands of frequencies

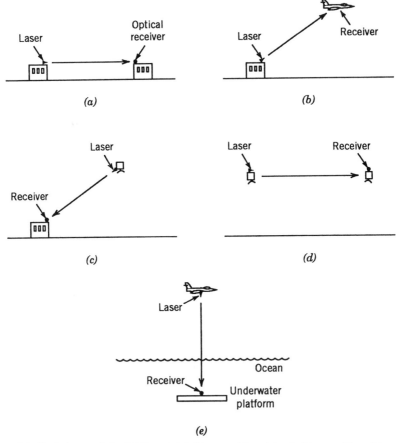

Figure 1.5. Optical space links. (*a*) Terrestrial, (*b*) ground based to air, (*c*) spaced based to ground, (*d*) intersatellite, (*e*) air to underwater.

and reject others. The latter filters determine the bandwidth of the resulting optical field subsequently photodetected.

Photodetectors convert the focused optical field into an electrical signal for processing. Although there are several types of detectors available, all behave according to quantum mechanical principles, utilizing photosensitive materials to produce current or voltage responses to changes in impinging optical field power. The basic model defining this interaction for all photodetectors is well accepted, although detectors may differ in their output response characteristics. This basic model, which is examined in detail in Chapter 3, is extremely important to the communication engineer because it generates the inherent statistics that must be utilized in design of the postdetection processing. The most common type of photodetectors are the phototubes, photodiodes, and photomultipliers.

The detection of optical fields is hampered by the various noise sources

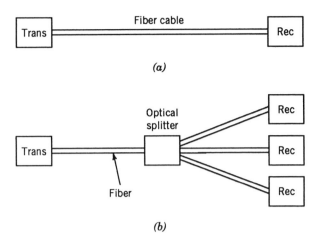

Figure 1.6. Fiber optic links. (*a*) Single fiber link, (*b*) fiber optic distribution.

present throughout the receiver. The most predominant in long-distance space communication is the background light or stray radiation that is collected at the receiver lens along with the desired optical field. Although this radiation may be reduced by proper spatial filtering, it still represents the most significant interference in the detection operation. The background effect can be eliminated when direct-coupled fiberoptic waveguides can be used for the transmission path. A second noise source is the photodetector itself, which, not being a purely ideal device, produces internal interference during the photodetection operation. This induced noise is referred to as detector noise. The last noise source is the circuit and electronic thermal noise generated in the processing operations following photodetection. The thermal noise is accurately modeled as additive white Gaussian noise, whose spectral level is directly related to the receiver temperature, just as in any RF or microwave communication system. Each of these noise sources must be properly accounted for in any receiver analysis.

The models in Figures 1.3 and 1.4 are common to any optical communication system. In a space system, the transmitted optical field is focused into a beam of light and transmitted as a propagating electromagnetic field through a medium. Examples of these are shown in Figure 1.5. The system can be a terrestrial (ground-based) link, a ground-to-space (atmospheric) link, a space-to-space crosslink, or even a space-to-underwater link. All such systems use optical beams transmitted as unguided fields and are susceptible to the effects of the medium (atmosphere, clouds, water, etc.) over the communication path. The effects of atmospheric propagation in space links are examined in Chapter 9.

A fiberoptic system (Fig. 1.6*a*) confines the transmitted field to an optical waveguide (fiber) during its propagation. The system, therefore, is operated as is any cable-connected link. Because the field is guided, only the properties of the fiber itself affect the field transmission. In particular, atmospheric and background noise effects are no longer important to system performance. The enormous improvement in fiber quality has permitted long communication links and fiberoptic distribution systems (Fig 1.6b) to be readily established. Today, fiberoptic systems are rapidly replacing the more traditional cable and wireline systems of the past.

1.2   OPTICAL SOURCES, MODULATORS, AND BEAM FORMERS

The key element in any optical communication system is the availability of a light source that can be easily modulated. Such a source should produce energy concentrated in a narrow wavelength band. The primary sources of light in modern optical systems are the light-emitting diodes (LED), the laser, and the laser diode (LD) [2–4]. Although the physical descriptions of these devices are beyond our scope here, their output properties and characteristics will be important in assessing the performance when used in an optical communication system.

An LED is formed from semiconductor junctions that interact when subjected to external current so as to radiate light energy, as indicated by the diagram in Figure 1.7a. A detailed theory of band energy is needed to describe this interaction and is not pursued here. The choice of the junction materials determines the emitted wavelength. Light-emitting diodes are typically formed from compounds of gallium arsenide, and produce light in the 0.8–0.9-μm wavelength bands. An LED is small in size (centimeters), relatively inexpensive, and can produce radiation with low-current drive levels. However, they are limited in output power (1 to 10mW), and the emitted light tends to be unfocused. Table 1.1 summarizes these basic characteristics.

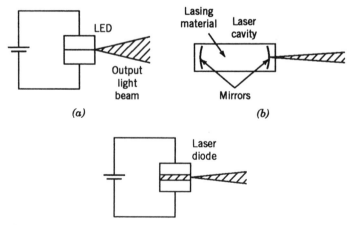

Figure 1.7. Light sources. (a) Light emitting diode (LED), (b) laser cavity, (c) laser diode (LD).

A laser tube is constructed as an optical cavity filled with light amplification material (gas or solid) and mirrored facets at each end as shown in Figure 1.7b. If the propagation gain of the material overcomes the reflection losses of the mirrors, then an initiated optical field reflected back and forth by the mirrors, will be self-sustaining. We say the cavity "lases," and optical energy is produced within the cavity. By placing a small aperture in one mirror, some internal energy will escape as radiated light. As a result the laser can produce high power levels (0.1 to 1 W) with output radiation that can be more focused than an LED.

Many different materials can be used in the cavity to produce lasing (Table 1.1), each having specific atomic structures that form particular wavelength bands. In addition, the cavity length must be such that a propagating internal field will exactly reinforce (be phase aligned) after two-way mirror reflections. This means the cavity length must be precisely an integer multiple of a half wavelength of the internal field. That is, the lasing wavelength λ must be related to the cavity length L by $\lambda = 2L/n$, for some integer n. The lasing material sustains those wavelengths that are commensurate with its propagation gain profile. Laser tubes are therefore high-power devices, but are much bulkier than diode sources.

Laser diodes are semiconductor junction devices that contain substrates that are etched, or cleaved, to act as reflecting facets for field reinforcements over the junctions. They therefore combine the properties of an LED and the

TABLE 1.1 Optical Sources

Laser material	Wavelength (μm)	
Solid state		
GaAs	0.87	
InGaAs	1.0–1.3	
InGaAsP	0.9–1.7	
AlGaAs	0.8–0.89	
Ruby	0.694	
Nd-Yag	1.06	
Gas		
CO_2	10.6	
HeNe	0.63	

Laser Types	Output Power	Linewidth (nm)
Diodes		
LED	0.1–10 mW	20–100
laser diode	1–40 mW	1–5
Distributed feedback laser	1–40 mW	0.1
Tubes		
CO_2	1–5 W	0.01–1.0
HeNe	50–100 mW	0.01–1.0

cavity reflector, producing an external light radiation that is higher in power (10 to 50 mW) and better focused than a simple LED.

The important communication characteristics of any optical source are its modulation bandwidth (the rate at which the source can be modulated), its input–output power curve, and its frequency spectrum. Light sources, generated from extremely narrow cavities, can be modulated at bandwidths up to 1 to 4 GHz. This provides an enormous potential advantage over RF communications, where modulation bandwidths of only hundreds of megahertz are available.

Figure 1.8a sketches typical diode and laser power characteristics, plotting output light power versus external bias current. The threshold current is that needed to produce output light. The linear range defines the modulation range, where input current converts proportionally to output power. The saturation level limits the maximum available output power. Light-emitting diodes have low thresholds and can operate at low-current values, but they have limited peak powers. Laser diodes require more drive current, but have higher peak power. Laser tubes generally have to be pumped above threshold and are difficult to stabilize in the linear range. Hence, high-power lasers are usually operated as continuous-wave devices at their peak power capability.

The frequency spectrum of an optical source, as sketched in Figure 1.8b, indicates the spectral extent, or purity, of the light source. The spreading of spectrum around the desired wavelength indicates the presence of unwanted frequencies, or undesired noise modulations, superimposed on the output wavelength. This spectral spreading can hinder the ability to recognize desired information modulated on the source. Light-emitting diodes have relatively wide spectral extent (hundreds of Angstroms), whereas lasers significantly improve the light purity. The effect of source spreading on modulation performance is examined further later in this volume.

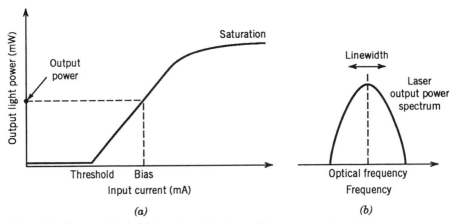

Figure 1.8. Characteristics of light sources. (a) Output light power vs. input drive current; (b) power spectrum of the output light.

Modulators superimpose the information signals (analog or digital) on to the source. Optical modulators [4] are of two basic types: internal or external (Fig. 1.9). An internal modulator is one in which the source itself is directly modified by the information signal to produce a modulated optical field. Amplitude or intensity modulation can be imposed by varying the bias current, as is evident from Figure 1.9a. Frequency or phase modulation can be inserted on a laser tube by varying its cavity length. Pulse modulation is easily applied to a diode by driving it above and below threshold. Such modulations are generally limited to the linear range of the power characteristic.

In external modulation (Fig. 1.9b), the source light is focused through an external device, whose propagation characteristics are altered by the modulating signal. Such systems have the advantage of utilizing the full power capability of the source. Modulation is achieved via the electrooptic or acoustooptic effect of the material, in which external currents can modify the transmission properties (index of refraction, polarization, direction of flow, etc.) of the inserted light. These effects produce delay variations (phase modulation) or polarization changes (intensity modulation) on the excited beam. Pulsed outputs can be achieved by blocking or deflecting the light path. Unfortunately, external modulators insert significant coupling losses, limit the modulation range, and generally require relatively higher modulation drive power.

Light fields from the radiating surfaces of optical sources and modulators are emitted with varying degrees of focusing, usually described by its emission angle, as shown in Figure 1.10. The light emission is further characterized by the source brightness function, $\mathscr{B}(\theta)$, which is in units of watts/steradian-area, that describes the normalized light power emitted in a given direction angle θ out from the source. The power is usually normalized to a unit solid angle per unit of source area. Hence the source brightness indicates the distribution of power radiated out from the source. A uniformly radiating source will have the same brightness at all angles within its emission solid angle Ω_s, as shown in Figure 1.10. The total power in watts emitted from a uniform source with area A_s and emission angle Ω_s is then

$$P_s = = \mathscr{B}A_s\Omega_s \tag{1.2.1}$$

For symmetrically radiating sources, the solid angle Ω_s can be related to the

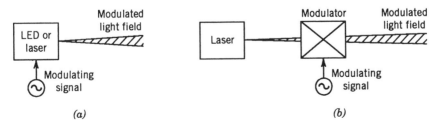

Figure 1.9. Optical modulators. (a) Internal, (b) external.

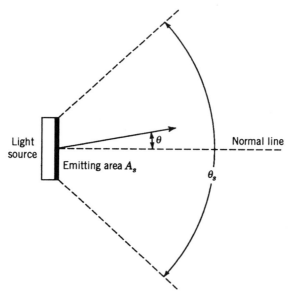

Figure 1.10. Light emission angles.

planar emission angle θ_s in Figure 1.10 by

$$\Omega_s = 2\pi[1 - \cos(\theta_s/2)] \tag{1.2.2}$$

A Lambertian source is a uniformly radiating source that emits in all forward directions $|\theta| < \pi/2$. For the Lambertian source, $\Omega_s = 2\pi$ steradians, and $P_s = \mathscr{B}2\pi A_s$. The assumption of uniformly radiating surfaces is generally used in describing most optically emitting sources.

Light fields from radiating sources can also be collected and refocused by means of beam-forming optics. The latter are usually combinations of various

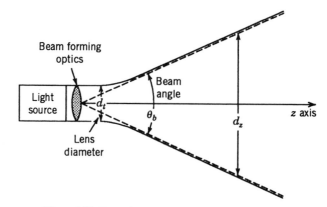

Figure 1.11. Beam forming and focused light.

types of lenses placed at the source or modulator output that orient the light into particular directions. Although light focusing will be considered in Chapter 2, Figure 1.11 shows a simple type of beam formation commonly used in long-range space links. A combination of a converging and diverging lens placed at the source is used to produce a collimated beam. Ideally, the converging lens focuses the source field light to a point, and the diverging lens expands it to a perfect beam. In practice, the source field is instead focused to a spot, and the expanded beam spreads during propagation with a planar beam diameter of approximately [5]

$$d_z = d_t \left[1 + \left(\frac{\lambda z}{d_t^2} \right)^2 \right]^{1/2}$$ (1.2.3)

where λ is the wavelength, d_t is the output lens diameter, and z is the distance from the lens. At points in the near field ($\lambda z/d_t^2 < 1$), the emerging light is collimated with a diameter equal to the lens diameter. That is, the light appears to uniformly exit over the entire lens. In the far field ($\lambda z/d_t^2 > 1$) the beam diameter expands with distance, and appears as if the light is emerging from a single point with a planar beam angle of approximately

$$\theta_b \cong \frac{\lambda}{d_t} \text{ rad}$$ (1.2.4)

The angle θ_b is called the diffraction-limited transmitter beam angle. The expanding field far from the source is therefore confined to a two-dimensional solid angle of approximately

$$\Omega_b = 2\pi[1 - \cos(\theta_b/2)]$$

$$\simeq \left(\frac{\pi}{4} \right) \theta_b^2$$ (1.2.5)

Figure 1.12 shows a plot of Eq. (1.2.4) as a function of diameter d_t for several optical wavelengths. For example, a 6-in. lens at an optical wavelength of $0.5\,\mu$m has a beamwidth on the order of 3×10^{-6} rad or approximately 0.16 mdeg. This is a spectacular advantage compared to RF transmitters where antenna beams are usually on the order of degrees. This ability to concentrate field flow to small beam angles with relatively small size optics is a significant advantage in long-range space communications.

The optical advantage of this source focusing can be further emphasized by converting to an effective antenna gain. From antenna theory [6], a transmitter with the beamwidth in Eq. (1.2.4) will have an effective antenna gain of

$$G_t = \frac{4\pi}{\Omega_b} \approx \left(\frac{4d_t}{\lambda} \right)^2$$ (1.2.6)

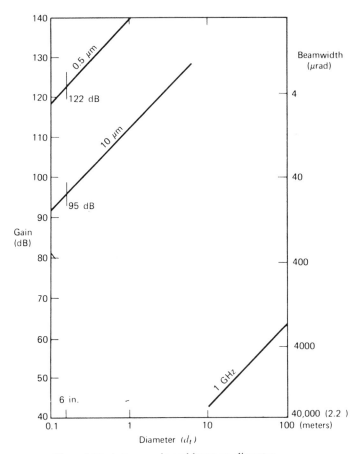

Figure 1.12. Antenna gain and beam vs. diameter.

This corresponding gain is also shown in Figure 1.12. The same 6-in. lens would have an effective gain of 122 dB, a sizable improvement over an RF antenna. (A 210-ft antenna generates only 60 dB of gain at RF.) It should be pointed out, of course, that with these thin beams, difficult pointing problems (aiming the transmitter at a particular spot) are now generated. This latter problem, which immediately becomes an integral part of the overall system design, is considered in Chapter 10.

1.3 TRANSMITTED OPTICAL FIELDS

The field from an optical source is transmitted as a propagating optical field. It is necessary to properly describe this emanating field to determine power flow. Because the transmitted field is an electromagnetic wave, it is described at any spatial point by solutions to Maxwell's equations. Let ξ represent a point of a selected coordinate system in which the field source is located at the origin, as

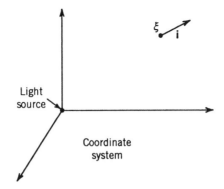

Figure 1.13. Three-dimensional coordinate space.

shown in Figure 1.13. At any time t, the radiation field is described at ξ by a unit vector $\mathbf{i}$ in the direction of field flow and its electrical field function

$$\text{Electric field} = \text{Real}\{f(t, \xi)\} \qquad \text{V/length} \qquad (1.3.1)$$

where $\text{Real}\{\cdot\}$ means "real part of" and $f(t, \xi)$ is referred to as the complex field. At each point ξ, the field has an intensity given by

$$I(t, \xi) = \frac{|f(t, \xi)|^2}{Z_w} \qquad \text{W/area} \qquad (1.3.2)$$

where Z_W is the wave impedance (i.e., the impedance of the medium; in free space, $Z_W = 377$ ohms). The *instantaneous power* of the complex field over a planar area A at time t is given by the surface integral of the intensity over the area,

$$P_A(t) = (\cos\theta) \int_A I(t, \xi)\, d\xi \qquad \text{W} \qquad (1.3.3)$$

where the integration is over all ξ in A, and θ is the angle between the normal line to the area and the unit vector $\mathbf{i}$ in the direction of power flow.* We point out that without loss of generality, the medium impedance Z_w can be directly incorporated into the field definition $f(t, \xi)$, so that the field intensity can be simply written as $I(t, \xi) = |f(t, \xi)|^2$. This normalized field $f(t, \xi)$ now has units of watts$^{1/2}$/length.

Equations (1.3.1) to (1.3.3) are key equations relating optical field, intensity, and power and are important to subsequent communication analysis. This is because optical detectors inherently respond to the intensity of the impinging fields. The most important optical field is the plane wave field that propagates in direction $\mathbf{i}$ and has its complex field defined over a normal planar area A

*The vector $I(t, \xi)\mathbf{i}$ through ξ, is called the Poynting vector. Equation (1.3.3) is the integration of the Poynting vector over the surface area.

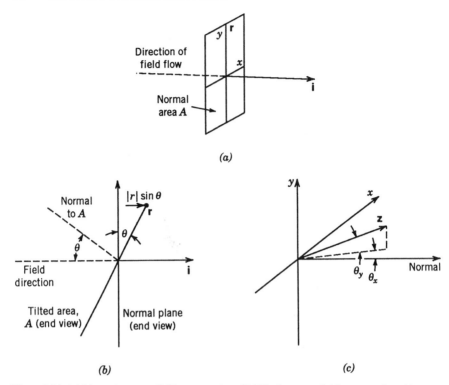

Figure 1.14. (*a*) Normal area to field propagation. (*b*) Tilted area to field propagation. (*c*) Subangles of the vector **z**.

(Fig. 1.14*a*) containing points **r** as

$$f(t, \mathbf{r}) = a(t)e^{j\omega_o t} \qquad \mathbf{r} \in A \tag{1.3.4}$$

where ω_o is the optical radian frequency of the transmitter and $a(t)$ is a complex amplitude function (normalized by Z_w) describing the field polarization in the plane (x, y) normal to the direction of propagation. This can be expanded as

$$a(t) = a_x(t)\mathbf{1}_x + a_y(t)\mathbf{1}_y \tag{1.3.5}$$

where $a_x(t)$ and $a_y(t)$ are the complex polarization components and $(\mathbf{1}_x, \mathbf{1}_y)$ are unit coordinate vectors in the (x, y) plane. The polarization components determine the polarization state of the plane wave. If $|a_x(t)| = |a_y(t)|$ and are 90 degrees out of phase with the other, the field is circularly polarized. If both are in phase, or one is zero, the field is linearly polarized. Unless otherwise stated, we assume linearly polarized fields throughout subsequent discussions, and the polarization notation can be dropped. In this case $a(t)$ represents the complex modulation applied to the optical carrier. If we write

$$a(t) = |a(t)|e^{j\phi(t)} \tag{1.3.6}$$

then $|a(t)|$ describes the amplitude modulation, and $\phi(t)$ the phase modulation ($d\phi/dt$ would describe the frequency modulation) occurring over the normal plane. The intensity modulation would then be given by $|a(t)|^2$ over the plane. Note that all points $\mathbf{r}$ in A vary in phase synchronism, and we say the plane wave is *phase coherent* over A. As the field propagates in the direction $\mathbf{i}$ through A, the coherent wavefront varies in time at each point in the plane.

Suppose we wish to describe a plane wave over an area A tilted from the normal, as shown in Figure 1.14b. Since the points in a normal plane are in phase, the points $\mathbf{r}$ on the tilted plane are phase shifted due to the field propagation. Let $f(t, 0) = a(t)e^{j\omega_o t}$ be the complex field at the point $\mathbf{r} = 0$ on A. A point $\mathbf{r}$ will either lag or lead this point as the plane wave propagates through it. Thus we can write for any point $\mathbf{r}$ in A,

$$f(t, \mathbf{r}) = a(t)e^{j\omega_o t}\exp\left[\left(-j\left(\frac{2\pi}{\lambda}\right)|\mathbf{r}|\sin\theta\right)\right] \tag{1.3.7}$$

where again θ is the angle between $\mathbf{i}$ and the normal area. The exponential phase term accounts for the additional phase shift at point $\mathbf{r}$. This phase shift can be written in other ways. Define the propagation vector

$$\mathbf{z} = (2\pi/\lambda)\mathbf{i} \tag{1.3.8}$$

which is a vector passing through the origin of A with the direction of the field vector $\mathbf{i}$ and with amplitude $2\pi/\lambda$. We first note that we can write $(2\pi/\lambda)|\mathbf{r}|\sin\theta = |\mathbf{z}||\mathbf{r}|\cos(90° - \theta) = \mathbf{z}\cdot\mathbf{r}$, denoting the vector dot product between the vectors $\mathbf{r}$ and $\mathbf{z}$. Thus we have the equivalent of Eq. (1.3.7),

$$f(t, \mathbf{r}) = a(t)e^{j\omega_o t}e^{-j\mathbf{z}\cdot\mathbf{r}} \tag{1.3.9}$$

The above defines the field at any point $\mathbf{r}$ in A due to a plane wave field with complex envelope $a(t)$ arriving from angle θ relative to the normal. Note that the optical carrier is phase shifted from one point in the plane to another by an amount depending on the dot product at the point location.

We can also write the phase shift directly in terms of the coordinates of the point $\mathbf{r} = (x, y)$ by expanding the dot product, using the x and y coordinates of the vector $\mathbf{z}$ in Eq. (1.3.8). Denote $\mathbf{z} = (z_x, z_y)$ and let (θ_x, θ_y) be the subangles of the arrival angle θ, as shown in Figure 1.14c. Specifically, θ_y is the angle between $\mathbf{z}$ and the plane formed by the normal axis and the x axis of A (elevation angle). The subangle θ_x is the angle between the normal axis and the vector formed by projecting $\mathbf{z}$ on to this plane (azimuth angle). Then, in terms of these subangles,

$$\begin{aligned} z_y &= (2\pi/\lambda)\sin\theta_y \\ z_x &= (2\pi/\lambda)\cos(\theta_y)\sin(\theta_x) \end{aligned} \tag{1.3.10}$$

This expansion allows us to substitute $\mathbf{z} \cdot \mathbf{r} = x z_x + y z_y$, so that

$$f(t, \mathbf{r}) = a(t)e^{j\omega_o t}e^{-j(xz_x + yz_y)} \tag{1.3.11}$$

The phase factor contains only the arrival subangles and the coordinates of $\mathbf{r}$. This further simplifies if the arrival subangles are small ($\mathbf{z}$ arrives close to the normal) so that $\cos\theta_y \cong 1$, $\sin\theta_y \cong \theta_y$ and $\sin\theta_x \cong \theta_x$. In this case, $z_x \cong (2\pi/\lambda)\theta_x$, $z_y = (2\pi/\lambda)\theta_y$, and

$$f(t, \mathbf{r}) \cong a(t)e^{j\omega_o t} \exp\left[-j\left(\frac{2\pi}{\lambda}\right)\left(x\theta_x + y\theta_y\right)\right] \tag{1.3.12}$$

These phase relations between arrival angles of a plane wave and points $\mathbf{r}$ in the plane will be important in later focusing and detection analysis.

1.4 OPTICAL SPACE CHANNELS

In the optical space systems in Figure 1.5 the laser source produced an optical field that is intended for a receiving area that is located a significant distance away. This distance is usually far enough so that the laser source appears as a point source when viewed from the receiver. Systems in which this point source assumption can be made can be analyzed similar to RF communications.

If the source producing the optical field is a point source, operating in a free space medium and modulated with the field function in Eq. (1.3.6), then the field at a remote point ξ due to this source is given by

$$f(t, \xi) = \frac{\sqrt{G(\xi)}\, a\left(t - \dfrac{|\xi|}{c}\right) e^{j(\omega_o t - 2\pi|\xi|/\lambda)}}{\sqrt{4\pi|\xi|}} \tag{1.4.1}$$

where $|\xi|$ is the distance to the point ξ, λ is the source wavelength corresponding to the frequency ω_0, c, is the speed of light, and $G(\xi)$ is the transmitter power gain in the direction ξ. Equation (1.4.1) describes an optical field that propagates out from the source as a spherical wave with diminishing amplitude. However, to an area A located a distance z away, where $z \gg A^{1/2}$, all points on A are approximately the same distance from the source. Thus $|\xi|$ is the same over A and the field arriving at A from the point source is basically a plane wave field. That is, the field emminating from a point source appears as a plane wave when observed by a small area from a remote distance. The field intensity at a point ξ in A is then

$$I(t, \xi) = G_t \frac{\left| a\left(t - \frac{z}{c}\right) \right|^2}{4\pi z^2} \tag{1.4.2}$$

where G_t is the transmitting optics gain in the direction of A. Defining

$$P_s(t) = |a(t)|^2 \tag{1.4.3}$$

as the power variation of the point source, Eq. (1.4.2) reduces to

$$I(t, \xi) = G_t \frac{P_s(t - t_d)}{4\pi z^2} \tag{1.4.4}$$

where $t_d = z/c$ is the propagation delay from the source to the point ξ. Thus the field intensity produced by an optical point source with envelope function $a(t)$ propagates out from the transmitter according to the gain functions of the transmitting optics and decreases in strength as the square of the distance.

As an example, consider a point source that transmits a constant power P_s through beam-forming optics over a solid angle Ω_s, as shown in Figure 1.15. At a distance $|\xi| = z$ away, the field intensity within the beam will be

$$I(t, \xi) = \frac{G_t P_s}{4\pi z^2} \tag{1.4.5}$$

where G_t is the transmitter gain in Eq. (1.2.6). A normal receiving area A within the beam collects the field power, using Eq. (1.3.3), as

$$P_r = \left[\frac{G_t P_s}{4\pi z^2} \right] A \tag{1.4.6}$$

Note this corresponds to the standard power flow equation associated with any electromagnetic field produced from a source with the gain G_t. It is common practice to define a receiving gain related to the receiving area as

$$G_r = \left(\frac{4\pi}{\lambda^2} \right) A \tag{1.4.7}$$

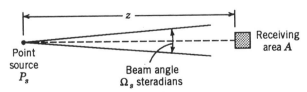

Point
source
P_s

Beam angle
Ω_s steradians

Receiving
area A

Figure 1.15. Space link model.

Equation (1.4.6) now becomes

$$P_r = P_s G_t L_p G_r \tag{1.4.8}$$

where

$$L_p = \left(\frac{\lambda}{4\pi z}\right)^2 \tag{1.4.9}$$

is the propagation loss associated with the transmission of an electromagnetic field of wavelength λ over a distance z. Thus the power flow equation (1.4.6) is identical to the link budget equations used in all standard communication link analysis. Note that the use of Eq. (1.4.7) in Eq. (1.4.6) inserts the optical wavelength λ into the power flow equation, giving the somewhat erroneous conclusion that collected power in a free space channel depends on wavelength.

If the receiving area A is tilted relative to the normal (i.e., the point source field arrives from an offset angle), Eqs. (1.3.3) must be used with the angle to determine received power. This inserts an additional power loss from Eq. (1.4.8) due to the cosine of the angle offset. This loss is negligible if the arrival angle is small (i.e., the field arrives close to the area's normal).

When the propagating medium is not free space, additional effects must be included on the propagating field. This is because when propagating through a nonfree space link, such as the atmosphere, the wave undergoes effects that tend to further alter the power levels and the field phase across the wavefront. These alterations depend on the wavelength of the wave and are due primarily to interactions with inhomogeneities and foreign particles comprising the medium. These effects become most predominant as the wavelength approaches the physical size of these particulates. For this reason, atmospheric distortion tends to be quite severe in the optical range, where wavelengths are commensurate with particulates as small as molecules. Thus the atmospheric space channel presents a major communication hurdle that must be of concern to optical system engineers.

The atmosphere primarily causes field absorption and scattering, the degree of which depends on the type of atmospheric conditions (clear air, clouds, rain, etc.) [7]. This absorption manifests itself in the propagating wavefront as both amplitude and spatial phase effects. Amplitude effects are exhibited in the time variation of the field and involve power loss, power fluctuations, and frequency filtering. Spatial effects appear as variations in the beam direction, beam spreading, or phase variations across the wavefront. In particular, it alters the phase coherency that would normally exist across the wavefront in standard plane waves.

Since the particulate structure of the atmosphere tends to be random in nature, these amplitude and phase effects evolve as random disturbances that can only be stochastically described. A detailed analysis of the atmospheric effects in optical space channels is considered in Chapter 9.

1.5 THE FIBER OPTICAL CHANNEL

An optical fiber is a thin pipe of glass, through which one can insert light and have it shine out at the output end. Hence a fiber can transmit a modulated light field from its input to its output. The basic construction of a fiber is shown in Figure 1.16, and corresponds to a central glass core encased in a plastic shielding. The core is made of high-quality glass through which the light can propagate. Glass, or a similar silicon compound, is used, because it offers the least attenuation for optical transmission. The core is supported by the shielding, or cladding, which provides mechanical protection, isolates the core from external radiation, and aids in confining the light to only the glass core. The cladding is opaque, and basically reflects any escaping light back into the core. Typically the core diameter is on the order of micrometers, whereas the fiber cross-section (core plus cladding) is on the order of several millimeters. Thus the fiber is literally a thread of transmission path over which the modulated light field can be propagated. This ability to pack large amounts of modulated data over an extremely small spatial area is an overwhelming advantage for optical fiber communications [8, 9].

The key parameters describing the propagation properties of the fiber light field are the refractive index of the core and cladding. The cladding is usually designed to have an index a few percent smaller than that of the core. (Glass typically has an index of approximately 1.5). These indices determine the propagation angles of the field in the fiber.

Consider the ray diagram in Figure 1.17, which shows an optical ray (optical field propagating in the ray direction) being inserted into the fiber core at angle θ. If θ is too large, the ray will be absorbed into the cladding, and will not propagate. Propagation will occur down the fiber only if the incident ray angle at the cladding boundary is shallow enough to be reflected back into the core. The ray will then continue to propagate down the core as it continually reflects off the core walls. The application of Snells law at the boundary requires that $\theta \leqslant \theta_p$, where

$$\theta_p = \cos^{-1}\left(\frac{n_2}{n_1}\right) \tag{1.5.1}$$

for the reflections to occur, n_1 is the core index and n_2 is the cladding index.

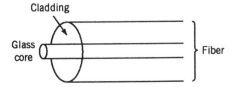

Figure 1.16. Fiber construction.

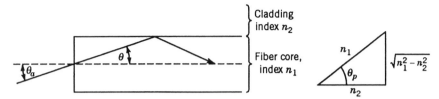

Figure 1.17. Fiber ray geometry.

The right-hand side of Eq. (1.5.1) is called the *critical angle* of the fiber and represents the maximum angle at which light will propagate down the fiber. Note that the angle depends only on the index ratio. By adjusting these indices one can control these light flow angles. In particular, by having $n_2 \gtrsim n_1$ the propagation angles can be made quite small, so the light that propagates does so at small angles, and the light will flow only down the center of the core. This means fiber boundary losses will be reduced, and less leakage will occur from the core into the cladding. Since n_2 will necessarily be close to the value of n_1, we often deal with the fractional difference,

$$\Delta = \frac{n_1 - n_2}{n_1} \tag{1.5.2}$$

For small propagation angles, we generally desire a Δ of a few percent.

To propagate light into the core at angle θ_p, it is necessary that the light be properly inserted (because the light ray may bend at the fiber input). Let θ_a be the fiber insertion angle (i.e., the angle at which light is fed into the fiber from an external source), as shown in Figure 1.17. Again, by Snells law applied to the index change from the external medium (assumed to be free space with index of unity) to the core glass, we have

$$n_1 \sin \theta_p = (1) \sin \theta_a = \sin \theta_a \tag{1.5.3}$$

The numerical aperture of the fiber is defined as

$$\text{NA} \stackrel{\Delta}{=} \sin \theta_a$$
$$= n_1 \sin \theta_p \tag{1.5.4}$$

We see that NA is an indication of the allowable insertion angle. For a given core index, we see that a small NA corresponds to a small propagation angle θ_p, and therefore a highly collimated field. Equations (1.5.3) and (1.5.4) suggest the triangle relationship in Figure 1.17 and, from simple geometry, we obtain

$$NA = n_1 \sin \theta_p$$

$$= \frac{n_1(n_1^2 - n_2^2)^{1/2}}{n_1}$$

$$= (n_1^2 - n_2^2)^{1/2} \tag{1.5.5}$$

$$= n_1 \left[\left(1 + \frac{n_2}{n_1} \right) \left(1 - \frac{n_2}{n_1} \right) \right]^{1/2}$$

Thus, the numerical aperture NA is obtained directly from the indices. Since $n_2 \approx n_1$, we can substitute from Eq. (1.5.2) and use the approximation

$$NA \cong n_1 \left[2 \left(1 - \frac{n_2}{n_1} \right) \right]^{1/2} \tag{1.5.6}$$

$$\cong n_1 \sqrt{2\Delta}$$

When the core index is fixed, NA is actually a measure of the fractional difference of the indices. Because Δ is a few percent, NA typically takes on values between 0.05 and 0.2.

When light is inserted into the fiber, it will propagate at many of the ray angles θ satisfying Eq. (1.5.1). Each individual propagating ray line in the fiber is called a field mode, and each has its own spatial E-field satisfying the necessary boundary conditions. Each mode propagates as an individual fiber field with orthogonally polarized E-field vectors (corresponding to the various transpose electric fields similar to those found in microwave waveguides) and at slightly different angles. From Maxwells equations for cylindrical guides, the number of such modes [8, 9] is given by

$$\text{number of fiber modes} = \left(\frac{d^2}{2} \right) \left[\left(\frac{2\pi n_1}{\lambda} \right)^2 - \left(\frac{2\pi n_2}{\lambda} \right)^2 \right] \tag{1.5.7}$$

where d is the core diameter. By direct substitution from Eq. (1.5.5), we see that

$$\text{number of fiber modes} = 2 \left(\frac{\pi d}{\lambda} \right)^2 [n_1^2 - n_2^2]$$

$$= 2 \left(\frac{\pi d}{\lambda} \right)^2 (NA)^2 \tag{1.5.8}$$

Thus the numerical aperture NA also determines the number of propagating fiber modes. A fiber can therefore have hundreds of modes if the core diameter is many times the light wavelength.

A fiber will propagate only a single mode if

$$2 \left(\frac{\pi d}{\lambda} \right)^2 (NA)^2 \leqslant 1 \qquad (1.5.9)$$

or if

$$d \leqslant \frac{\lambda}{\sqrt{2}\,\pi\,NA} \approx (0.22) \left(\frac{\lambda}{NA} \right) \qquad (1.5.10)$$

Single-mode fibers, therefore, require core diameters to be only several times the wavelength. A single-mode fiber will have the least loss during propagation (since the field flows primarily down the center of the fiber), but the extremely narrow core diameter will lead to obvious problems in fiber coupling and interconnections.

For large-core diameters the fiber becomes a multimode fiber with modes propagating at the various angles within θ_p. From a communication point of view having many modes is generally undesirable. This is due to the fact that power is usually distributed somewhat uniformly over the modes. (Even if the power is inserted at a single ray line direction, the internal reflections due to fiber bends will redistribute the power over all the modes.) The modes at the larger angles θ_p will be more lossy, and the overall attenuation will be greater than for a single-mode fiber. In addition, since modes propagate at different angles, their group velocities are different, and they are therefore delayed with respect to each other at the fiber output. This delay dispersion can cause an inherent filtering on the modulated light field on each mode. This effect will be considered in more detail in Chapter 7.

As in the space system, the key element is the source power that can be delivered to the fiber output. The important parameters are the efficiency of inserting source power into the fiber, and the power loss that occurs during the fiber propagation. Power insertion from an optical source into a fiber can be described by the diagram in Figure 1.18. The fiber can be butted against the source-emitting surface, or can be interconnected via lensing elements. Only the source power coupled into the acceptance angle of the fiber will produce propagating fields. Let the source have an emitting area of A_s, a brightness of $\mathscr{B}$ W/steradian-area and divergence cone angle Ω_s. If the fiber is connected directly to the source (Fig. 1.18a) the power into the fiber is

$$
\begin{aligned}
P_f &= \mathscr{B}\Omega_p A_f \qquad \text{if } A_f < A_s \\
&= \mathscr{B}\Omega_p A_s \qquad \text{if } A_s < A_f
\end{aligned}
\qquad (1.5.11)
$$

where the solid angle Ω_p is related to the planar angle θ_p by Eq. (1.2.5). From Eq. (1.2.1), the total source power is $P_s = \mathscr{B}\Omega_s A_s$. The fractional power loss

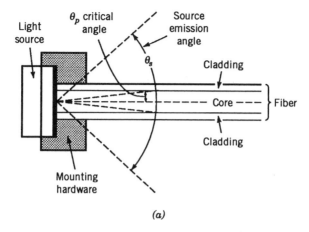

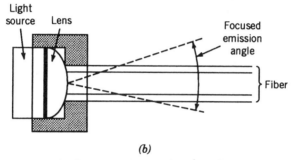

Figure 1.18. Power coupling with fibers. (a) Direct butting of the fiber to source. (b) Lens focusing into fiber.

while coupling into the fiber is then

$$\frac{P_f}{P_s} = \left(\frac{\Omega_p}{\Omega_s}\right)\left(\frac{A_f}{A_s}\right) \qquad \text{if } A_f < A_s$$

$$= \left(\frac{\Omega_p}{\Omega_s}\right) \qquad \text{if } A_s < A_f \tag{1.5.12}$$

Thus the coupling loss may have an area mismatch term and an angle mismatch term. The area mismatch term will depend on the relative areas of the source and fiber core, with the loss contribution being most severe for the narrow-core fibers. If the light source produces unfocused Lambertian light in Eq. (1.5.12), then the angle coupling loss becomes

$$\frac{\Omega_p}{\Omega_s} = \frac{2\pi(1 - \cos\theta_p)}{2\pi} \simeq \frac{\theta_p^2}{2} \tag{1.5.13}$$

Thus the angle coupling loss is related to the square of the numerical aperture, and will be greater for low NA fibers.

By inserting a focusing lens between the source and the fiber (Fig. 1.18*b*), the source light can be focused into smaller angles Ω_s in Eq. (1.5.12), which improves the coupling efficiency. For coupling into single-mode fibers it is almost mandatory that some type of field focusing be used for power insertion. Various types of lenses and index matching fluids (which provide some degree of focusing) have been designed in pigtail formats to permit reduced fiber coupling losses.

The propagating fields within the fiber will be attenuated due to the absorption and scattering of the glass impurities, with the scattering effect decreasing at the higher wavelengths. Figure 1.19 shows a typical plot of the overall attenuation in decibels per kilometer of a typical single-mode fiber as a function of wavelength. Minimal attenuation windows occur in the vicinity of wavelengths of approximately 0.8, 1.3, and 1.5 μm, where most of the fiber systems tend to operate. The latter two bands indicate that propagation losses less than 1 dB/km are possible.

As more modes are produced in the fiber, the power in the fiber is distributed over these modes with the outermost modes having higher attenuation factors. As a result, when there are many modes propagating, the attenuation effects tend to be averaged over all modes. The multimode fibers, therefore have an attenuation loss higher than that for single-mode fibers and tend to be somewhat constant with wavelength, as indicated in Figure 1.19.

The total propagation loss in a fiber is therefore directly related to its length, and can be obtained from appropriate curves similar to Figure 1.19. The fiber attenuation α_f in decibels/length is read off at the proper wavelength, and, knowing the fiber length Z, the total attenuation in decibels is $\alpha_f Z$. If P_f is the source power coupled into the fiber in Eq. (1.5.11), the fiber output power then follows as

$$P_r = P_f 10^{-\alpha_f Z/10} = P_f e^{-0.23\alpha_f Z} \tag{1.5.14}$$

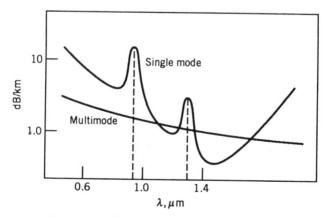

Figure 1.19. Fiber attenuation vs. wavelength.

Thus, fiberoptic power flow can be determined in this manner. Additional losses may have to be included for fiber splicings, cable interconnections, and fiber bending.

Other effects in fiber propagation, such as dispersion and filtering, will be considered in Chapter 7.

1.6 FIELD EXPANSIONS

In the analysis of communication systems, it is often convenient to use orthogonal expansions of signals to formulate a more general procedure, and sometimes to gain further insight into results [10]. Orthogonal expansions produce mathematically valid waveform representations that permit signals to be treated in pieces (multiple dimensions) rather than as a single entity. Likewise, in optical communications it is convenient to represent optical fields in terms of similar orthogonal expansions as an alternative to field description.

Consider again a general complex optical field defined over an area A as

$$f(t, \mathbf{r}) = a(t, \mathbf{r})e^{j\omega_o t} \qquad \mathbf{r} \in A \tag{1.6.1}$$

where we have denoted $a(t, \mathbf{r})$ as the complex envelope and ω_o as the optical source frequency. Let us now consider a spatial orthogonal expansion of this complex envelope as

$$a(t, \mathbf{r}) = \sum_{i=1}^{\infty} a_i(t)\Phi_i(\mathbf{r}) \tag{1.6.2}$$

where $\{\Phi_i(\mathbf{r})\}$ are a set of complex two-dimensional orthogonal space functions defined over A. That is,

$$\int_A \Phi_i(\mathbf{r})\Phi_j^*(\mathbf{r}) \, d\mathbf{r} = 0 \qquad \text{if } i \neq j \tag{1.6.3}$$

with the asterisk denoting complex conjugates. The corresponding expansions functions are given by

$$a_i(t) = \frac{\displaystyle\int_A a(t, \mathbf{r})\Phi_i^*(\mathbf{r}) \, d\mathbf{r}}{\displaystyle\int_A |\Phi_i(\mathbf{r})|^2 \, d\mathbf{r}} \tag{1.6.4}$$

That is, each $a_i(t)$ is the complex time function corresponding to the spatial function $\Phi_i(\mathbf{r})$. Equation (1.6.2) corresponds to the orthogonal expansion of the field envelope into the spatial orthogonal set $\{\Phi_i(\mathbf{r})\}$. Combining Eqs. (1.6.1) and (1.6.2) produces an overall field description defined over A as

$$f(t, \mathbf{r}) = \sum_{i=1}^{\infty} a_i(t)e^{j\omega_o t}\Phi_i(\mathbf{r}) \qquad (1.6.5)$$

The equality in Eq. (1.6.5) is in a squared integrable sense, and the convergence of the sum on the right to the field function on the left requires only a bounded energy constraint on the radiation field [10]. The advantage of an expansion of this form is primarily for mathematical convenience, although such expansions often yield physical insight into the related optical processing. It is natural to consider the functions $\{\Phi_i(\mathbf{r})\}$ as defining the spatial "modes" of the field over the area A. The time functions $\{a_i(t)\}$ then become the modal functions describing the temporal process in each mode. Because different orthogonal sets may be available, mode descriptions of a given field are not necessarily unique, each corresponding to a different expansion set.

For the fiber link, where the area A can represent the cross-sectional area of the fiber, the orthogonal expansion modes $\Phi_i(\mathbf{r})$ appear naturally as the orthogonal field modes that propagate in the fiber as the transverse E-fields, each satisfying the necessary boundary conditions. Hence the fiber modes discussed in Section 1.5 become the mathematical modes in Eq. (1.6.5), and the mode functions $a_i(t)$ are the time variations of these modes described over the area A. Thus the orthogonal expansion in Eq. (1.6.5) can represent the internal fiber field at any point along the fiber. In particular, it can be used to represent the field at the fiber output.

For an area collecting an arriving optical field in a space system, the concept of a mode is no longer as obvious. As an example, consider the complex function

$$\Phi_i(\mathbf{r}) = e^{-j(\mathbf{z}_i \cdot \mathbf{r})} \quad \mathbf{r} \in A \qquad (1.6.6)$$

corresponding in Eq. (1.3.9) to a constant amplitude plane wave arriving from direction $\mathbf{z}_i$. Now consider two such functions with arrival vectors $\mathbf{z}_i$ and $\mathbf{z}_q$, and assume a square area A of dimension $l \times l$, as shown in Figure 1.20. The spatial integral of these functions become

$$\int_A \Phi_i(\mathbf{r})\Phi_q^*(\mathbf{r}) \, d\mathbf{r} = \int_{-l/2}^{l/2} \int_{-l/2}^{l/2} e^{-j[\mathbf{r}\cdot(\mathbf{z}_i - \mathbf{z}_q)]} \, dx \, dy$$
$$= \iint_{-l/2}^{l/2} e^{-j(x\Delta z_x + y\Delta z)} \, dx \, dy \qquad (1.6.7)$$

where we have used $(\mathbf{z}_i - \mathbf{z}_q) = (\Delta z_x; \Delta z_y)$, with Δz_x and Δz_y the x and y components of the difference vector. Letting $(\theta_{ix}, \theta_{iy})$ be the subangles (azimuth

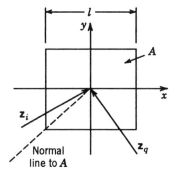

Figure 1.20. Receiver area geometry with impinging ray lines.

and elevation) of each vector z_i, we can use Eq. (1.3.10) to write

$$\Delta z_y = \left(\frac{2\pi}{\lambda}\right)(\sin\theta_{iy} - \sin\theta_{qy})$$

$$\Delta z_x = \left(\frac{2\pi}{\lambda}\right)(\cos\theta_{iy}\sin\theta_{ix} - \cos\theta_{qy}\sin\theta_{qx})$$

(1.6.8)

Because

$$\int_{-l/2}^{l/2} e^{-j\gamma x}\,dx = [l\,\sin(\gamma l/2)]/(\gamma l/2)$$

(1.6.9)

the integral in Eq. (1.6.7) is then

$$l^2\left[\frac{\sin(l\Delta z_x/2)}{(l\Delta z_x/2)}\right]\left[\frac{\sin(l\Delta z_y/2)}{(l\Delta z_y/2)}\right]$$

$$= \begin{cases} A & \text{if } \Delta z_x = 0,\ \Delta z_y = 0 \\ 0 & \text{if } \Delta z_x = k_1(2\pi/l) \text{ or } \Delta z_y = k_2(2\pi/l) \end{cases}$$

(1.6.10)

where k_1 and k_2 are any nonzero integers. Thus the plane wave fields corresponding to arrival vectors z_i and z_q are orthogonal fields over the area A if their angles produce coordinates satisfying $\Delta z_x = 2k\pi/l$ or $\Delta z_y = 2k\pi/l$ for any nonzero integer k. That is, the vectors will be orthogonal if their arrival angles satisfy the proper conditions. To expand on this, consider the small angle assumption, as in Eqs. (1.3.12), producing the requirement $\Delta z_y \simeq (2\pi/\lambda)(\theta_{iy} - \theta_{qy})$ and $\Delta z_x = (2\pi/\lambda)(\theta_{ix} - \theta_{qx})$. Under this assumption, the vectors z_i and z_q are orthogonal if

$$(\theta_{ix} - \theta_{qx}) \approx k(\lambda/\sqrt{A}) \quad \text{or}$$

$$(\theta_{iy} - \theta_{qy}) \approx k(\lambda/\sqrt{A})$$

(1.6.11)

Thus complex plane waves arriving close to the normal at a squared area A having arrival angles separated by multiples of $\lambda/A^{1/2}$ are spatially orthogonal functions. The set of all such arrival vectors satisfying Eq. (1.6.11) will therefore constitute an orthogonal set and can be used as the modes of the expansion. We conclude, therefore, that any field function defined over A can be expanded into sets of these arriving plane waves properly separated in angle of arrival. The expansion in Eq. (1.6.5) then corresponds to describing the field $f(t, \mathbf{r})$ as sums of plane waves, each with its own amplitude time functions in Eq. (1.6.4). We find this rather simple interpretation quite useful in later analyses. Although the result was here derived specifically for a square area, the plane wave orthogonality concept can in fact be extended to more general area shapes [11], with approximately the same conditions as in Eq. (1.6.11). We use these generalizations in later studies.

1.7   RANDOM FIELDS

In an optical system, we are often forced to deal with stochastic or random fields. Such fields arise when dealing with atmospheric or weather effects (which cause random variations to occur in the transmitted field), or in describing background or stray light that may appear in the system. In addition, the laser itself may produce random contributions (self-emission noise) to its own emitted fields. These random variations lead to statistical fluctuations in the optical field, which can only be analyzed after associating proper statistics with the field itself. We wish to develop some basic properties for describing these general random fields.

The complex envelope of a stochastic field must be considered random at each point t and $\mathbf{r}$ describing the field over a designated area. As such, random fields are completely described by their probability densities, that is, the probability densities associated with each point, or set of points, of the field. These densities are often difficult to model exactly, and often assumptions must be imposed upon these statistics. For example, a Gaussian random field is one whose probability density at any point $(t, \mathbf{r})$ is taken as a complex Gaussian variable.

Stochastic field analysis is often confined to second-order statistics associated with the field; in particular, its coherence function [12]. In this regard, the time-space (mutual) coherence function of a stochastic field $f(t, \mathbf{r})$ at points $(t_1, \mathbf{r}_1)$ and $(t_2, \mathbf{r}_2)$ is formally defined as

$$R_f(t_1, t_2, \mathbf{r}_1, \mathbf{r}_2) = \overline{f(t_1, \mathbf{r}_1)f^*(t_2, \mathbf{r}_2)} \tag{1.7.1}$$

where the overbar represents the expectation operator over the joint field densities at the points involved, and the asterisk denotes the complex conjugate. The field coherence therefore describes the degree of correlation from one point to another.

Stochastic fields are often described by their inherent coherence properties. A stochastic field is said to be *temporally stationary* if the time dependence in its coherence function depends only on the time difference $t_1 - t_2$. The field is *spatially homogeneous* if the spatial dependence in the coherence function depends only on the spatial distance $(\mathbf{r}_1 - \mathbf{r}_2)$. A field is *completely homogeneous* if it is both temporally stationary and spatially homogeneous. A stochastic field is said to be *coherence-separable* if its coherence function factors as

$$R_f(t_1, t_2, \mathbf{r}_1, \mathbf{r}_2) = R_t(t_1, t_2)R_s(\mathbf{r}_1, \mathbf{r}_2) \tag{1.7.2}$$

The factor $R_s(\mathbf{r}_1, \mathbf{r}_2)$ describes the spatial coherence variation, whereas $R_t(t_1, t_2)$ defines the temporal correlation of the field.

The average intensity of a field at a point $(t, \mathbf{r})$ is defined as the mean squared value of the field at time t and point $\mathbf{r}$. Thus

$$\overline{|f(t, \mathbf{r})|^2} = R_f(t, t, \mathbf{r}, \mathbf{r}) \tag{1.7.3}$$

and therefore can be obtained directly from the mutual coherence. When applied to stationary coherence-separable fields, the average intensity at $\mathbf{r}$ becomes

$$\begin{aligned} \bar{I}(\mathbf{r}) &= R(t, t, \mathbf{r}, \mathbf{r}) \\ &= R_t(0)R_s(\mathbf{r}, \mathbf{r}) \end{aligned} \tag{1.7.4}$$

and depends only on the spatial coherence function. The value of $\bar{I}(\mathbf{r})$ is often called the field irradiance at the point $\mathbf{r}$.

A stochastic field is space coherent over an area A if $R_s(\mathbf{r}_1, \mathbf{r}_2) = R_s(\mathbf{r}_1, \mathbf{r}_1)$ for all $\mathbf{r}_1, \mathbf{r}_2$ in A. Note that in this case only the temporal randomness is exhibited over A. In this sense, the field is identical to a plane wave, except it has its field amplitude $a(t)$ as a random process in t. Hence, random fields coherent over an area behave as random plane waves over that area. A stochastic field is space incoherent if $R_s(\mathbf{r}_1, \mathbf{r}_2) = 0$, $\mathbf{r}_1 \neq \mathbf{r}_2$. Otherwise, it is partially space coherent.

Stochastic field envelopes also have infinite series expansions into orthogonal functions as in Eq. (1.6.5). That is, we can represent a random field as

$$f(t, \mathbf{r}) = \sum_{i=1}^{\infty} a_i(t)e^{j\omega_o t}\Phi_i(\mathbf{r}) \tag{1.7.5}$$

where the component functions $\{a_i(t)\}$ are now random processes. Equation (1.7.5) describes a field whose spatial mode functions have a random time envelope variation imposed. For example, the expansion can represent randomly varying arriving offset plane waves, or random spatial modes in a fiber, at the frequency ω_o. The convergence of the sum is now a mean square [13], and requires bounded average energy in the field over the expansion area. Again, any set of complete spatially orthogonal functions can be used for the expansion.

It is common practice to assume that the individual modal time processes $\{a_i(t)\}$ are pairwise uncorrelated random processes. This assumption has an implicit condition that the spatial orthogonal functions satisfy a set of required conditions (called the *Karhunen-Loeve*, or KL, conditions; see Problem 1.18). If the random field is assumed to be a Gaussian field with uncorrelated modal functions, then the $a_i(t)$ processes are in fact independent random processes, and the field is said to have independent random modes, as defined by Eq. (1.7.5). The random analysis presented here is useful later to better understand stochastic field effects in our optical receivers.

1.8   OPTICAL AMPLIFIERS

One of the more important advances in fiber technology is in the area of optical amplifiers. As with their electronic counterparts, optical amplifiers can increase the power level of inserted optical fields, thereby overcoming fiber propagation power losses and directly improving the detection performance in a fiberoptic communication system.

Optical amplifiers amplify light levels through various types of forced stimulated emission [14, 15]. Although a detailed discussion of the various amplification mechanisms are beyond the scope of our study here, the communication characteristics of such devices are important in assessing system performance and understanding potential applications. The most basic optical amplifier is the semiconductor laser amplifier (SLA) in which optical power gain is achieved by biasing a laser cavity below threshold, and inserting the light field at one end. The internal reflections from the mirrored facets build field intensity, similar to the lasing operation described in Section 1.2. Achievable power gain values in SLA devices will depend on the lasing material and the reflection coefficients of the cavity [14].

A fiberoptic amplifier (Fig. 1.21a) uses Raman or Brillouin scattering in the fiber core to transfer energy from an external optical pump field to the inserted field. The two fields are coupled together in the fiber amplifier, and energy transfer occurs during the joint propagation. The pump field is used to pump the core ions into an excited state, and the energy is transferred to the input signal field as it propagates as a traveling wave. At the amplifier output, the pump field is removed and a higher intensity signal field appears. Gain values depend on the pumping power and on the fiber amplifier length.

Improvements in fiber amplification have occurred with the insertion of core doping to improve the conversion from pump to field. Various dopant materials can be used, but erbium has the advantage of producing maximum gain at the center wavelength of 1.55 μm, which matches the fiber loss window,

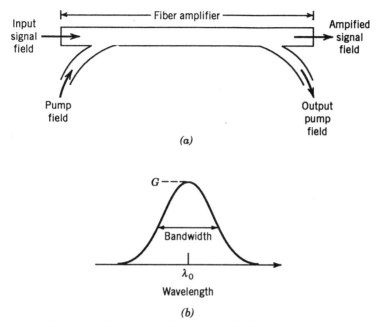

Figure 1.21. (*a*) Fiber amplifier diagram. (*b*) Gain function.

as shown in Figure 1.19. Hence the erbium-doped fiber amplifier (EDFA) has emerged as a critical fiberoptic element.

Any optical amplifier produces a wavelength-dependent power-gain transfer function, similar to electronic amplifiers. A typical gain function is sketched in Figure 1.21*b* and is characterized by its peak gain G, its optical bandwidth, and its center wavelength, all of which depend on the various design parameters of the device itself. Reported results [14–16] of present day amplifiers have gain values in the 20 to 40-dB range, with bandwidths of approximately 10 to 20 nm (approximately 10^{12} Hz). Hence optical amplifiers are extremely wideband devices with potentially high gain. However, as with any amplifier, gain levels may become saturated if the amplifier input signal power is increased. Thus peak gains may be higher when amplifying weaker optical signals than when amplifying a stronger signal.

In the process of generating the energy transfer that provides the power gain, the pump-signal coupling also generates a random spontaneous emission noise that is superimposed on the output signal field [17, 18]. This spontaneous noise field acts as a random independeent field in the fiber, similar to the random optical fields described in Section 1.7. This noise is added to all the signal field modes that are being amplified [19].

From a communication point of view, the optical amplifier can be represented by a direct multiplication of the input optical field over all its modes and the insertion of the output spontaneous noise field. If we represent the general input fiber field $f_1(t, \mathbf{r})$ as in Eqs. (1.6.5), the output field after amplification with power gain G is then

$$f_o(t, \mathbf{r}) = \sum_{i=1}^{\infty} [G^{1/2}a_i(t) + b_i(t)]e^{j\omega_o t}\Phi_i(\mathbf{r}) \qquad (1.8.1)$$

Here G is the power gain at the wavelength λ corresponding to the optical frequency ω_o, and $b_i(t)$ is the envelope of the spontaneous noise in the ith mode. Studies have shown that the complex process $b_i(t)e^{j\omega_o t}$ will have a flat power distribution over the amplifier bandwidth [18, 19], with spectral level

$$N_{sp} = (G - 1)(h f_o)n_{sp} \qquad \text{W/Hz} \qquad (1.8.2)$$

where h is Planks constant, f_o is the optical frequency in hertz, and n_{sp} is the spontaneous emission coefficient, related to the population states of the gain material [14]. Values of n_{sp} typically range from slightly above 1 to approximately 5.

It is evident from Eq. (1.8.1) that amplification has produced an output signal field whose intensity is

$$\left| \sum_{i=1}^{\infty} G^{1/2}a_i(t)e^{j\omega_o t}\Phi_i(\mathbf{r}) \right|^2 = G|f_1(t, \mathbf{r})|^2 \qquad (1.8.3)$$

and the amplifier has directly multiplied the input field intensity by the gain factor G. In later analyses, in Chapter 4 we examine the combined effects of the amplified field in Eq. (1.8.3) and the added field noise on the overall communication performance.

PROBLEMS

1.1 Communication engineers deal with bandwidth in terms of frequency. Physicists deal with bandwidth in terms of wavelength.

   (a) Determine the conversion between these two bandwidths. That is, find the frequency band Δf corresponding to a given wavelength band $\Delta\lambda$.

   (b) Use this result to sketch a plot of frequency band Δf in hertz versus optical wavelength bandwidth $\Delta\lambda$ in angstroms, assuming an optical center wavelength of $\lambda = 1\ \mu m$.

1.2 A laser cavity produces a lightwave by sustaining an optical field between two reflecting surfaces located a distance L apart, such that the fields reinforce after each reflection.

(a) Derive the relation between the field wavelength produced (there may be more than one) and the cavity length.

(b) Relate the change in field frequency to a change in cavity length.

1.3   A laser cavity 1-m long is to provide light at wavelength $\lambda = 0.5\,\mu m$.

(a) What is the nearest wavelength to $0.5\,\mu m$ at which the laser can also produce light?

(b) What is the required wavelength bandwidth of the lasing material propagation gain to guarantee light at only $0.5\,\mu m$?

1.4   An LED emits $10\,mW$ of power when injected with $2.5\,mA$ of current from a 5-V bias voltage. What is the power efficiency (power out/power in) of this LED?

1.5   A laser diode has the power characteristic shown in Figure 1.8a. Write an expression for the fraction of the output average power P_o that can be linearly modulated (equal positive and negative variations) in terms of P_o, the threshold power, and the saturation power level of the device.

1.6   A laser with power P_s transmits with a circular symmetric Gaussian beam pattern such that the source power transmitted at angle θ off boresight is given by

$$P(\theta) = \frac{P_s \pi}{\theta_b^2}\, e^{-(\theta/\theta_b)^2}$$

(a) What is the effective antenna gain on boresight $(\theta = 0)$?

(b) How much is the field intensity reduced from boresight intensity at transmission angle $\theta = \theta_b$? Express in decibels.

(c) Repeat for $\theta = n\theta_b$.

1.7   Assume all optical power is confined to the diffraction-limited beamwidth of the transmitter. Approximately how large must a receiver lens diameter be to collect all the source power from a transmitting lens of diameter d meters located L meters away?

1.8   A 10-mW laser transmits a red light beam through a 3-in. lens to a circular receiving lens having a 6-in. diameter located 10^7 m away.

(a) In free space how much power will fall on the receiver lens?

(b) Recompute the receiver lens power if the laser is located only 10^5 m away.

1.9   A stationary satellite located 22,000 miles above Earth is to transmit a green light beam to cover a 1-mile radial distance on Earth. What is the approximate transmitting lens size needed on the satellite, and what is its associated antenna gain?

1.10   An Earth–Moon link is to communicate at $10.6\,\mu m$. The Earth beam is to have a solid angle just encompassing the Moon. The Moon beam must illuminate the United States (3000 miles diameter).

(a) Design the transmitting and receiving lens that satisfy these specifications.

(b) What is the overall antenna gain (transmit + receive) of the system.

1.11   Determine the propagation loss L_p in Eq. (1.4.9) of an optical field at $1\,\mu m$ when transmitted over a 22,000-mile path. Express in decibels.

1.12   A fiber uses a core diameter of 1 mm, an index of 1.5, and a cladding with index 1.485. For an optical wavelength of $0.6\,\mu m$, find

(a) the maximum propagation angle,

(b) the number of propagating orthogonal modes, and

(c) how small the core diameter should be for a single propagating mode.

(d) Repeat (a), (b), and (c) with a cladding index of 1.0.

1.13   A 1-mW Lambertian laser source is to be coupled to an equal diameter core fiber. Two fibers are available: (a) fiber with $NA = 0.1$ and a loss factor of $3\,dB/km$, or (b) a fiber with $NA = 0.4$ and a loss of $50\,dB/km$. For each fiber, estimate the power at the output for a 10-m fiber length and a 10-km fiber length.

1.14   A fiber has a glass core ($n_1 = 1.5$), an NA of 0.2, a diameter of $100\,\mu m$, and a loss coefficient of $2\,dB/km$. It is butted up against a uniformly emitting source of diameter 1 cm, power output 1 mW, and a total emitting angle of 30 degrees.

(a) Determine the area mismatch loss.

(b) Determine the angle mismatch loss.

(c) If the fiber is 2 km long, what is the power at the fiber output?

1.15   Assume arriving orthogonal plane waves from offset angles satisfying Eq. (1.6.11) define field modes for a collecting area A. How many signal modes are contained in an arriving field from an optical source that extends over an area A_s located Z units away? *Hint*: assume the source area can be modeled as an array of point sources distributed over the area A_s.

1.16   Consider the set of orthogonal plane waves at wavelength λ defined over a square area $l \times l$. Now consider a general signal field from a source z-units away that produces a field expansion containing Q plane wave components. Estimate the effective transmitting area of the source as seen by the receiver.

1.17 Show that if the real and imaginary parts of a zero mean complex random field $f(t, \mathbf{r})$ are each uncorrelated fields, then its mutual coherence function is always real.

1.18 Show that the complex mode processes $a_i(t)$ and $a_j(t)$ generated from an orthogonal spatial expansion of a zero mean random field are uncorrelated time processes if the orthogonal function set satisfies the KL condition

$$\int_A R_f(t_1, t_2, \mathbf{r}_1, \mathbf{r}_2)\Phi_i(\mathbf{r}_2) \, d\mathbf{r}_2 = \gamma(t_1, t_2)\Phi_i(\mathbf{r}_1)$$

for some deterministic function set $\gamma(t_1, t_2)$.

1.19 **(a)** Show that a single plane wave field defined over an area A with a random envelope function $a(t)$ is always a coherence-separable field over this area.

   (b) Find the condition required for an arbitrary random field having the general orthogonal expansion in Eqs (1.7.5), with $\{a_i(t)\}$ as uncorrelated random processes, to be coherence separable.

REFERENCES

1. R. Gagliardi, *Introduction to Communication Engineering*, 2nd ed. Wiley, New York, 1988.
2. B. Lengyel, *Lasers*, 2nd ed., Wiley, New York, 1971.
3. B. Wherrett, *Laser Advances and Applications*, Wiley, New York, 1980.
4. A. Yariv, *Quantum Electronics*, 2nd ed., Wiley, New York, 1975.
5. M. Born and E. Wolf, *Principles of Optics*, 4th ed., Pergammon Press, London, 1970.
6. J. Kraus, *Antennas*, McGraw Hill, New York, 1950.
7. S. Karp, R. Gagliardi, S. Moran, and L. Stotts. *Optical Channels*, Plenum Press, New York, 1988.
8. J. Palais, *Fiber Optic Communications*, 2nd ed., Prentice Hall, Englewood Cliffs, NJ, 1988.
9. H. Killen, *Fiber Optic Communications*, Prentice Hall, Englewood Cliffs, NJ, 1991.
10. R. Courant and D. Hilbert, *Methods of Mathematical Physics*, Vol. 1, Wiley, New York, 1953.
11. C. Helstrom, Mode decomposition in aperture fields, *J. Opt. Soc. Am.*, 60(4), April, 1970.
12. E. O'Niell, *Introduction to Statistical Optics*, Addison-Wesley, Reading, MA, 1967.
13. R. Gagliardi and S. Karp, *Optical Communications*, 1st ed., Wiley, New York, 1978. Appendix A.
14. G. Agrawal, *Fiberoptic Communication Systems*, McGraw Hill, New York, 1992. Chapter 8.

15. Y. Yamamoto and T. Mukai, Fundamentals of optical amplifiers, *Opt. Quant. Electron.*, 27, pp. S1-S14, May 1989).

16. N. Olsson, Lightwave systems with optical amplifiers, *J. Lightwave Technol.*, 9, 505–513, March 1989.

17. W. Louisell, *Radiation and Noise in Quantum Electronics*, McGraw Hill, New York, 1964.

18. R. Loudon, Theory of noise in linear optical amplifier chains, *IEEE J. Quant. Electron.* QE-21, 766–773, April 1985.

19. B. Saleh and M. Teich, *Fundamentals of Photonics*, Wiley, New York, 1991.

2

OPTICAL FIELD RECEPTION

In optical receivers, the incoming light field in either a space or fiber system is generally collected or focused by front end optics on to the detecting surface, as was shown in Figures 1.4. In space systems, these front ends usually involve combinations of lenses that act to image the received field onto the photodetector, thereby creating a focused field that represents the actual field being observed. In fiber systems, the fiber field is generally coupled directly to the detector surface. To determine the performance of the overall link, it is first necessary to accurately describe these received fields and the manner in which they are converted to the photodetector field. This requires a review of basic lens and focusing equations.

2.1 FIELD FOCUSING

The focusing of fields by an optical lens can be described using the diagram in Figure 2.1. The field collected at the lens input is defined in the aperture (receiver) plane, and the focused field is defined in the focal (detector) plane. The focal plane is located at distance f_c behind the aperture plane, where f_c is the *focal length* of the lens. The optical lens, physically placed in the aperture plane, redirects the incoming light field on to the focal plane, where the photodetector is located. The field produced in the focal plane is often called the diffracted field.

A well-designed receiver lens allows for Fraunhofer diffraction [1, 2] in the focal plane. Thus, if $f_r(t, \mathbf{r})$ is the received field over the aperture lens and if $f_d(t, u, v)$ is the diffracted field in the focal plane, then the two are related by

$$f_d(t, u, v) = \frac{\Gamma(u, v)}{\lambda f_c} \int_A f_r(t, x, y) \exp\left[-j\frac{2\pi}{\lambda f_c}(xu + yv)\right] dx\, dy \quad (2.1.1)$$

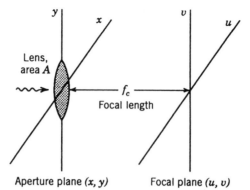

Figure 2.1. Imaging geometry of the optical receiver.

where

$$\Gamma(u, v) = \frac{1}{j} \exp\left[j \left(\frac{\pi}{\lambda f_c}\right) (u^2 + v^2) \right] \tag{2.1.2}$$

is a phase factor, $\mathbf{r} = (x, y)$ are the field coordinates in the aperture plane, and (u, v) are the field coordinates in the focal plane, as shown in Figure 2.1. Equation (2.1.1) describes the manner in which the received and focal plane fields are related. Note that $f_d(t, u, v)$ is also related to the two-dimensional Fourier transform of $f_r(t, \mathbf{r})$. That is, if we denote

$$F_r(t, \omega_1, \omega_2) = \int_A f_r(t, x, y) \exp[-j(x\omega_1 + y\omega_2)] \, dx \, dy \tag{2.1.3}$$

as the two-dimensional spatial Fourier transform of $f_r(t, x, y)$, then

$$f_d(t, u, v) = \frac{\Gamma(u, v)}{\lambda f_c} F_r\left(t, \frac{2\pi u}{\lambda f_c}, \frac{2\pi v}{\lambda f_c}\right) \tag{2.1.4}$$

Thus, diffraction patterns in optical receivers can be generated by simply resorting to transform theory. In communication analyses, this is an extremely useful result, because it means much of our optical receiver analysis reduces to straightforward linear system theory [3].

Consider a normal plane wave impinging on a receiver lens of area A. From our discussion in Section 1.3, the received field is then

$$\begin{aligned} f_r(t, x, y) &= a(t)e^{j\omega_o t} \quad (x, y) \in A, \\ &= 0 \qquad\qquad \text{elsewhere} \end{aligned} \tag{2.1.5}$$

The resulting diffraction pattern in the focal plane is then obtained directly from Eq. (2.1.1) as

$$f_d(t, u, v) = a(t)e^{j\omega_o t}\Gamma(u, v)f_{do}(u, v) \tag{2.1.6}$$

where $f_{do}(u, v)$ is the spatial integral

$$f_{do}(u, v) = \frac{1}{\lambda f_c}\int_A \exp\left[-j\left(\frac{2\pi}{\lambda f_c}\right)(xu + yv)\right]dx\,dy \tag{2.1.7}$$

Note that the diffraction pattern is simply the product of the time-varying envelope function, and the spatial function defined by the phase factor $\Gamma(u, v)$ and the two-dimensional transform $f_{do}(u, v)$. Thus the focal plane field distribution produced by an arriving normal plane wave field can be determined by evaluating the two-dimensional integral in Eq. (2.1.6). The integration, however, depends on the form of the receiver aperture area A. Consider the following examples.

Rectangular Aperture Lens   If the aperture lens area is assumed rectangular with dimensions (d, b), the limits of integration in Eq. (2.1.7) becomes $|x| \leqslant d/2$, $|y| \leqslant b/2$. The integral factors into a product of similar x and y integrals, producing the integrated result

$$
\begin{aligned}
f_{do}(u, v) &= \frac{1}{\lambda f_c}\int_{-d/2}^{d/2}\int_{-b/2}^{b/2}\exp\left[-j\left(\frac{2\pi}{\lambda f_c}\right)(xu + yv)\right]dy\,dx \\
&= \left(\frac{bd}{\lambda f_c}\right)\left[\frac{\sin(\pi du/\lambda f_c)}{(\pi dv/\lambda f_c)}\cdot\frac{\sin(\pi bv/\lambda f_c)}{(\pi dv/\lambda f_c)}\right]
\end{aligned}
\tag{2.1.8}
$$

The resulting magnitude for the u coordinate is sketched in Figure 2.2a. A similar plot exists along the v coordinate, with the two combining to produce a single "hump" centered on the origon of the (u, v) plane.

Circular Aperture Lens   If a circular lens of diameter d is used, the transform in Eq. (2.1.7) can be evaluated by first converting to polar coordinates. This yields

$$
\begin{aligned}
f_{do}(u, v) &= \left(\frac{1}{\lambda f_c}\right)2\pi\int_0^{d/2}rJ_0\left(\frac{\pi r\rho}{\lambda f_c}\right)dr \\
&= \left(\frac{\pi d^2/4}{\lambda f_c}\right)\left[\frac{2J_1(\pi d\rho/\lambda f_c)}{(\pi d\rho/\lambda f_c)}\right]
\end{aligned}
\tag{2.1.9}
$$

where $\rho = (u^2 + v^2)^{1/2}$ and $J_0(x)$ and $J_1(x)$ are Bessel functions. The diffraction pattern magnitude is sketched in Figure 2.22b as a function of the circular

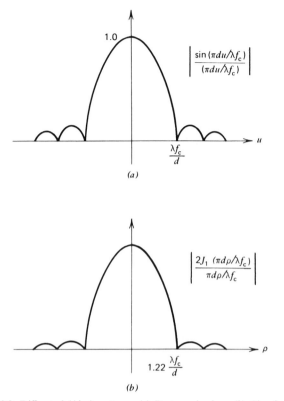

Figure 2.2. Diffracted (Airy) patterns. (*a*) Rectangular lens. (*b*) Circular lens.

radial distance ρ and corresponds to a pattern that is similar in shape to the diffracted pattern of the rectangular lens.

The diffracted fields in Figure 2.2 are the familiar Airy patterns in optical diffraction theory. Note that in both cases (2.1.8) and (2.1.9), the diffracted pattern occupies a height of approximately $A/\lambda f_c$ and a width of approximately $2\lambda f_c/d$ (i.e., the width encompassed by the largest hump) in the focal plane. Because this width is extremely small, we see that the incoming plane wave field has been imaged to an infinitesimal spot in the focal plane. In particular, the focused field pattern is many times smaller than the receiver lens size. In general, most lenses are designed with a focal length f_c about the size of the lens width d (f_c/d is the lens f-number), so that the Airy pattern occupies a width of about 2λ, and therefore is on the order of microns in size.

It is helpful to consider this lens imaging in terms of the overall diagram in Figure 2.3. A remote point source produces the modulated plane wave field in Eq. (2.1.5) that impinges on the receiver lens aperture. The lens focuses the field into the Airy spot, according to the previous equations and containing the same time varying envelope variations as the received field. One can consider

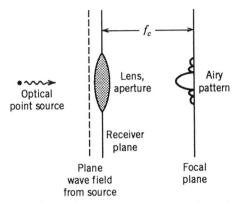

Figure 2.3. Imaged point source at the receiver.

the spot to be the focal plane image of the point source. In other words, the focal plane is reproducing a spatial pattern of the source to within the diffraction pattern of the lens. We say the point source has been *imaged* in the detector plane. Note that any detector placed in the focal plane need only collect the Airy pattern to "see" the point source and its envelope modulation. Hence focal plane detecting areas can be much smaller than the receiver lens aperture.

Suppose the plane wave arrives off the normal at the receiver with the arrival vector **z**, as in Figure 2.4a. From Eq. (1.3.9), the received field over the receiver lens is now described by

$$f_r(t, x, y) = a(t)e^{j\omega_o t} e^{-j\mathbf{z} \cdot \mathbf{r}}$$
$$= a(t)e^{j\omega_o t} \exp[-j(xz_x + yz_y)] \tag{2.1.10}$$

where z_x and z_y are the x and y coordinates of **z**. Using the small angle approximation, we can write $z_x = (2\pi/\lambda)\theta_x$, $z_y = (2\pi/\lambda)\theta_y$, where (θ_x, θ_y) are the arrival subangles of the field vector **z** relative to the normal, as was shown in Figure 1.14c. The resulting spatial diffraction pattern is now

$$f_d(t, u, v) = a(t)e^{j\omega_o t} \left(\frac{\Gamma(u, v)}{\lambda f_c}\right)$$

$$\times \int_A \exp\left[-j\left(\frac{2\pi}{\lambda}\right)(x\theta_x + y\theta_y)\right] \exp\left[-j\left(\frac{2\pi}{\lambda f_c}\right)(xu + yv)\right] dx\, dy$$

$$= a(t)e^{j\omega_o t}\Gamma(u, v)f_{do}(u + u_o, v + v_o) \tag{2.1.11}$$

where $f_{do}(u, v)$ is given in Eq. (2.1.6) and $u_o = f_c\theta_x$, $v_o = f_c\theta_y$. Thus off-angle incident plane waves generate position-shifted diffraction patterns in the focal

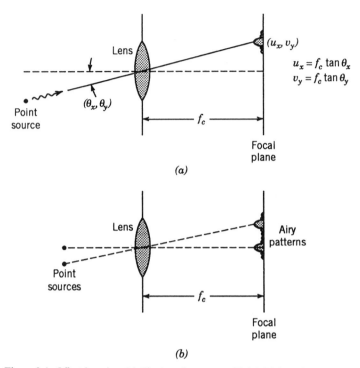

Figure 2.4. Offset imaging. (*a*) Single point source. (*b*) Multiple point sources.

plane. Note that the location of the shifted pattern can be determined by extending the line of arrival direction of the plane wave field through the center of the lens aperture to the point where it intersects the focal plane.* Because the field originates from a point source along the line, we again see that the lens is effectively imaging the point source to a shifted position in the focal plane.

Two such point sources (Fig. 2.4*b*) will each produce individual Airy patterns that will superimpose in the focal plane (from the linearity of the lens transformation). Their patterns can be resolved as long as they are sufficiently separated. Two patterns are considered unresolvable if the pattern of one is located within the Airy width of the other. Because the patterns have a width of approximately 2λ, two patterns will be unresolved if they are within λ of each other. This translates to point source fields arriving with angles separated by less than

$$\theta_{dL} \approx \frac{\lambda}{f_c} \qquad (2.1.12)$$

---

*The angles (θ_x, θ_y) are limited by the Fraunhofer equations to small offset angles for the transform theory to apply. At large offsets, the shifted Airy pattern will be distorted.

We can therefore define a solid angle of unresolvable arrival angles as

$$\Omega_{dL} \cong \frac{\pi}{4} \left(\frac{\lambda}{f_c} \right)^2 \tag{2.1.13}$$

Under the condition $f_c \cong d$, where d is a circular lens diameter,

$$\Omega_{dL} \approx \left(\frac{\pi}{4} \right)^2 \left(\frac{\lambda^2}{A} \right) \tag{2.1.14}$$

where A is the receiver aperture area. This is called the *diffraction-limited* field of view of the aperture optics and defines the solid angle in which all arriving plane wave angles will superimpose their Airy patterns so as to be indistinguishable (Fig. 2.5). Thus Ω_{dL} defines the resolution capability of the aperture optics at wavelength λ and depends only on the aperture area. Note that Ω_{dL} effectively divides the received field into separate distinguishable regions, each of which produces a distinct Airy pattern in the focal plane. The ability to distinguish at such narrow angles in Eq. (2.1.13) is of course an indication of the high-resolution capability of optical imaging.

We can relate the diffraction-limited angle in Eq. (2.1.13) to our earlier concept of orthogonal modes. Recall from Eq. (1.6.11) that arrival angles separated by approximately $\lambda/A^{1/2}$ produced orthogonal fields over A. We now see that this angle separation is roughly equivalent to the diffraction-limited angles in Eq. (2.1.12) that produced resolvability. Thus the diffraction-limited angle is the minimum angle separation for orthogonal plane waves. Conversely, arriving orthogonal plane waves produce resolvable diffraction patterns in the focal plane.

When the received field is not a simple plane wave, the resulting focused field is no longer a basic Airy pattern as before, and we must resort to two-dimensional transform theory to determine the focal plane field. Consider a general received field $f_s(t, x, y)$ defined in the aperture plane. The lens focuses the portion of this on its surface onto the focal plane. If we define the lens pupil function

$$w(x, y) = 1 \qquad (x, y) \in A$$
$$= 0 \qquad \text{elsewhere} \tag{2.1.15}$$

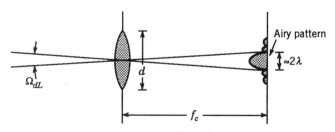

Figure 2.5. Diffraction-limited field-of-view geometry.

then the received field collected by the lens is

$$f_r(t, x, y) = f_s(t, x, y)w(x, y) \qquad (2.1.16)$$

The focal plane field is then obtained by substituting Eq. (2.1.16) into Eq. (2.1.1). However, it can also be written using the complex convolution of the transforms of each term in Eq. (2.1.16). The Fourier transform of $f_s(t, x, y)$ is a function $F_s(t, \omega_1, \omega_2)$. The transform of the aperture function $w(x, y)$ in Eq. (2.1.15) is precisely the Airy pattern $f_{do}(u, v)$ of the lens. Thus, the focal plane field is the complex field

$$f_d(t, u, v) = F_s(t, u, v) \otimes f_{do}(u, v) \qquad (2.1.17)$$

where $\otimes$ denotes spatial convolution. That is, the focal plane spatial field is determined by convolving the transform of the received field distribution with the lens Airy pattern. This means that the received field is imaged in the focal plane through the windowing effect of the Airy pattern convolution. For a received field with a widely distributed transform, $f_{do}(u, v)$ appears as a delta function in Eq. (2.1.17), and the received field transform is exactly reproduced in the focal plane. On the other hand, if the field transform is very narrow (i.e., a delta function, corresponding to an infinitely wide plane wave), the focused field appears as $f_{do}(u, v)$, and the delta function is effectively spread out into the Airy pattern by the aperture. For this reason Airy patterns are also referred to as the *point spread function* of the receiver.

The lens system in Figure 2.1 indicates a single-lens field-focusing operation at the receiver. In some receivers, the same focusing is often accomplished with folded optics in which the combinations of primary and secondary reflecting lenses are used to generate the imaged field, as shown in Figure 2.6. The

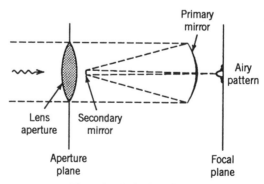

Figure 2.6. Folded optics.

focusing is accomplished in several stages, resulting in an overall assembly that can be shorter than the actual focal length. This, of course, is commonly done in telescopes and binoculars.

In this section, we have presented several basic equations from lens theory that are useful in our subsequent analysis of receivers. We choose not to digress further at this time, because the material presented is sufficient for a basic understanding of optical communication systems. The reader interested in pursuing the areas of diffraction theory, optics, and imaging is referred to any of the many representative texts in this area [4–7].

2.2 POWER DETECTION AND RECEIVER FIELD OF VIEW

In the optical receiver using lenses, the aperture field is imaged on to the focal plane, where the photodetector is placed. The photodetector, therefore, responds to the imaged field on its collecting surface. Two important parameters are the amount of power in the detector field and the receiver field of view.

Detector field power can be determined by direct application of Parceval's theorem in two-dimensional transform theory. Let $f_1(x, y) \leftrightarrow F_1(u, v)$ and $f_2(x, y) \leftrightarrow F_2(u, v)$ be two transform pairs. Then Parceval's theorem states

$$\int_{-\infty}^{\infty} \int_{-\infty}^{\infty} f_1(x, y) f_2^*(x, y) \, dx \, dy = \left(\frac{1}{2\pi}\right)^2 \int_{-\infty}^{\infty} \int_{-\infty}^{\infty} F_1(u, v) F_2^*(u, v) \, du \, dv \tag{2.2.1}$$

When $f_1 = f_2$, this reduces to

$$\int_{-\infty}^{\infty} \int_{-\infty}^{\infty} |f_1(x, y)|^2 \, dx \, dy = \left(\frac{1}{2\pi}\right)^2 \int_{-\infty}^{\infty} \int_{-\infty}^{\infty} |F_1(u, v)|^2 \, du \, dv \tag{2.2.2}$$

This can be related to the integrals of the aperture and to the focused fields of an optical lens, which are themselves transforms. In particular, it can be stated that a direct relationship exists between spatial integrals of focused fields after lens transformations. Let us apply the field power definitions in Eq. (1.3.3) with

$$\begin{bmatrix} \text{power in the focal} \\ \text{plane field at time } t \end{bmatrix} = \int_{-\infty}^{\infty} \int_{-\infty}^{\infty} |f_d(t, u, v)|^2 \, du \, dv \tag{2.2.3}$$

Then, by use of Eq. (2.1.4), the integral in Eq. (2.2.3) becomes

$$\int_{-\infty}^{\infty}\int_{-\infty}^{\infty} |f_d(t, u, v)|^2 \, du \, dv = \left(\frac{1}{\lambda f_c}\right)^2 \int_{-\infty}^{\infty}\int_{-\infty}^{\infty} \left|F_r\left(t, \frac{2\pi u}{\lambda f_c}, \frac{2\pi v}{\lambda f_c}\right)\right|^2 du \, dv$$

$$= \left(\frac{1}{2\pi}\right)^2 \int_{-\infty}^{\infty}\int_{-\infty}^{\infty} |F_r(t, \omega_1, \omega_2)|^2 \, d\omega_1 \, d\omega_2$$

$$(2.2.4)$$

where $F_r(t, \omega_1, \omega_2)$ is the inverse Fourier transform of $f_r(t, x, y)$ over the aperture area A. By applications of Eq. (2.2.2), the last integral is identical to:

$$\int_A |f_r(t, x, y)|^2 \, dx \, dy = \left[\begin{array}{c}\text{power in the field over the}\\ \text{aperture area at time } t\end{array}\right] \quad (2.2.5)$$

This means that the power collected over the focal plane from the focal plane field is identical to the power collected over the receiver aperture area from the received field. Hence, power is preserved in focusing the received field onto the focal plane. A detector placed in the plane will collect the aperture power, provided that the detector is large enough to encompass the focused field. Therefore, detector power levels can be computed directly at the receiver lens without the necessity of computing the actual diffracted field. We need only specify the lens aperture area as a receiving-collecting area, and directly apply the power flow equations developed in Section 1.4.

This focal-plane power detection is based on the fact that a focal-plane detector encompasses the entire focused field no matter the direction from which it arrives. For a given detector, we can determine exactly which directions of the arriving received field is actually detected. Consider the diagram in Figure 2.7, showing a circular detector of diameter d_d in the focal plane of a lens of circular area A and focal length f_c. Recalling our discussion of arrival angles and shifted Airy patterns in Figure 2.4, we can clearly see that we will only image on the detector surface (neglecting end effects) field arrival angles within the solid angle

$$\Omega_{fv} = \frac{\pi}{4}\left(\frac{d_d}{f_c}\right)^2$$

$$= \frac{A_d}{f_c^2} \quad (2.2.6)$$

where A_d is the detector area. The parameter Ω_{fv} defines the range of receiver arrival angles observed by the detecting surface. Thus, Ω_{fv} determines how much of the incoming light field will actually be detected, and is referred to as the receiver *field of view*. The latter is therefore the solid angle, looking out from the receiver, within which all arriving plane waves must occur in order to

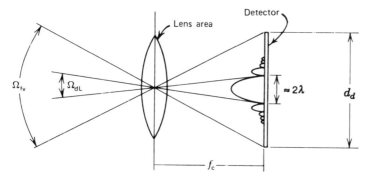

Figure 2.7. Receiver field of view and its relation to lens and detector areas.

project their diffraction pattern onto the detector. Recall that all plane waves that superimpose their diffraction patterns are in essence indistinguishable in terms of direction of arrival. Thus, the absolute minimal field of view that we can have is the diffraction-limited angle in Eq. (2.1.13). This would require a detecting surface whose area is on the order of $\pi\lambda^2$.

In practice detector areas are much larger than $\pi\lambda^2$, so that $\Omega_{fv} \gg \Omega_{dL}$. Thus optical receivers collect incoming fields over angles much wider than the diffraction angles. If we consider each Ω_{dL} as a distinguishable arrival direction, or spatial orthogonal mode, then the optical receiver will have

$$\begin{bmatrix} \text{number of received} \\ \text{field modes} \end{bmatrix} = \frac{\Omega_{fv}}{\Omega_{dL}} = \frac{(A_d/f_c^2)}{(\pi/4)^2(\lambda^2/A)}$$

$$\approx \left(\frac{4}{\pi}\right)\left(\frac{A_d}{\lambda^2}\right) \tag{2.2.7}$$

Since $A_d \gg \lambda^2$, an optical receiver collects many field modes. Even with a detector area as small as 1 cm^2, an optical receiver can have as many as 10^6 to 10^8 such modes. Contrast this with RF antennas that typically have feed horns that are on the order of the wavelength itself, producing only approximately 1 to 10 such modes. Thus optical fields have a much higher dimension in their detector characterization.

The above discussion pertains to receiving systems that use lensing receivers. In a fiber link, the fiber can be directly connected to the photodetector by basic mounting hardware, as shown in Figure 2.8. The detector is usually mounted to a circuit board, and the fiber is coupled directly to it. Because the detector area is much larger than the typical fiber core area, theoretically all the light power emerging from the fiber output will impinge on the detector area without focusing. Hence the detector power is generally equal to the fiber output power; the latter is determined by the fiber propagation analysis in Section 1.5. Mathematically, the orthogonal modes of the fiber field become the impinging orthogonal modes on the detector surface.

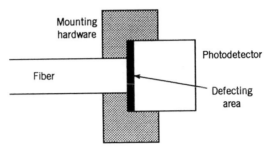

Figure 2.8. Interconnecting fibers to detectors.

2.3 DETECTOR FIELD EXPANSIONS

Earlier in Section 1.6, the notion of an orthogonal spatial expansion of an optical field over a collecting area was introduced. This led to the idea of considering fields as sums of modulated orthogonal spatial functions. It can now be shown that a similar expansion can be made for the focused detector field. As might be expected, the latter can be directly related to the received field expansion.

Let the received field $f_r(t, \mathbf{r})$ be expanded into orthonormal spatial functions as in Eq. (1.6.5),

$$f_r(t, \mathbf{r}) = \sum_{i=1}^{\infty} a_i(t) e^{j\omega_o t} \Phi_i(\mathbf{r}) \qquad \mathbf{r} \in A \qquad (2.3.1)$$

Then the focused field, via Fourier transformations, must have an associated expansion

$$f_d(t, u, v) = \sum_{i=1}^{\infty} a_i(t) e^{j\omega_o t} \phi_i(u, v) \qquad (2.3.2)$$

where $\phi_i(u, v)$ is the focused field corresponding to $\Phi_i(x, y)$. That is,

$$\phi_i(u, v) = \frac{\Gamma(u, v)}{\lambda f_c} \int_A \Phi_i(x, y) \exp\left[-j\left(\frac{2\pi}{\lambda f_c}\right)(xu + yv) \right] dx\, dy \qquad (2.3.3)$$

The focal-plane functions $\{\phi_i(u, v)\}$ are the diffracted versions of the receiver functions $\{\Phi_i(x, y)\}$ and therefore are obtained by its Fourier transform. Using Parceval's theorem in Eq. (2.2.1), we can easily show that these focal-plane functions are themselves orthogonal, since the $\{\Phi_i(x, y)\}$ are orthogonal. Hence Eq. (2.3.2) itself represents an orthogonal expansion of the focused field over the focal plane, having the same coordinate time functions as the receiver field expansion. This means the focal-plane field expansion can be derived from the received field expansion by using the same coefficients and modifying the orthogonal functions according to Eq. (2.3.3). Note that since the focal-plane orthogonal functions differ from those in the aperture plane, the associated field intensity will be spatially distributed differently in the focal plane from that in the receiver plane, even though its power (integral) has been preserved. The principal point here is that the optical field can be expanded either at the receiver plane or at the focal plane, with a straightforward transform conversion between the two. The former is obviously more convenient when discus-sing the received field, whereas the latter is better suited for analysis at the detector surface.

Consider specifically the case of the orthogonal-plane wave expansion using the functions in Eq. (1.6.6), corresponding to a set of arrival angles properly spaced. Because off-normal arriving plane waves transform to shifted Airy patterns, the resulting focal-plane expansion corresponds to separated Airy patterns over the focal plane. Each Airy pattern has a time variation identical to the plane wave from which it came, and the plane-wave field modes become Airy-pattern focal-plane modes. This permits us to consider the focal-plane field as a set of time-varying Airy patterns distributed over the plane. It is of interest to note that the preservation of orthogonality during Fourier transformation shows that properly spaced Airy patterns are themselves orthogonal.

This Airy-pattern expansion leads naturally to the concept of focal-plane arrays, a topic of significant importance in optical signal processing. A focal-plane array is an array of individual, micron-size detecting surfaces placed in the receiver focal plane (Fig. 2.9). Each such detector responds only to the field over its surface. Hence the array elements detect the focused field in pieces, effectively partitioning the field into individual pixels. Ideally, if each detector area can be made small enough to encompass a single Airy pattern, then each detector responds to a particular field mode (direction of arrival). The array is, therefore, spatially sampling the arriving receiver field and, by processing all detectors in parallel, can perform separate operations on each element of the field expansion. This affords tremendous potential in the area of

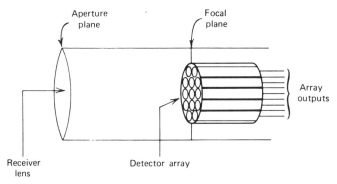

Figure 2.9. Receiver with focal plane processing.

parallel plane processing of optical images, and has application to communications, as we shall see in Chapter 4. The development of optically sensitive mosaic surfaces, composed of extremely small integrated detecting areas distributed over the surface, and arrays of charge-coupled devices (CCD), become important optical elements in this technology.

2.4   FOCUSING RANDOM FIELDS

Let us extend the concept of aperture–focal-plane focusing to the case of random arriving fields. This is of importance in communications since received optical noise is almost always present and, in many cases, the signal field itself may become random. Consider again the system in Figure 2.1, except assume that the received field $f_r(t, \mathbf{r})$ is random. We know the resulting imaged field will also be random and is given by Eq. (2.1.11), which we rewrite as

$$f_d(t, \mathbf{q}) = \frac{\Gamma(u, v)}{\lambda f_c} \int\int\limits_{-\infty}^{\infty} w(\mathbf{r}) f_r(t, \mathbf{r}) e^{-j(\mathbf{q} \cdot \mathbf{r})} \, d\mathbf{r} \tag{2.4.1}$$

where $\mathbf{r} = (x, y)$ and $\mathbf{q} = (2\pi/\lambda)(u, v)$, and $w(\mathbf{r})$ represents the lens pupil function in Eq. (2.1.15),

$$w(\mathbf{r}) = 1 \qquad \text{for } \mathbf{r} \in \mathbf{A}$$
$$= 0 \qquad \text{elsewhere} \tag{2.4.2}$$

The random field $f_d(t, \mathbf{q})$ will have a mutual coherence function

$$R_{f_d}(t_1, t_2, \mathbf{q}_1, \mathbf{q}_2) = \overline{f_d(t_1, \mathbf{q}_1) f_d^*(t_2, \mathbf{q}_2)} \tag{2.4.3}$$

Since $\Gamma(u, v)\Gamma^*(u, v) = 1$, this becomes

$$R_{f_d}(t_1, t_2, \mathbf{q}_1, \mathbf{q}_2) = \left(\frac{1}{\lambda f_c}\right)^2 \int\int\int\int\limits_{-\infty}^{\infty} w(\mathbf{r}_1) w(\mathbf{r}_2) \overline{f_r(t_1, \mathbf{r}_1) f_r^*(t_2, \mathbf{r}_2)}$$

$$\times \exp^{-j(\mathbf{r}_1 \cdot \mathbf{q}_1)} e^{j(\mathbf{r}_2 \cdot \mathbf{q}_2)} \, d\mathbf{r}_1 \, d\mathbf{r}_2$$

$$= \left(\frac{1}{\lambda f_c}\right)^2 \int\int\int\int\limits_{-\infty}^{\infty} w(\mathbf{r}_1) w(\mathbf{r}_2) R_{f_r}(t_1, t_2, \mathbf{r}_1, \mathbf{r}_2) e^{j(\mathbf{r}_2 \cdot \mathbf{q}_2 - \mathbf{r}_1 \cdot \mathbf{q}_1)} \, d\mathbf{r}_1 \, d\mathbf{r}_2$$

$$\tag{2.4.4}$$

where R_{f_r} denotes the mutual coherence function of the receiver field. Thus the received field coherence function is converted to the corresponding focal plane coherence function by Eq. (2.4.4).

Let us concentrate on the average field intensity at a time t and point q in the focal plane, defined by the irradiance

$$\bar{I}_d(\mathbf{q}) = \overline{f_d(t, \mathbf{q}) f_d^*(t, \mathbf{q})}$$
$$= R_{f_d}(t, t, \mathbf{q}, \mathbf{q}) \tag{2.4.5}$$

From Eq. (2.4.4),

$$\bar{I}_d(\mathbf{q}) = (\lambda f_c)^{-2} \iiiint_{-\infty}^{\infty} w(\mathbf{r}_1) w(\mathbf{r}_2) R_{f_r}(t, t, \mathbf{r}_1, \mathbf{r}_2) e^{j[\mathbf{q} \cdot (\mathbf{r}_2 - \mathbf{r}_1)]} \, d\mathbf{r}_1 \, d\mathbf{r}_2 \tag{2.4.6}$$

If the received field is coherence separable and homogeneous,

$$R_{f_r}(t, t, \mathbf{r}_1, \mathbf{r}_2) = R_t(0) R_s(\mathbf{r}_1 - \mathbf{r}_2) \tag{2.4.7}$$

and Eq. (2.4.6) becomes

$$\bar{I}_d(\mathbf{q}) = \frac{R_t(0)}{(\lambda f_c)^2} \iiiint_{-\infty}^{\infty} w(\mathbf{r}_1) w(\mathbf{r}_2) R_s(\mathbf{r}_1 - \mathbf{r}_2) e^{j[\mathbf{q} \cdot (\mathbf{r}_2 - \mathbf{r}_1)]} \, d\mathbf{r}_1 \, d\mathbf{r}_2 \tag{2.4.8}$$

Substituting $\boldsymbol{\rho} = (\mathbf{r}_1 - \mathbf{r}_2)$ allows us to rewrite this as

$$\bar{I}_d(\mathbf{q}) = \frac{R_t(0)}{(\lambda f_c)^2} \iint_{-\infty}^{\infty} R_s(\boldsymbol{\rho}) H(\boldsymbol{\rho}) e^{-j(\mathbf{q} \cdot \boldsymbol{\rho})} \, d\boldsymbol{\rho} \tag{2.4.9}$$

where

$$H(\boldsymbol{\rho}) = \iint_{-\infty}^{\infty} w(\mathbf{r}_2) w(\mathbf{r}_2 + \boldsymbol{\rho}) \, d\mathbf{r}_2 \tag{2.4.10}$$

The function $H(\boldsymbol{\rho})$ is the spatial convolution of the aperture function in Eq. (2.4.2) with itself, and is called the *optical transfer function* (OTF) of the receiver. Hence the average field intensity in the focal plane is distributed according to the scaled integral in Eq. (2.4.9). This integral can be recognized as the Fourier transform of the product of the two-dimensional functions $R_s(\boldsymbol{\rho})$

and the OTF $H(\mathbf{\rho})$. This means that $\bar{I}_d(\mathbf{q})$ involves the convolutional of their individual transforms. That is, we can symbolically write

$$\bar{I}_d(\mathbf{q}) = [R_t(0)/(\lambda f)^2][\mathscr{F}[R_s(\mathbf{\rho})] \otimes \mathscr{F}[H(\mathbf{\rho})]] \qquad (2.4.11)$$

where $\mathscr{F}$ denotes the two-dimensional Fourier transform, and $\otimes$ denotes convolution. For the idealized aperture function in Eq. (2.4.2), the resulting OTF appears as in Figure 2.10 and is effectively spread over a range equal to the aperture diameter. Hence, its Fourier transform in Eq. (2.4.11) is a relatively narrow spatial function in the transform domain. This means $\bar{I}_d(\mathbf{q})$ will have a spatial variation approximately equivalent to the Fourier transform of the spatial portion of the received coherence function. That is, average intensity in the focal plane is distributed approximately according to the transform of the input spatial coherence function. This produces a transformation similar to that sketched in Figure 2.11. In particular, we see that narrow coherence functions (received random fields with small coherence areas) will produce a widely distributed average intensity in the focal plane. On the other hand, fields with wide coherence areas produce focal-plane fields with narrow spatial intensity distributions. We emphasize that these comments refer to average intensities only (the actual imaged field is randomly varying in time and space). Because average intensity is directly related to average power, these intensity distributions are helpful in indicating how average field power is distributed in the focal plane, when random fields impinge on the input lens.

Earlier it was stated that the focal-plane field effectively images the source. When the focal-plane field intensity is spread over a wide area, it appears as if the receiver is collecting light from a widely distributed source field. Hence random receiver fields with narrow spatial coherence functions appear to be arising from a source that is distributed over a wide angle. This gives an interesting interpretation to the effect of a random medium on a propagating field from a point source observed by a receiver. In free space, the source

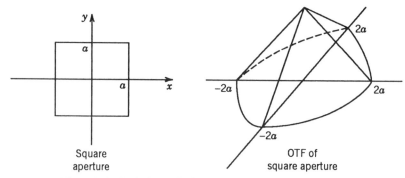

Figure 2.10. Optical transfer function (OTF) for square apertures.

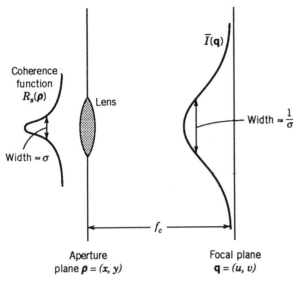

Figure 2.11. Example of coherence function transformation with random fields.

produces a plane wave that is imaged to a spot in the focal plane. If the medium randomly "breaks up" the source plane wave, producing a wavefront with small coherence areas at the receiver, the imaged intensity spreads, and is no longer a single spot. It is said that the medium has "spread" the source from a point to an extended area source. Note that this is a fictitious source area (the transmitting source is still a point) and its size can only be defined through the field coherence function. We make use of these notions in Chapter 4.

Background or ambient random light fields produced by diffuse light arriving at an optical receiver from all directions tend to have narrow coherence widths. As a result, background light produces an average intensity distributed widely over the focal plane. Over most detecting areas, this will appear as a spatially flat intensity distribution, uniformly spread over the detector surface, and effectively adds a spatial noise flow to the focused source field. In Section 2.6, this model is used to directly estimate the incident background light power at a space optics receiver.

While the previous discussion refers to intensity distributions, the actual focal field statistics (field probability densities) can best be studied via orthogonal expansions. Assume the received random field is represented by the orthogonal expansion, as in Eq. (2.3.1)

$$f_r(t, \mathbf{r}) = \sum_{i=1}^{\infty} a_i(t) e^{j\omega_o t} \Phi_i(\mathbf{r}) \tag{2.4.12}$$

where now the $\{a_i(t)\}$ corresponds to the random mode processes. As stated in Eq. (2.3.2), the imaged field will likewise be a random field, with a similar expansion, in which the new mode functions will each have the same random time processes superimposed. Thus the temporal statistics of the input field are preserved in the focal plane. In particular, Gaussian fields (fields in which $a_i(t)$ is a complex Gaussian random process) will produce Gaussian fields in the focal plane.

Note that if Eq. (2.4.12) corresponds to the expansion of a random field in which the $\{a_i(t)\}$ are uncorrelated processes, we automatically have uncorrelated processes in the focal plane, because the mode time functions are preserved. When the expansion functions are plane waves, the Airy patterns produced in the focal plane now have randomly varying time functions that are in fact uncorrelated. When the fields are also Gaussian, each Airy pattern mode varies randomly and independently of all others. This type of argument tends to justify the use of statistically independent field modes in the detector plan, an assumption that greatly simplifies detector analysis.

In fibers, the orthogonal modes often become random from self-emission noise of the laser itself or from undesired mode coupling during fiber propagation that appears as crosstalk noise. These effects produce an inherent additive noise process to the fiber mode signal component. The modal time functions $\{a_i(t)\}$ within the fiber now evolve as a summation of desired signal modulation plus additive random noise appearing on each mode component. This type of signal-plus-noise model is often used to characterize fiber fields.

2.5 OPTICAL FILTERS

In most space communication receivers, optical filters are used in the front end during the light focusing, as was indicated in Figure 1.4. An optical filter is a material or element placed in the path of the optical beam to control the transmissivity of various wavelengths. As with any front end filter used in communications, its objective is to remove as much unwanted light as possible during the reception of the source light field. Thus optical filters are designed with properly selected passbands (high transmissivity) and reject bands (low transmissivity) with respect to optical wavelengths. In fiber systems, receiver filters are used primarily in frequency-division multiplexed systems, where multiple signals are separated over a wavelength band and transmitted simultaneously over the fiber. The filters are used to remove the undesired portions of the signal spectrum, as well as the noise. These systems are considered in Chapter 8.

The basic types of optical filters are the absorption filters, interference filters, birefringent filters, and atomic resonance filters [8]. Absorption filters are generally coatings of material connected to or painted on lens or photodetector surfaces to absorb (or scatter or reflect) incident light at certain wavelengths while passing others. The filtering characteristics depend on the type of material and thickness of coating. Figure 2.12 shows some basic filter materials and their transmissivity. Note that the filter bands tend to be somewhat wide,

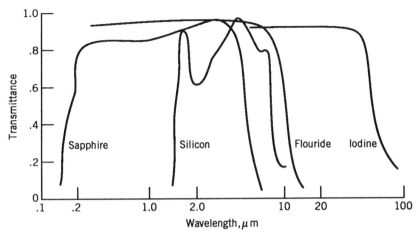

Figure 2.12. Transmissivity of several photomissive materials used in absorption filters.

with bandwidths on the order of tenths of microns (approximately 1000 Å). Thus these filters are primarily for wideband filtering, although a cascade of such filters can sometimes be combined with overlapping passbands to achieve narrower filtering.

Interference filters are devices placed in front of the detector into which the focused field is propagated. Interference filters make use of reflected wavelengths reinforcing or cancelling to provide selective filtering, as shown in the sketch in Figure 2.13a. The device could be designed in the form of an optical cavity that only sustains internal reflections at certain wavelengths, while producing out-of-phase cancellation at other wavelengths. A popular cavity filter is the Fabret-Perot device using reflecting metallic mirrors to form the cavity. Such devices have the ability to be tuned over a wavelength range for centering the passband.

Interference filters can also be designed as multilayer dielectric filters, composed of layers of materials of different indices deposited on parallel glass substrates. With the indices properly set, reflected fields from the substrates recombine to reinforce at desired wavelengths while interfering at other wavelengths. The overall effect is to produce light at the output at the desired wavelength, with relatively high attenuation at unwanted wavelengths, producing the desired passband. Because interference filters use phase shift combining, relative sharp filtering can be achieved, especially when multiple sections similar to Figure 2.13a can be cascaded. These filters can have bandwidths on the order of 1 to 10 Å, as shown in some typical filter curves in Figure 2.13b.

Sharply tuned interference filters however are designed for focused light entering the filter along the boresight axis only, where the proper phase shifts can be maintained by the material thickness. Light beams entering off-axis travel longer paths in the cavity or dialectric and, therefore, have shifted phases that reduce the wavelength reinforcement. As a result, off-axis beams have

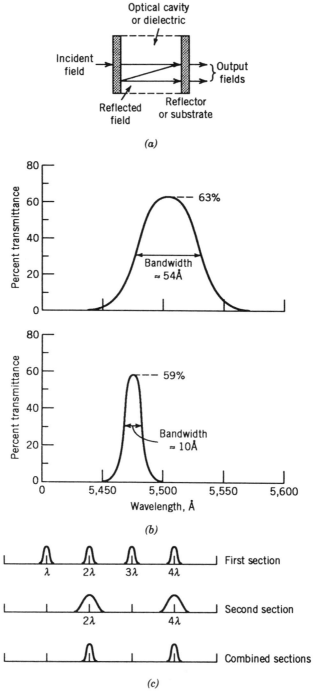

Figure 2.13. Optical filters. (*a*) Single section. (*b*) Interference filter function. (*c*) Combined sections.

reduced peak gains and wider filter bandwidths than boresight beams. Interference filters, therefore, have defined input field-of-view angles into which the light must be inserted to achieve the design bandwidth. This requires careful control of beam alignment.

A single section of interference filter in Figure 2.13a will always have a comb-filter response, since multiple passbands will exist at all the harmonics of the desired wavelength, as shown in Figure 2.13c. By cascading sections of different thicknesses tuned to specific harmonics of the first section, nearby bands can be eliminated, as shown. The result is a tuned filter with larger separation between adjacent bands than the original section. In general, increasing the number of sections of an interference filter also increases the power loss (attenuation) of the desired band.

Another type of filter is the birefringent, or *Lyot*, filter. These filters use polarization shifts occurring in birefringent crystals as light waves pass through. The amount of polarization shift depends on the field wavelength and crystal width. If the crystal is followed by a polarizer (only light polarized in one direction is allowed to propagate) a passband is created for only those wavelengths that are properly polarized, producing a tuned filter. Lyot filters can achieve bandwidths as low as 1 to 10 Å, similar to interference filters.

Atomic resonance filters use the atomic line widths of certain materials to resonate with input light fields to produce passbands on the order of the line width. As a result, bandwidths of fractions of an angstrom are theoretically possible. However, these passbands are only available at specific optical wavelengths associated with the atomic line. Because of the narrowness of the filter bands, atomic filters require careful control of the input optical wavelength when used as a communication filter. Also, atomic filters have low transmissivities, usually only 20 to 30 percent.

2.6 BACKGROUND RADIATION

In addition to the desired source power, a receiving space system viewing a lighted background also collects undesirable strong background radiation falling within the spatial and frequency ranges of the detector. The collected background radiation is processed along with the desired signal background, and presents a basic degradation to the overall system performance. Of particular importance is the actual amount of background radiation power that is collected. The determination of the power, however, requires an accurate model for the source of this radiation. The accepted model is to consider the background to be generated from uniformly radiating sources. These sources divide into two basic types: (1) the extended background, assumed to occupy the entire background, and therefore is present in any receiver field of view, and (2) discrete or point sources that are more localized but are more intense and may or may not be in the receiver field of view. In a space system, the sky is the primary extended background, and the localized sources may correspond to stars, planets, moon, sun, and so on. In an indoor environment, reflecting walls become the extended background, and localized sources may be room lights, reflecting surfaces, and the like. In this section, we review the analysis of general background optical noise.

Extended background radiators are most often described by their *spectral radiance function*, $W(\lambda)$, defined as the power radiated at wavelength λ per cycle of bandwidth into a unit solid angle per unit of source area. If the receiving lens occupies an area A at distance Z from the source (Fig. 2.14), it represents a solid angle, measured from the source, of approximately $A/Z^2 sr$. If the radiation source has area A_s, then the total power collected depends on the portion of source area lying within the receiver field of view, Ω_{fv}. Thus the background power collected at the receiver in a wavelength bandwidth of $\Delta\lambda$ around a wavelength λ is

$$
\begin{aligned}
P_b &= W(\lambda)(\Delta\lambda)(\Omega_{fv}Z^2)(A/Z^2) \quad &\text{if } A_s > \Omega_{fv}Z^2 \\
&= W(\lambda)(\Delta\lambda)A_s(A/Z^2) \quad &\text{if } A_s < \Omega_{fv}Z^2
\end{aligned}
\tag{2.6.1}
$$

Defining Ω_s as the source solid angle viewed from the receiver, then $\Omega_s \cong A_s/Z^2$, and we can rewrite Eq. (2.6.1) as

$$
\begin{aligned}
P_b &= W(\lambda)(\Delta\lambda)\Omega_{fv}A \quad &\text{if } \Omega_{fv} < \Omega_s \tag{2.6.2a} \\
&= W(\lambda)(\Delta\lambda)\Omega_s A \quad &\text{if } \Omega_s < \Omega_{fv} \tag{2.6.2b}
\end{aligned}
$$

Thus, if the background source is extended to encompass the receiver field of view, the background power is given by Eq. (2.6.2a) and depends only on the

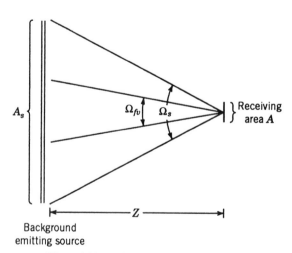

Background emitting source

Figure 2.14. Background noise source model.

receiver area, field-of-view, and bandwidth. In particular, note that P_b does not depend on the range Z. We also see that P_b in Eq. (2.6.2a) increases linearly with Ω_{fv} until the source is entirely encompassed, at which point Eq. (2.6.2b) is valid, and P_b remains constant with Ω_{fv}.

For a localized point source producing noise fields for which Eq. (2.6.2b) is true, it is convenient to define the source *irradiance* at wavelength λ as the product

$$\mathscr{W}(\lambda) = W(\lambda)\Omega_s \qquad (2.6.3)$$

Power can then be computed directly from $\mathscr{W}(\lambda)$ as

$$P_b = \mathscr{W}(\lambda)(\Delta\lambda)A \qquad \text{if } \Omega_s < \Omega_{fv} \qquad (2.6.4)$$

without having to specify the source angle Ω_s. Thus background power levels from either extended or localized sources can be obtained from knowledge of their radiance and irradiance functions.

Most background sources are described by a blackbody radiation model in which the radiance is given by

$$W(\lambda) = \frac{c^2 h}{\lambda^5}\left[\frac{1}{e^{hc/\lambda \kappa T^\circ} - 1}\right] \qquad (2.6.5)$$

where c is the speed of light, h is Planck's constant, κ is Boltzmann constant,

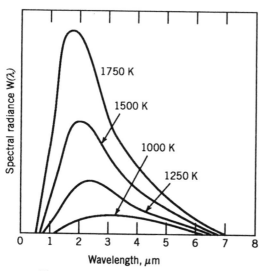

Figure 2.15. Blackbody spectral radiance.

and $T°$ is the temperature of the radiation in degrees Kelvin. Figure 2.15 shows a plot of Eq. (2.6.5) for different temperatures $T°$ over the range of optical wavelengths.

The most important point source of light in a space environment is of course the sun itself. Figure 2.16a shows a plot of the sun irradiance measured outside the Earth's atmosphere, and closely resembles a blackbody of approximately 6,000°K. As this radiation passes through the atmosphere, the particulates of the atmosphere cause a wavelength-dependent attenuation that alters the irradiance profile. Figure 2.16b shows the sun irradiance collected at sea level when looking at the sun directly overhead. (When observed at lower elevation angles the irradiance is proportionally attenuated.)

The moon, planets, and stars also appear as point sources, and their irradiance values, as observed at sea level, are shown in Figure 2.17. In terms of background contribution, the sources are obviously of much lower magnitude than the sun itself.

Also important in any space link is the light from the extended daytime sky.

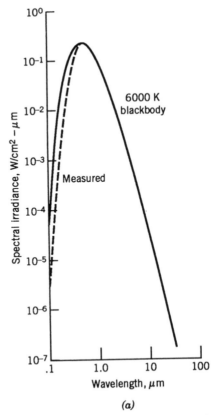

(a)

Figure 2.16. Sun's spectral irradiance (a) from above the atmosphere and (b) from Earth.

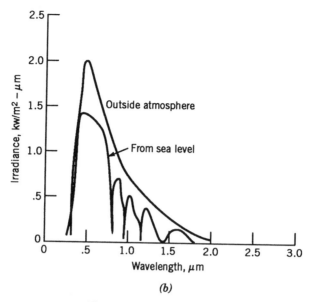

(b)

Figure 2.16. (*Continued*).

The diffuse daytime sky (portion of the sky observed when looking away from the sun) is produced primarily by the scattering of the sunlight by the atmosphere particulates. Since the scattering is least for the shorter wavelengths, the daytime sky often looks blue. Figure 2.18 shows a typical radiance plot of the daytime sky observed from sea level. The radiance profile is typically multimodel, having a basic (lower) mode exhibiting directly the sun's scattered energy. The secondary mode is from the re-radiation of the Earth's energy by the atmosphere, and closely resembles a blackbody at approximately 300°K. Table 2.1 summarizes some basic examples of how these radiance and irradiance values are used in the earlier equations to compute background power levels.

At night time the atmospheric scattering is excited primarily by the moon and galactic sources and is always several orders of magnitude less, as shown in Figure 2.19. Clearly the background advantages of operating a space optical link at night rather than in the daytime are apparent.

A down-looking airborne optical receiver will likewise be affected by background due to Earth-shine. Figure 2.20 illustrates typical Earth-shine irradiance for zenith angle viewing through a clear sky.

The background effects in Eq. (2.6.2a) can be normalized to that of a diffraction-limited receiver. If we set

$$\Omega_{fv} = \Omega_{dL} = \frac{\lambda^2}{A} \qquad (2.6.6)$$

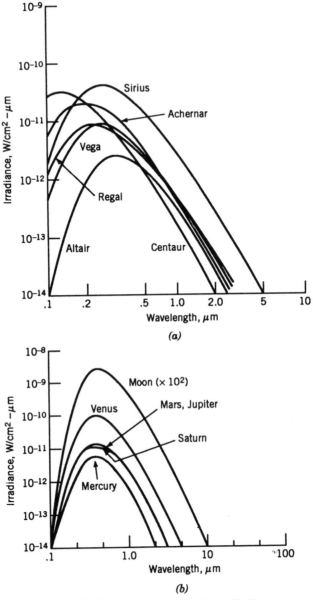

Figure 2.17. Spectral irradiance. (*a*) Stars. (*b*) Planets.

then the background power collected from an extended source is

$$P_{bo} = W(\lambda)(\Delta\lambda)A\left(\frac{\lambda^2}{A}\right)$$

$$= W(\lambda)(\Delta\lambda)\lambda^2$$

(2.6.7)

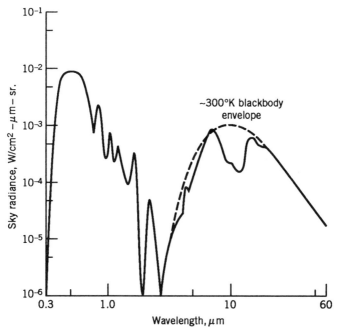

Figure 2.18. Background angle radiance of diffuse sky from sea level, elevation angle 45, no clouds. Reprinted from Karp et al. [9].

For the blackbody radiance in Eq. (2.6.5),

$$W(\lambda)\lambda^2 = \frac{(hc/\lambda)(c/\lambda^2)}{e^{hc/\lambda\kappa T^\circ} - 1} \tag{2.6.8}$$

Noting that $(c/\lambda^2)\Delta\lambda = \Delta f$ (the equivalent bandwidth in hertz corresponding to $\Delta\lambda$), Eq. (2.6.7) becomes

$$P_{bo} = N_0(\Delta f) \tag{2.6.9}$$

where

$$N_0 = \frac{hf}{e^{hf/\kappa T^\circ} - 1} \tag{2.6.10}$$

and f is the frequency in hertz corresponding to the wavelength λ. Thus N_0 plays the role of a one-sided noise spectral level at the receiver surface caused by a blackbody background at temperature T° in a diffraction-limited receiver.

When the field of view is not diffraction limited, as is usually the case, we must resort to Eq. (2.6.2a) to determine noise power. However, it is convenient to write this result in terms of the diffraction-limited result in Eq. (2.6.9).

TABLE 2.1   Example of a Background Noise Calculation

Receiver Parameters

Area 10^{-2} m²

$\Delta\lambda$ 10 Å

λ 0.5 μm

Field-of-view angle   100 μrad

Sky background	*Sun background*
Radiance (Fig. 2.18)	Irradiance
$W(0.5) = 10^{-2}$/cm²-sr-μm	$\mathscr{W}(0.5) = 2 \times 10^{-1}$ W/cm²-μm
Convert receiver parameters	Convert receiver parameters
$A = 10^{-2}$ m² $= 100$ cm²	$A = 100$ cm²
$\Delta\lambda = 10$ Å $= 10^{-3}$ μm	$\Delta\lambda = 10^{-3}$ μm
$\Omega_{fv} \cong (\pi/4)(10^{-4})^2 = 0.78 \times 10^{-8}$	
Power calculation	Power calculation
$P = W \cdot A \cdot \Delta\lambda \cdot \Omega_{fv}$	$P = \mathscr{W} \cdot A \cdot \Delta\lambda$
$= (10^{-2})(100)(10^{-3})(0.78 \cdot 10^{-8})$	$= (2 \times 10^{-1})(100)(10^{-3})$
$= 0.78 \times 10^{-11}$ W	$= 0.02$ W

Multiplying and dividing in Eq. (2.6.2a) by Ω_{dL} allows us to write the total collected power as

$$P_b = P_{bo} \left(\frac{\Omega_{fv}}{\Omega_{dL}} \right) \qquad (2.6.11)$$

The ratio Ω_{fv}/Ω_{dL} appears as a multiplying factor by which the diffraction-limited power is multiplied to get total power. The diffraction-limited power P_{bo} is often called the power per spatial "mode," and the ratio Ω_{fv}/Ω_{dL} is the number of modes of the optical receiver, as we discussed in Eq. (2.2.7). Thus, background power is collected by multiplying the diffraction-limited power P_{bo} by the number of receiver field modes being observed. This is equivalent to assuming the background noise has a uniform distribution in the focal plane, and therefore collected background power will increase directly as the detector area increases.

In some types of space links, the background light may be from man-made light sources instead of the sky background. This may occur, for example, in

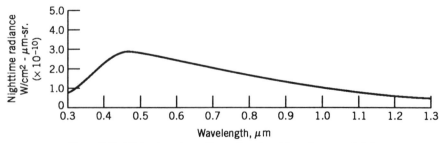

Figure 2.19. Night time sky radiance. Reprinted from Karp et al. [9].

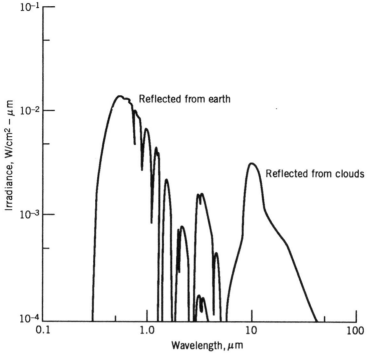

Figure 2.20. Spectral radiance of Earth from space. Reprinted from Karp et al. [9].

indoor optical links or short-range terrestial links, where stray light from light bulbs, street lamps, urban illumination, and so on may produce scattered radiation that could appear as the primary interference to an optical receiver. For systems of this type, it is necessary to build up an interference light model that can be used similar to the sky background radiance function. Consider the system depicted in Figure 2.21, showing an optical receiver operating in the presence of reflected light from a lamp in the vicinity of a reflecting wall. The lamp radiates P_s watts over an optical bandwidth of $\Delta\lambda$ and is located a distance z from the wall. Assume the lamp isotropically illuminates the room in all directions, and assume a reflection factor of η_w for the wall (depending on the wall texture). The lamp therefore transmits $P_s/(4\pi z^2)(\Delta\lambda)$ watts/area-bandwidth over the wall. The wall can be considered a Lambertian source that uniformly reradiates its incident light, with the reflection loss η_w, over 2π steradians. Hence the reflecting wall appears to an optical receiver as a diffuse background with effective radiance

$$W(\lambda) = \frac{P_s \eta_w}{8\pi^2 z^2 (\Delta\lambda)} \tag{2.6.12}$$

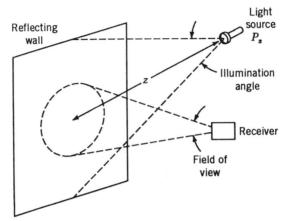

Figure 2.21. Model for indoor reflected background light.

in watts/area-steradians-micron. An optical receiver containing the wall in its field of view will then collect background light from the wall according to Eq. (2.6.2) with the radiance in Eq. (2.6.12) inserted. If the lamp light was directed to only a specific section of the wall of area A_s, then the above equations must be modified, with the term $P_s/4\pi z^2$ replaced by P_s/A_s.

2.7 EXTENDED SIGNAL SOURCES

This mode concept has an interesting interpretation when related to optical communication links in which the source signal itself takes on the appearance of an extended background noise source. This would occur, for example, if a point source were behind an extended scattering medium (cloud) or perhaps observed after reflection off a scattering surface (like the ocean or mountains), producing diffuse light throughout. A receiver would observe optical transmissions from the entire surface, the latter appearing as an extended source to the receiver, as shown in Figure 2.22. If the source radiance of the scattering mechanism, $W(\lambda)$, can be properly described, Eq. (2.6.2a) can be used to determine the amount of signal source power collected at the receiver. In this case, however, the collected power is the desired signal power rather than background. (The interference light would have to be computed separately.) It would now be important to collect as much source radiance as possible, which now means selecting the receiver field of view to match the subtended solid angle of the extended source, that is, observing as many source modes (directions of arrival) as possible. This notion is particularly important in atmospheric and terrestial space links, and will be considered again in Chapters 4 and 5.

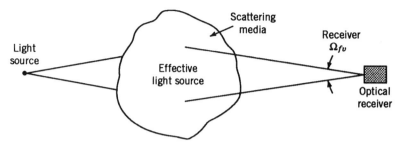

Figure 2.22. Extended source due to scattering.

PROBLEMS

2.1   A point source transmitter at $\lambda = 1\mu m$ transmits toward a circular lens of diameter 5 cm, as shown in Figure P2.1. The source field at the lens has an intensity of 10^{-6} W/cm².

(a) Show approximately where the source will image in the focal plane shown.

(b) What is the peak intensity and diameter of the imaged Airy pattern?

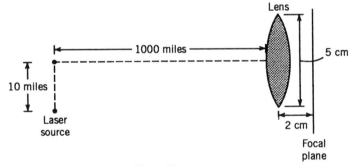

Figure P2.1.

2.2   A field at $\lambda = 1\ \mu m$ in a 1-mm diameter fiber core is a plane wave at the output. A lens matching the fiber diameter is inserted at the output to focus the fiber output field on to a detector located 2 mm from the fiber output. What is the minimal diameter of the detecting area to collect the main hump Airy pattern of the focused fiber field?

2.3   A square array of individual detectors, each 1 mm in diameter, is placed in the focal plane of a 50-cm receiver aperture.

(a) Estimate the resolution angle (radian field of view) of each detector.

(b) How large must the array be to cover an overall 2 degree by 2 degree field of view?

2.4 An optical receiver at $\lambda = 1\,\mu m$ uses a 3-in. aperture lens and a 1 cm detector.

(a) Estimate its diffraction-limited field of view.

(b) Estimate its receiver field of view.

(c) How many field modes (directions of arrival) can the receiver resolve?

(d) Approximately how far off the normal axis in angle can a point source be before it is not detectable?

2.5 Five lasers transmit a plane wave to an optical receiver, as snown in Figure P2.5. Each laser produces an intensity of 10^{-6} W/m at the receiver.

(a) Estimate the total power collected by the detector.

(b) What is the power collected if the receiver lens is removed and the detector is placed directly in the receiver plane?

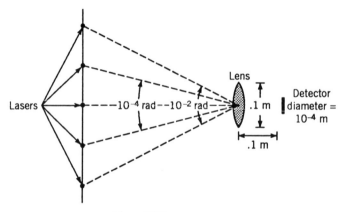

Figure P2.5.

2.6 Use the identity

$$\int_{-\infty}^{\infty} \frac{\sin[\pi(x-a)]}{[\pi(x-a)]} \cdot \frac{\sin[\pi(x-b)]}{[\pi(x-b)]} \, dx = \frac{\sin[\pi(a-b)/]}{[\pi(a-b)]}$$

to determine the required angle separation for arriving plane waves at a square aperture to produce orthogonal Airy patterns. *Hint*: Use Eqs. (2.1.6) and (2.1.8).

2.7   Repeat Problem 2.6 for a circular lens, using the identity

$$\frac{1}{\pi} \int_{\text{plane}} \frac{J_1(|\mathbf{u}|)}{|\mathbf{u}|} \cdot \frac{J_1(|\mathbf{u} - \mathbf{u}_0|)}{|(\mathbf{u} - \mathbf{u}_0)|} \cdot d\mathbf{u} = \frac{2J_1(|\mathbf{u}_0|)}{|\mathbf{u}_0|}$$

Hint: $J_1(x) = 0$ at $x = 3.83$.

2.8   A random field with spatial coherence

$$R_s(x, y) = e^{-(x^2 + y^2)/2\sigma^2}$$

occurs at a receiver. The field is imaged by a receiver lens with diameter much greater than σ and with a focal length of f_c. Estimate the 3-db width of the average intensity distribution in the focal plane. *Hint*: The Fourier Transform of a Gaussian function is another Gaussian function.

2.9   A random field at wavelength λ has the idealized coherence function,

$$R_f(t, t + \tau, \mathbf{r}_1, \mathbf{r}_2) = R_t(0)\delta(\boldsymbol{\rho}) \qquad \boldsymbol{\rho} = \mathbf{r}_1 - \mathbf{r}_2$$

with $\delta(\boldsymbol{\rho})$ a delta function, appears at a lens of area A. The field is focused onto a focal plane having a detector area A_d. Show that with $f_c = A^{1/2}$, the power collected at the detecting area is

$$P_d = \left(\frac{R_t(0)}{\lambda^2}\right) A_d$$

2.10   Determine the background power produced from a sunlit sky that also contains the moon in the field of view. Assume the following receiver parameters: 10 cm lens, 100 Å bandwidth, 100 μrad field of view angle, and operating at $\lambda = 10$ μm.

2.11   A 4 in. telescope with bandwidth 100 Å observes the Earth from outer space at $\lambda = 0.5$ μ. How much background power is collected from the Earth by the telescope?

2.12   (a) How much light power can be expected from the star Cirius if it is in the field of view of an 8-in. telescope with a bandwidth of 100 Å at $\lambda = 0.5$?

(b) How much additional power will be added if Saturn is also in the field of view?

2.13   The ocean is illuminated by the sun. The sun has an irradiance of 10^{-4} W/m-Hz. The light is reflected as a Lambertian source (0-dB loss) from the ocean surface. How much power is collected by a down-looking 6 in. receiver at an altitude of 10^4 m with an optical bandwidth of 10^{12} Hz?

2.14   A laser with power P_s and wavelength λ transmits with a d_t meter aperture. A receiver at distance Z_1, as shown in Figure P2.14, has a diffraction limited field of view, aperture area A_1, and optical bandwidth B_o Hz.

(a) Determine the ratio of laser power to background power collected at the receiver.

(b) Repeat if the receiver is moved to distance Z_2 ($Z_2 < Z_1$), the field of view is increased to Ω_{fv}, and the aperture area set to A_2, with the same bandwidth B_o.

(c) Compare the two ratios, and determine which parameters determine the relative values of the ratios.

2.15 Consider an optical surface A_1 transmitting to a receiving surface A_2

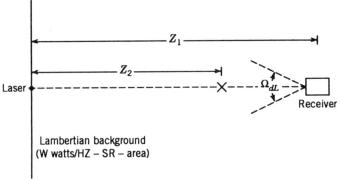

Figure P2.14.

located a distance Z away. Let the transmitting surface be defined by a Lambertian radiance function W (watts/area steradian-micron) at wavelength λ.

(a) What is the maximum power that can be transferred from A_1 to A_2?

(b) What fraction of the total power available from A_1 will be collected by A_2 if the acceptance angle (field of view) of A_2 is Ω_2.

(c) Now assume a lens is placed in front of A_1 so that its radiation field is expanded to an angle $\Omega_1 = 2A_2/Z^2$. How much power is collected at A_2?

2.16 A 100-W fluorescent lamp in a room illuminates a wall 3 m away with a circular beam having a beam angle of 90 degrees. Assume all the lamp radiation is in a wavelength band from 0.2 to 2 μm. The wall has a reflection loss of 6 dB and scatters the light as a Lambertian source. The scattered light is collected by a receiver with area 1 cm, a 45 degree angle field of view (looking at the wall), and a bandwidth of 1 nm around $\lambda = 1$ μm. Estimate the reflected background light power collected by the receiver.

REFERENCES

1. J. Goodman, *Introduction to Fourier Optics*, McGraw Hill, New York, 1968.
2. E. Steward, *Fourier Optics, An Introduction*, Wiley, New York, 1983.
3. A. Papoulis, *System and Transform Theory in Optics*, McGraw Hill, New York, 1968.
4. M. Born and E. Wolf, *Principles of Optics*, 4th ed., Pergamon Press, London, 1970.
5. R. Kingslake, *Lens Design Fundamentals*, Academic Press, New York, 1978.
6. D. O'Shea, *Elements of Modern Optical Design*, Wiley, New York, 1985.
7. W. Smith, *Modern Optical Engineering*, 2nd ed., McGraw Hill, New York, 1990.
8. G. Agrawal, *Fiberoptic Communications Systems*, Wiley, New York, 1992.
9. S. Karp, R. Gagliardi, S. Moran, and L. Stotts, *Optical Channels*, Plenum Press, New York, 1988.

3

PHOTODETECTION

Photodetection of the light field represents the key operation in the optical receiver, converting the collected field to a current or voltage waveform for subsequent postdetection processing. This interfacing of the optics and the electronics represents an area commonly referred to as *photonics*. For optimal design of the photonic processing, it is important that the system designer be cognizant of the characteristics of the photodetecting operation. This becomes particularly significant when the actual statistics of the detector waveform are necessary for optimal design procedures. This occurs in detection and decoding systems where performance depends explicitly on the mathematical model of the photodetector.

3.1 THE PHOTODETECTION PROCESS

Photodetection is achieved by having a photosensitive material respond to incident light by producing free electrons. These electrons are then susceptible to an externally applied electric potential that forces the free electrons to drift in a given direction. This electron flow is then exhibited at the detector output as a current flow. Hence photodetection converts impinging light fields to output current. In a vacuum tube, the free electrons are produced from a photosensitive surface material and released into a vacuum cavity, where they are collected by a charged anode plate. In a solid-state detector, the light excites electrons at a positive-negative (PN) junction, and the current corresponds to an electron flow across the junction gap. The general photodetection operation can therefore be represented by the simplified diagram in Figure 3.1 showing the incident light and the external bias circuitry producing the output current flow. This current can then be converted to a voltage by passing through a load resistor.

Photomultiplication is achieved during photodetection by having the excited electrons regenerate additional free electrons so that the resulting accumulated current flow is many times higher than for the primary electrons alone. This electron enhancement can be obtained by using multiple anode plates to

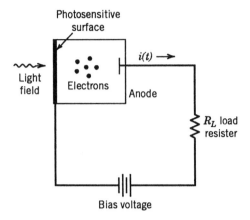

Figure 3.1. The basic photodetection model.

produce secondary electron emissions in a vacuum tube or by using avalanching effects in doped semiconductor compounds placed in the junction gap in solid-state detectors.

The conversion of an optical field to electron flow and the reproduction of electrons by photomultiplication are both probabilistic in nature, and the photodetector output always evolves as a random current process in time. The overall effect is to induce an inherent randomness on the photodetection operation when responding to any optical field, whether stochastic or not. This detector randomness must be properly accounted for in system models.

The photodetected output current process is described mathematically by the superposition of the individual current effects of each released electron. As a single electron moves within the photosensitive material, it produces a current response function $h(t)$, as shown in Figure 3.2. (This would be the time response observed in an ideal ammeter connected to the output, when a single electron is released at $t = 0$.) Physically, each electron moves for a short time period, ideally coming to rest after a fixed transit time. Hence, its current response $h(t)$ will also be of a finite time duration, existing only while the electron is in motion. The actual shape of $h(t)$ depends on the electron velocity

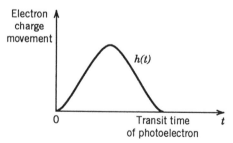

Figure 3.2. Typical electron response.

during transition. In all cases, the area under the response function is a fixed constant, since the integral of $h(t)$ is the change in electric charge during electron motion. Thus,

$$\int_0^\infty h(t)dt = \text{charge of a single electron}$$

$$= 1.6 \times 10^{-19} \text{ coulombs} \tag{3.1.1}$$

In a photomultiplier, the release of an electron produces the movement of many other electrons. If the photomultiplier is ideal, exactly g electrons are collected for each photoemissive electron released, and each $h(t)$ effectively becomes $gh(t)$. Thus the output current is increased by the factor g; the latter is called the *gain* of the photomultiplier. In nonideal photomultiplication, the number of secondary electrons g for each released electron is itself random, and g must be treated as a random variable. As a result g will have a specific probability density, and in particular a mean $\bar{g}$, as well as other moments. The ratio of the mean squared gain $\overline{g^2}$ to the squared mean $(\bar{g})^2$ is called the *excess noise factor* of the photomultiplier and is denoted by

$$F = \overline{g^2}/\bar{g}^2 \tag{3.1.2}$$

The excess noise factor is an indication of the randomness of the photomultiplier and, as we shall see, appears as a key parameter in communication analysis. The excess noise factor can also be written in terms of the variance of the gain (var g), since $\overline{g^2} = (\bar{g}^2) + \text{var } g$. This means that

$$F = \frac{\bar{g}^2 + \text{var } g}{\bar{g}^2} = 1 + \frac{\text{var } g}{\bar{g}^2} \tag{3.1.3}$$

and clearly $F \geq 1$. An ideal (nonrandom) photomultiplier of gain g would have $\bar{g} = g$ and $\overline{g^2} = (g)^2$, so that $F = 1$.

An electron released at time, say t_m, produces the response function $h(t - t_m)$. (That is, the response function $h(t)$ is shifted to the point $t = t_m$.) If the optical field is impressed at time $t = 0$, the total cumulative response at time t is the combined response of all electrons released during the interval $(0, t)$. This produces the photodetector output current function

$$i(t) = \sum_{m=0}^{k(0,t)} g_m h(t - t_m) \tag{3.1.4}$$

where t_m is the time of release of the mth electron, g_m is the photomultiplier gain of that electron, and $k(0, t)$ is the number of electrons released during the time interval $(0, t)$. Note that only electrons released prior to time t can contribute to the output current at time t. The number $k(0, t)$ is often called the *count* of the electrons, and $k(0, t)$ is considered to be a function of t and is often called the *counting process* of the photodetection operation. Because photodetection is statistical in nature, the location times $\{t_m\}$, the electron count $k(0, t)$, and sometimes the gain g are random variables when used to model the optical detector output. Thus, Eq. (3.1.4) represents a sum of a random number of randomly located response functions $g_m h(t)$. Such processes are called *shot noise processes* and have been used to model more general types of burstlike noise phenomena, such as shot noise in vacuum tubes (from which its name was derived). The functions $h(t)$ are called the *component functions* of the shot noise and, in photodetection, are determined by the electron transit behavior, as stated before.

Since the $\{t_m\}$ are random variables, and the count $k(0, t)$ is a random count process in time, the shot noise current $i(t)$ in Eq. (3.1.4) is itself generated as a somewhat complicated random process. Clearly, any analysis to determine its inherent statistics (e.g., probability densities, averages, moments) requires specification of the statistics of the location times, gain, and counting process. This is our objective in this chapter. Note that the output $i(t)$ does not explicitly contain the received field in any obvious way, yet we expect the photodetector to exhibit the properties of the received field. Indeed, the optical communication system is based on the ability to faithfully recover the modulation imposed on the optical beam. Thus, we would expect the optical field to be embedded within the shot process in some manner.

Besides the response functions and gain, there are other parameters used in the description of practical photodetectors. The most important are the quantum efficiency, dark current, and bandwidth. The *quantum efficiency* indicates the fraction of the incident field power that will be actually detected. This parameter depends on the type of material used and is generally wavelength dependent. *Dark current* is current flow that occurs with no incident light on the detector and is due to inherent thermal agitation of electrons in the material. As such, dark current is itself a random current process that appears as an inherent detector noise. This current is typically described by its mean current flow, the latter being temperature dependent and proportional to the size of the detector collecting area. The *detector bandwidth* is an indication of the rate at which input light variations can be detected as output current variations and therefore measures the frequency response of the photodetectors. This bandwidth is approximately inversely related to the transit time (time of flight) of the free electrons or, equivalently, to the time length of the component functions $h(t)$. This point will be made more definitive in Section 3.4. Note that this detector bandwidth is different from the optical bandwidth of the front-end filters, the latter defining the wavelength bandwidth of input light impinging on the detector.

3.2 PHOTODETECTORS

Practical photodetectors are of two basic types: phototube devices using vacuum tube construction and solid-state devices using junction effects [2–5]. The phototube, shown in Figure 3.3a uses a photosensitive material as its receiving surface. Excited electrons are released into a vacuum as a space charge and are collected at an anode plate to produce the current flow. A photomultiplier tube (PMT) is a phototube with multiple plates (dynodes) that each produce secondary emissions (Figure 3.3b), thereby multiplying up the current flow. Because secondary emissions can be made extremely high, a PMT can have mean gains in the 10^3 to 10^5 range. The gain variance is often described by a spreading factor ξ such that $\sqrt{\operatorname{var} g} = \xi\bar{g}$. This means the excess noise factor in Eq. (3.1.3) has the form $F = 1 + \xi^2$ for the PMT. Spreading factors of approximately 1 to 30 percent are typical.

The photosensitive materials used in phototubes are generally composed of silicon or germanium components. Figure 3.4 shows the quantum efficiency factor of some of these materials as a function of wavelength. These efficiency factors are peaked in the visible range, and fall off rapidly for wavelengths exceeding 2 to 3 μm. Hence, efficient optical photodetectors do not exist for the longer wavelengths.

A photodiode (Fig. 3.5) is a solid-state PN junction device (commonly called a *PIN* diode), in which the gap material is selected to be responsive to the wavelength of the light. PIN diodes have small collecting areas (cms), low dark current, and extremely high bandwidths ($\approx 10\,\mathrm{GHz}$), due to the shorter gap transit times. The *avalanche photodiode* (APD) is a solid-state diode with junction material providing current gain by repeated electron ionization. An APD can provide mean gain values in the range 50 to 300. A complete theory of the avalanching mechanism has been developed, and fairly reliable gain statistics can be modeled for the APD. This will be presented in Section 3.6.

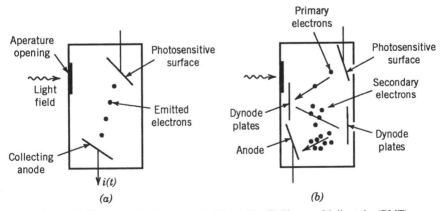

Figure 3.3. Vacuum tube detectors. (a) Phototube. (b) Photomultiplier tube (PMT).

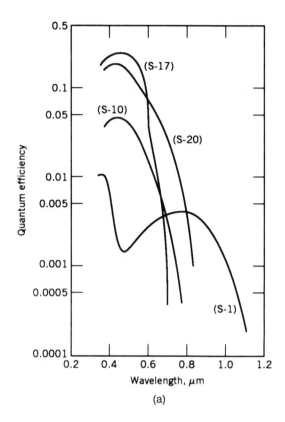

(a)

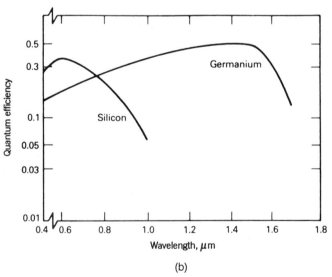

(b)

Figure 3.4. Quantum efficiency factors of several photosensitive materials. (*a*) Classes of silicon. (*b*) Germanium.

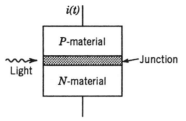

Figure 3.5. Solid-state photodetectors.

TABLE 3.1 Detector Parameter Values

Detector Type	Gain, $\bar{g}$	Excess Noise Factor, F^a	Bandwidth	Average Dark Current amp/cm^2
Phototube	1	1	500 MHz	10^{-12}
PMT	10^3–10^5	$1 + \xi^2$	100 MHz	10^{-11}
PIN diode	1	1	1–5 GHz	10^{-13}
APD	50–200	$\gamma\bar{g} + \left(2 - \dfrac{1}{\bar{g}}\right)(1 - \gamma)$	1–5 GHz	10^{-14}

$^a \xi$ = PMT spreading factor; γ = APD ionization coefficient.

Solid-state photodetectors use silicon, germanium, and gallium arsenide components for the junction material. These materials are primarily sensitive to visible light and have extremely short transit times, which lead to high bandwidth, fast reponse detectors. Table 3.1 summarizes some typical parameter values associated with these various detectors.

3.3 COUNTING STATISTICS

The relationship of the received electromagnetic field and the number of released electrons is governed by the interaction between the radiation field and electrons of the photosensitive material. There are two accepted ways to treat this relaltionship. In the purely quantum treatment, the field is quantized into photons, and each field photon gives rise to an electron with some probability. The electrons released are, therefore, a statistical measurement of the photon occupancy in the field, and electron counting is often called photon or photoelectron counting. The alternate treatment (and the one we use) is the semiclassical approach, which is actually a consequence of the quantum treatment. This model treats the field classically (i.e., as a wave) and prescribes a probabilistic relation to account for its interaction with the atomic structure of the detector surface. Although a complete description of the emission and absorption of light by an atom is well beyond our interest here, an outline of the approach is as follows [1]. The semiclassical procedure begins with a charged particle in an electromagnetic field. It is then assumed that the combined system of atom plus field begins in some initial state, and a set of coupling equations is derived for the state transition probabilities. From these, one determines the probability rate of finding the combined system in a particular state. Summing over all final states, and making some simplifying assumptions, one derives Fermi's rule for the probability per second P_t for a state transition over a differential area $\Delta\mathbf{r}$ located at point $\mathbf{r}$ on the detector surface. The probability rate satisfies the equation

$$\frac{dP_t}{dt} = \alpha I_d(t, \mathbf{r})\Delta\mathbf{r} \tag{3.3.1}$$

where P_t can be interpreted as the probability of an electron release from $\Delta\mathbf{r}$ at t, α is a proportionality constant, and $I_d(t, \mathbf{r})$ is again the field intensity at time t and point $\mathbf{r}$ on the detector surface. The primary consequence of Fermi's rule is that it implies that in a short time interval Δt, the probability of ejecting an electron from an atom at the elemental surface area $\Delta\mathbf{r}$ is proportional to the incident field intensity over $\Delta\mathbf{r}$ and Δt. That is,

$$\begin{bmatrix} \text{probability that an} \\ \text{electron is released} \\ \text{from the area } \Delta\mathbf{r} \\ \text{during the time } \Delta t \end{bmatrix} \cong \alpha I_d(t, \mathbf{r})\Delta t \Delta\mathbf{r} \tag{3.3.2}$$

for sufficiently small $\Delta\mathbf{r}$ and Δt. In addition, Eq. (3.3.2) implies that the probability of more than one electron being released must go to zero as $(\Delta\mathbf{r}\Delta t)^2$, which means that

$$\begin{bmatrix} \text{probability of no} \\ \text{electrons being released} \\ \text{from } \Delta\mathbf{r} \text{ during } \Delta t \end{bmatrix} \cong 1 - \alpha I_d(t, \mathbf{r})\Delta t \Delta\mathbf{r} \tag{3.3.3}$$

Note that Eq. (3.3.2) states that the release of an electron from any elemental area at point $\mathbf{r}$ at any time t depends only on the field intensity at that time and point. This implies that the release of electrons from disjoint differential areas on the surface, and from disjoint intervals in time, can be treated as independent events when given a particular intensity function. This assumption, along with Eqs. (3.3.2) and (3.3.3), describes the key mathematical model of the photodetecting surface.

We must first determine the probability density of the number of electrons produced during a specific time interval, say $(0, T)$ from the photodetecting surface area A_d. We denote this random counting variable as k. To facilitate this derivation, we find it convenient to make use of the concept of a time–space domain. This domain contains vectors whose components correspond to time and spatial coordinates associated with these regions. We denote these vectors as $\mathbf{v} = (t, \mathbf{r})$, where t is the scalar time component and $\mathbf{r}$ represents the two-dimensional spatial coordinates of the detector surface. Unless otherwise stated, the origin of the coordinate axis for $\mathbf{r}$ is taken to be the detector center. We define the volume $\mathbf{V}$ in this domain to be composed of all vectors $\mathbf{v} = (t, \mathbf{r})$ such that $0 \leqslant t \leqslant T$ and $\mathbf{r} \in A_d$. This volume is shown in Figure 3.6a. This notation allows us to denote the normal electromagnetic field intensity at point $\mathbf{r}$ on the detector surface at time t by $I_d(t, \mathbf{r}) = I_d(\mathbf{v})$. The volume $\mathbf{V}$ is therefore the set of all points in the time–space domain over which we observe the radiation field with a given detector area in a given time interval.

Now consider the partition of the volume $\mathbf{V}$ into disjoint subvolumes, or cells, $\Delta \mathbf{v} = \Delta \mathbf{r} \Delta t$, as shown in Figure 3.6. We assume $\Delta \mathbf{r}$ and Δt are smaller than the spatial and temporal variations in $I_d(t, \mathbf{r})$ so that within each $\Delta \mathbf{v}$, $I_d(\mathbf{v})$ is approximately constant. (This is certainly possible with continuous fields.) Let ΔV be the volume of the cells $\Delta \mathbf{v}$, and let q be the total number of cells in $\mathbf{V}$ after partitioning. The ensemble of q disjoint cells $\Delta \mathbf{v}$ can now be ordered to

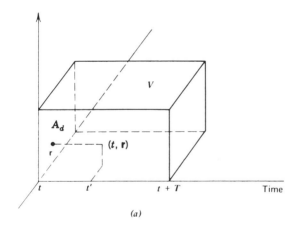

(a)

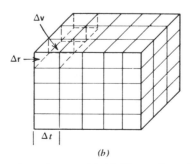

(b)

Figure 3.6. Photocounting volume model. (*a*) Observation volume (*b*) Observation cells.

form the sequence $\{\Delta\mathbf{v}_1, \Delta\mathbf{v}_2, \ldots, \Delta\mathbf{v}_q\}$, where each $\Delta\mathbf{v}_i$ is centered at some point $\mathbf{v}_i$ in $\mathbf{V}$. Each $\Delta\mathbf{v}_i$ can be interpreted as an observation cell corresponding to an elemental surface area and elemental time interval over which we observe the radiation field. In this notation, the probability model of electron emissions in Eq. (3.3.2) becomes, for $\Delta V \to 0$,

$$\begin{bmatrix} \text{probability of an electron} \\ \text{emitted from } \Delta\mathbf{v}_i \text{ at point } \mathbf{v}_i \end{bmatrix} = \alpha I_d(\mathbf{v}_i)\Delta V \qquad (3.3.4)$$

and

$$\begin{bmatrix} \text{probability of no} \\ \text{electrons emitted} \end{bmatrix} = 1 - \alpha I_d(\mathbf{v}_i)\Delta V \qquad (3.3.5)$$

where α is again the proportionality constant. We now consider the probability of the detector releasing exactly k total electrons from the total surface area A_d during the time interval $(0, T)$. This is equivalent to the compound probability that k electrons will be emitted from the totality of all cells $\{\Delta\mathbf{v}_i\}$ spanning $\mathbf{V}$. For small volumes ΔV, this can be written as

$$\begin{bmatrix} \text{probability of} \\ k \text{ emitted} \\ \text{electrons from } V \end{bmatrix} = \frac{1}{k!} \sum_{\substack{\text{all} \\ \text{orderings}}} \begin{bmatrix} \text{probability of} \\ \text{one electron} \\ \text{from } k \text{ different} \\ \text{ordered cells} \end{bmatrix} \begin{bmatrix} \text{probability of} \\ \text{no electrons in} \\ \text{the } q - k \text{ remaining} \\ \text{ordered cells} \end{bmatrix}$$

$$= \frac{\alpha^k}{k!} \sum_{\substack{\text{all} \\ \text{orderings}}} I_d(\mathbf{v}_{i_1}) \cdots I_d(\mathbf{v}_{i_k})(\Delta V)^k \prod_{j=k+1}^{q} [1 - \alpha I_d(\mathbf{v}_{i_j})\Delta V]$$

$$(3.3.6)$$

where $(i_1, i_2, \ldots, i_q)$ is a particular index ordering of the integers 1 to q. The summation considers all possible orderings, and therefore all possible arrangements of k and $(q - k)$ index groupings. The division by $k!$ is necessary since different arrangements for the same k cells need only be considered once. Note that Eq. (3.3.6) has used Eq. (3.3.4) and the assumption of independent electron emissions from disjoint cells. However, as long as q is finite, Eq. (3.3.6) must be considered only an approximation.

Now consider the limit (3.3.6) as $\Delta V \to 0$ and $q \to \infty$, so that the approximation approaches a true equality. Because the limits of sums and products are equal to the sum and products of the limits, we can investigate the limit of the individual terms in Eq. (3.3.6) and then recombine. We first show that the product term has the same limit for all orderings. This can be seen by considering the limit of the logarithm of the product. For finite q, this log is

$$\log \prod_{j=k+1}^{q} [I - \alpha I_d(\mathbf{v}_{i_j})\Delta V] = \sum_{j=k+1}^{q} \log[1 - \alpha I_d(\mathbf{v}_{i_j})\Delta V] \qquad (3.3.7)$$

Adding and subtracting the k terms not included in the summation allows us to write the right-hand side of Eq. (3.3.7) always as

$$\sum_{j=1}^{q} \log[1 - \alpha I_d(\mathbf{v}_j)\Delta V] - \sum_{j=1}^{k} \log[1 - \alpha I_d(\mathbf{v}_{i_j})\Delta V] \qquad (3.3.8)$$

Now, as $\Delta V \to 0$ in the limit, we have $q \to \infty$, and

$$\lim_{\Delta V \to 0} \log[1 - \alpha I_d(\mathbf{v}_j)\Delta V] \to -\alpha I_d(\mathbf{v}_j)\Delta V \qquad (3.3.9)$$

The first summation in Eq. (3.3.8), therefore, has the limit

$$\lim_{\substack{\Delta V \to 0 \\ q \to \infty}} \sum_{j=1}^{q} \log[1 - \alpha I_d(\mathbf{v}_j)\Delta V] \to \lim_{\substack{\Delta V \to 0 \\ q \to \infty}} \left[-\alpha \sum_{j=1}^{q} I_d(\mathbf{v}_j)\Delta V \right]$$

$$= -\alpha \int_V I_d(\mathbf{v}) \, d\mathbf{v} \qquad (3.3.10)$$

where the integration is over the volume V. However, the second summation in Eq. (3.3.8) involves only a finite number of terms and, therefore, has the limit

$$\lim_{\substack{\Delta V \to 0 \\ q \to \infty}} \sum_{j=1}^{k} -\alpha I_d(\mathbf{v}_{i_j})\Delta V = 0 \qquad (3.3.11)$$

for any ordering. Hence, in Eq. (3.3.6)

$$\lim_{\Delta V \to 0} \prod_{j=k+1}^{q} [1 - \alpha I_d(\mathbf{v}_{i_j})\Delta V] = \exp\left[-\alpha \int_V I_d(\mathbf{v}) \, d\mathbf{v} \right] \qquad (3.3.12)$$

Now consider the summation term in Eq. (3.3.6). We want to evaluate the limit

$$\lim_{\Delta V \to 0} \alpha^k \sum_{\substack{\text{all} \\ \text{orderings}}} [I_d(\mathbf{v}_{i_1}) \cdots I_d(\mathbf{v}_{i_k})(\Delta V)^k] \qquad (3.3.13)$$

Because each ordering above requires $i_1 \neq i_2 \neq \cdots \neq i_k$, the summation can be equivalently written as

$$\alpha^k \sum_{i_k = 1}^{q} \cdots \sum_{i_1 = 1}^{q} [I_d(\mathbf{v}_{i_1}) \cdots I_d(\mathbf{v}_{i_k})] \Delta V^k - \begin{bmatrix} \text{the sum over all} \\ \text{orderings in which at} \\ \text{least two } i_j \text{ are equal} \end{bmatrix} \qquad (3.3.14)$$

The second term involves q^{k-1} terms of order $(\Delta V)^k$. Because q behaves as $1/\Delta V$, the limit of the second term will be zero as $\Delta V \rightarrow 0$. The first term, on the other hand, has the limit

$$\lim_{\substack{\Delta V \rightarrow 0 \\ q \rightarrow \infty}} \left[\alpha \sum_{i=1}^{q} I_d(\mathbf{v}_i) \Delta V \right]^k = \left[\alpha \int_V I_d(\mathbf{v}) \, d\mathbf{v} \right]^k \qquad (3.3.15)$$

Therefore, using Eqs. (3.3.10), (3.3.11), and (3.3.15) in (3.3.6), we derive the final limiting form of the probability of k emissions over V as

$$P(k) = \frac{(m_v)^k e^{-m_v}}{k!} \qquad (3.3.16)$$

where

$$m_v = \alpha \int_V I_d(\mathbf{v}) \, d\mathbf{v} \qquad (3.3.17)$$

Recalling our definition of the volume $\mathbf{V}$, we can also write

$$m_v = \alpha \int_0^T \int_{A_d} I_d(t, \mathbf{r}) \, d\mathbf{r} \, dt \qquad (3.3.18)$$

The probability in Eq. (3.3.16) is the probability that exactly k electrons will be emitted over the observation volume $\mathbf{V}$, that is, from the spatial area A_d during the time interval $(0, T)$ defining $\mathbf{V}$. Note that the probability depends only on the integral of the field intensity over the volume $\mathbf{V}$. We have explicitly indicated this dependence by subscripting m. Because m_v depends on the spatial location and size of the detecting area A_d and the counting interval $(0, T)$, the probability is in general nonstationary in time and nonhomogeneous in space. That is, the probabilities may be different over different volumes.

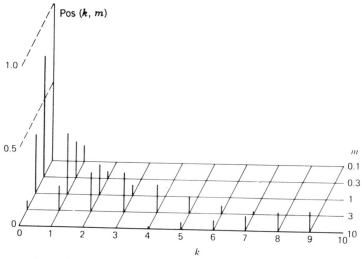

Figure 3.7. Poisson probabilities for various mean values $m_v = m$.

Because k can represent any nonnegative integer, Eq. (3.3.16) represents a probability over all nonnegative integers. This probability is called a Poisson probability, and m_v is its parameter. A sketch of this probability over the integers is shown in Figure 3.7 for several values of m_v. For convenience, we subsequently denote the Poisson probability with level m_v in Eq. (3.3.16) as $\text{Pos}(k, m_v)$. That is,

$$\text{Pos}(k, m_v) = \frac{m_v^k e^{-m_v}}{k!} \qquad (3.3.19)$$

The count probability density associated with these probabilities is then a discrete density over all nonnegative integers, with probability weight given by Eq. (3.3.19).

Table 3.2 summarizes some basic statistical parameters of the Poisson random variable. Note that the parameter m_v in Eq. (3.3.18) is also the mean value of the count. That is, $\bar{k} = m_v$. This fact allows us to evaluate specifically the proportionality constant α. This is accomplished by recalling that the photoelectric effect requires the absorption of an amount of energy hf from the field in order to release an electron, where f is the frequency of the field and h is Planck's constant. This means that the mean number of emitted electrons multiplied by hf must be equal to the average energy absorbed from the field at frequency f. Hence, we can equate

$$\bar{k}(hf) = \eta \int_0^T \int_{A_d} I_d(t, \mathbf{r}) \, d\mathbf{r} \, dt \qquad (3.3.20)$$

TABLE 3.2   Identities for the Poisson Variable $P(k) = \dfrac{m^k}{k!} e^{-m}$

Mean	$\sum\limits_{k} k P(k)$	m
Mean-square value	$\sum\limits_{k} k^2 P(k)$	$m^2 + m$
Characteristic function	$\sum\limits_{k} e^{j\omega k} P(k)$	$e^{(e^{j\omega} - 1)m}$
Moment-generating function	$\sum\limits_{k} (1 - z)^k P(k)$	e^{-zm}
qth moment	$\sum\limits_{k} k^q P(k)$	$\dfrac{\partial^q}{\partial z^q} [e^{-zm}]_{z=1}$

where η is the detector quantum efficiency. Substituting with $\bar{k} = m_v$, and using Eq. (3.3.18) yields the formal definition

$$\alpha = \frac{\eta}{hf} \tag{3.3.21}$$

in (watt-seconds)$^{-1}$. Note that α above has the units of (energy)$^{-1}$, and effectively converts field energy to a dimensionless count. We also point out that α is an extremely large number. For example, at $\lambda = 1\,\mu$m, with an efficiency of unity,

$$\alpha = \frac{1}{(6.6 \times 10^{-34})(3 \times 10^{14})} \approx 5 \times 10^{18} \tag{3.3.22}$$

Hence the mean count $\bar{k}$ generally is an extremely large number even for very low (10^{-9}) energy values.

Because m_v is the mean count over the detector A_d and time interval T, we also can write

$$m_v = \int_0^T n(t)\, dt \tag{3.3.23}$$

in counts with

$$n(t) = \alpha \int_{A_d} I_d(t, \mathbf{r})\, d\mathbf{r} \tag{3.3.24}$$

in counts/second. This function $n(t)$ in Eq. (3.3.24) now plays the role of the average count rate. That is, it can be considered as the average rate at which electrons are emitted at time t from the surface area A_d during photodetection. Furthermore, we see that $n(t)$ is directly proportional to the power collected over the detector area at time t. From our power collection discussion in Section 2.2, we can equate this power to that collected over the aperture and write

$$n(t) = \alpha \int_A I_r(t, \mathbf{r}) \, d\mathbf{r} \qquad (3.3.25)$$

where A is the aperture area and $I_r(t, \mathbf{r})$ is the aperture intensity corresponding to the detector intensity $I_d(t, u, v)$. Thus the count rate $n(t)$ can be computed at either the detector or the aperture lens.

Lastly, we point out that, in Table 3.2, the mean square value of the Poisson count variable is

$$\overline{k^2} = m_v + m_v^2 \qquad (3.3.26)$$

The variance of the count is then

$$\begin{aligned} \mathrm{var}(k) &= \overline{k^2} - (\bar{k})^2 \\ &= (m_v + m_v^2) - (m_v)^2 \qquad (3.3.27) \\ &= m_v \end{aligned}$$

Thus the parameter m_v of the Poisson probability also gives the variance of the count. Because the variance essentially describes the mean square variation from the mean value, we see that the Poisson count variable inherently has a variation that is itself related to the mean count. Later we relate this variation to an effective noise fluctuation associated with the ability of a photodetector to count electrons accurately. We emphasize again the dependence of these moments on the volume $\mathbf{V}$, further illustrating the general nonstationarity of the counting statistics.

3.4 PHOTOCOUNTING WITH RECEIVER FIELDS

When the impinging field on the photodetector surface is the focused field of the optical aperture, the previous counting theory can be directly applied. We consider the following examples.

3.4.1 Constant Point Source Fields

Let the received field be that due to the far field of a point source transmitting an unmodulated plane wave, producing the field at the receiver area A,

$$f_r(t, \mathbf{r}) = ae^{j\omega_o t} \qquad \mathbf{r} \in A \tag{3.4.1}$$

with receiver power $P_r = (|a|^2)A$. This field is focused to the detector Airy pattern $f_d(t, u, v)$ on the photodetector surface A_d, producing the resulting count rate process of the photodetector

$$n(t) = \alpha \int_{A_d} |f_d(t, u, v)|^2 \, du \, dv \tag{3.4.2}$$

or, from Eq. (3.3.25),

$$n(t) = \alpha \int_A |f_r(t, \mathbf{r})|^2 \, d\mathbf{r}$$
$$= \alpha P_r \tag{3.4.3}$$

The resulting mean count over any $(0, T)$ is then

$$m_v = \int_0^T n(t) \, dt$$
$$= \alpha P_r T \tag{3.4.4}$$

and the Poisson count probability is then

$$P(k) = \text{Pos}(k, \alpha P_r T)$$
$$= \frac{(\alpha P_r T)^k e^{-\alpha P_r T}}{k!} \tag{3.4.5}$$

Thus a point source plane wave with a constant envelope field produces a constant count rate in the photodetector, and a Poisson count probability with mean $\alpha P_r T$. Note that the mean count is directly proportional to the counting time length T, and is independent of when in time it occurs.

3.4.2 Intensity-Modulated Point Source Field

Consider again a point source field at the receiver with modulated envelope $a(t)e^{j\omega_o t}$ over the area A. The collected field power is then

$$P_r(t) = |a(t)|^2 A \tag{3.4.6}$$

and the resulting count rate is

$$n(t) = \alpha P_r(t) \tag{3.4.7}$$

The Poisson count over a time interval $(t, t + T)$ has mean

$$m_v = \int_t^{t+T} n(\rho)\, d\rho \tag{3.4.8}$$

and now depends on t through the integral in Eq. (3.4.8). This integral of field power over $(t, t + T))$ is the received field energy during this interval, which varies according to the modulation.

3.4.3 General Received Field

Let the received field be extended to an arbitrary optical field, described by the generalized orthogonal expansion

$$f_r(t, \mathbf{r}) = \sum_{i=1}^{D_s} a_i(t) e^{j\omega_o t} \Phi_i(\mathbf{r}) \qquad \mathbf{r} \in A \tag{3.4.9}$$

where D_s is the number of orthogonal modes, and each orthogonal function $\Phi_i(\mathbf{r})$ is normalized so that $\int_A |\Phi_i(\mathbf{r})|^2\, d\mathbf{r} = A$. Then

$$n(t) = \alpha A \sum_{i=1}^{D_s} |a_i(t)|^2 \tag{3.4.10}$$

and

$$m_v = \alpha \sum_{i=1}^{D_s} E_i \tag{3.4.11}$$

where

$$E_i = A \int_0^T |a_i(t)|^2\, dt \tag{3.4.12}$$

is the effective field energy of the ith spatial mode over $(0, T)$. Note that Eq. (3.4.10) is valid for any orthogonal function set used to describe the field $f_r(t, \mathbf{r})$. Thus, it holds for both space fields and fiber fields.

The count probability is again a Poisson count, with the m_v in Eq. (3.4.11). This has an alternative interpretation if we introduce the characteristic function in Table 3.2 of the count k as

$$
\begin{aligned}
\Psi_k(j\omega) &= \exp\left[(e^{j\omega} - 1)\alpha \sum_{i=1}^{D_s} E_i\right] \\
&= \prod_{i=1}^{D_s} \exp[(e^{j\omega} - 1)\alpha E_i]
\end{aligned}
\tag{3.4.13}
$$

Because Eq. (3.4.13) is the product of characteristic functions of individual Poisson variables, each with parameter αE_i, the resulting count probability is, equivalently, the sum of D_s independent Poisson counts. Hence the photodetector output count is developed as if each orthogonal field mode superimposes independently its own Poisson count, each with its own energy variable over the detector area and $(0, T)$ interval.

3.5 COUNTING WITH RANDOM FIELDS

It has been shown that, with deterministic receiver fields, the count variable is Poisson with parameter m_v; the latter is dependent on the field energy over the observation volume. When the received field is random, the parameter m_v becomes a random variable, and the true count probability must be obtained by subsequent averaging of the conditional Poisson count over the statistics of m_v. Thus if m_v has the probability density $p_{m_v}(m)$, $(0 < m < \infty)$, the count probability over **V** is obtained by

$$
\begin{aligned}
P(k) &= \int_0^\infty \text{Pos}(k, m) p_{m_v}(m)\, dm \\
&= \int_0^\infty \left[\frac{m^k e^{-m}}{k!}\right] p_{m_v}(m)\, dm
\end{aligned}
\tag{3.5.1}
$$

This counting probability for k is no longer a Poisson probability, and obviously depends on the probability density of m_v induced by the stochastic field. Since the conditional probability in the integrand is Poisson, we call the class of probabilities $P(k)$ generated from Eq. (3.5.1) *conditional Poisson* (CP) probabilities. In the literature they have also been called doubly stochastic Poisson probabilities. Thus, the count probability resulting from the photodetection of a random field always belongs to the class of CP probabilities. We call the count k a CP random variable. We emphasize again that the density of m_v, in general, depends on V, and, therefore, CP densities are generally functions of V and again are nonstationary in time and space.

Although higher moments of the CP count can be also related to the higher moments of m_v (see Problem 3.11), the integrations in Eq. (3.5.1) must still be performed to determine the exact CP count. This integration is difficult in general, since m_v is itself the integral of a random field, and usually has a somewhat unwieldy probability density. Sometimes this count probability is easier to determine by first computing its characteristic function. The latter is now obtained as the average of the Poisson characteristic function in Table 3.2,

$$\Psi_k(j\omega) = \int_0^\infty e^{(e^{j\omega} - 1)m} \, p_{m_v}(m) \, dm \tag{3.5.2}$$

The integral however is directly related to the characteristic function of the random variable m_v,

$$\Psi_{m_v}(j\omega) = \int_0^\infty e^{j\omega m} \, p_{m_v}(m) \, dm \tag{3.5.3}$$

Relating Eq. (3.5.2) to Eq. (3.5.3) shows the interesting fact that

$$\Psi_k(j\omega) = \Psi_{m_v}(j\omega)\big|_{j\omega \to e^{j\omega} - 1} \tag{3.5.4}$$

Thus the characteristic function of the CP count is obtained by merely substituting into the characteristic function of the random parameter m_v. Although this may still not lead to an explicit expression for P(k) (its characteristic function must still be inverted to get the true count probability), it often lends insight into the photodetection operation. For example, consider again the spatial orthonormal expansion of a generalized random field

$$f_r(t, \mathbf{r}) = \sum_{i=1}^{D_s} a_i(t) e^{j\omega_o t} \Phi_i(\mathbf{r}) \tag{3.5.5}$$

into D_s spatial modes, with the random modal envelopes $a_i(t)$ defined in Eq. (1.6.4). As in Eq. (3.4.11),

$$m_v = \alpha \int_0^T \int_A |f_r(t, \mathbf{r})|^2 \, d\mathbf{r}$$

$$= \alpha \sum_{i=1}^{D_s} E_i \tag{3.5.6}$$

except now the energy variables

$$E_i = A \int_0^T |a_i(t)|^2 \, dt \tag{3.5.7}$$

are random variables. To determine the statistics of m_v, it is first necessary to specify the field statistics of each $a_i(t)$ and, from this, determine the energy statistics of each E_i. In general, the $\{E_i\}$ set is not uncorrelated, and m_v in Eq. (3.5.6) is simply a sum of correlated random energy variables.

When the random field is a Gaussian field, with independent (uncorrelated and Gaussian) field modes, the m_v variable in Eq. (3.5.6) is now a sum of independent random energy variables. Its characteristic function follows as

$$\Psi_{m_v}(j\omega) = \prod_{i=1}^{D_s} \Psi_{E_i}(j\omega\alpha) \tag{3.5.8}$$

and from Eq. (3.5.4)

$$\Psi_k(j\omega) = \prod_{i=1}^{D_s} \Psi_{E_i}[\alpha(e^{j\omega} - 1)] \tag{3.5.9}$$

Thus, we only require the characteristic function of each E_i random variable to write the characteristic function of the total detector count. Since $\Psi_k(j\omega)$ is a product, the count variable k will always be a sum of a set of random independent counts, one from each independent mode of the received field. We again see the total detector count appearing as the accumulated effect of each random field component superimposing its own independent count at the output. As stated earlier, the resulting sum count is not Poisson, since the right-hand side of Eq. (3.5.9) is not a Poisson characteristic function.

For certain types of Gaussian fields, the energy variable E_i has a well-known characteristic function. Let the received $f_r(t, \mathbf{r})$ be

$$f_r(t, \mathbf{r}) = f_s(t, \mathbf{r}) + f_b(t, \mathbf{r}) \tag{3.5.10}$$

where $f_s(t, \mathbf{r})$ is a deterministic signal field and $f_b(t, \mathbf{r})$ is a random bandlimited Gaussian noise field with optical bandwidth B_o. When this noise field is due to blackbody background noise, as discussed in Section (2.6), each diffraction-limited mode produces noise power with the flat spectrum level N_0 in Figure 3.8a, and a total modal power of $P_{bo} = N_0 B_o$. For this field $f_r(t, \mathbf{r})$, the expansion in Eq. (3.5.10) has envelope functions

$$a_i(t) = s_i(t) + b_i(t) \tag{3.5.11}$$

corresponding to the sum of a deterministic complex signal term $s_i(t)$ and a

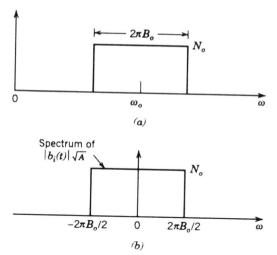

Figure 3.8. Spectrum of optical noise. (a)One-sided bandpass optical noise centered at ω_o. (b) Spectrum of the noise envelope of the process in (a).

Gaussian complex envelope modal process $b_i(t)$. The latter has an envelope magnitude process $|b_i(t)|$ with a low-pass spectrum obtained by shifting the bandpass mode spectrum in Figure 3.8a. In particular, the normalized process* $|b_i(t)|\,A^{1/2}$ has the low-pass spectrum shown in Figure 3.8b. The characteristic function of the energy variable E_i in Eq. (3.5.7) corresponding to this $a_i(t)$ can then be shown to be (see Problem 3.13)

$$\Psi_{E_i}(j\omega) = \left[\frac{1}{1 - j\omega N_0}\right]^{B_o T} \exp\left[\frac{j\omega E_{si}}{1 - j\omega N_0}\right] \tag{3.5.12}$$

where

$$E_{si} = A \int_0^T |s_i(t)|^2 \, dt \tag{3.5.13}$$

is the signal energy in the ith mode. Thus in Eq. (3.5.9)

$$\Psi_k(j\omega) = \left[\frac{1}{1 + \alpha N_0(1 - e^{j\omega})}\right]^{B_o T D_s} \exp\left[\frac{-\alpha E_s(1 - e^{j\omega})}{1 + \alpha N_0(1 - e^{j\omega})}\right] \tag{3.5.14}$$

with

$$E_s = \sum_{i=1}^{D} E_{si} \tag{3.5.15}$$

*Recall that $b_i(t)$ is the modal envelope of the field $f(t, \mathbf{r})$ and $|b_i(t)|^2$ is in units of watts/area. Hence the normalized process $|b_i(t)|A^{1/2}$ will have power units of watts.

Equation (3.5.14) is a general expression for the count characteristic function due to an impinging field composed of the sum of a deterministic signal field plus a band-limited white Gaussian noise field. It is interesting to note that Eq. (3.5.14) depends only on the total signal energy E_s, and not on the way it is distributed over the modes.

Equation (3.5.14) can be inverse Fourier transformed directly to obtain P(k). After some algebra (the computational details are outlined in Problem 3.16), the count probability can be shown to be

$$P(k) = \frac{(\alpha N_0)^k}{(1 + \alpha N_0)^{k+M}} \exp\left[\frac{-\alpha E_s}{1 + \alpha N_0}\right] L_k^{M-1}\left[\frac{-\alpha E_s}{\alpha N_0(1 + \alpha N_0)}\right] \quad (3.5.16)$$

where

$$M = B_o T D_s \quad (3.5.17)$$

and $L_k^M(x)$ is the Laguerre polynomial of order M, index k, and argument x,

$$L_k^M(x) = \sum_{j=0}^{k} \binom{M+k}{k-j} \frac{(-x)^j}{j!} \quad (3.5.18)$$

with $\binom{N}{n}$ the binomial coefficient,

$$\binom{N}{n} = \frac{N!}{n!(N-n)!} \quad (3.5.19)$$

This count probability is referred to as the *Laguerre probability*, and is an exact expression for the bandpass Gaussian field case. The Laguerre probability reduces to the special cases listed in Table 3.3 when $E_s = 0$ (noise only). The latter produce Bose–Einstein and negative binomial probabilities, dependent on the value of M in Eq. (3.5.17).

Note that P(k) in Eq. (3.5.16) depends on the parameter

$$\begin{aligned} \alpha N_0 &= \frac{\eta}{hf}\left[\frac{hf}{\exp(hf/\kappa T^\circ) - 1}\right] \\ &= \frac{\eta}{\exp(hf/\kappa T^\circ) - 1} \end{aligned} \quad (3.5.20)$$

which is negligibly small at optical frequencies. Likewise the parameter M in Eq. (3.5.17), which is the product of the number of space modes (D_s) and

TABLE 3.3 Limiting Forms of the Laguerre Count Probability when $E_s = 0$

Cases	$P(k)$	Mean
$M = 1$ (Bose–Einstein)	$\left(\dfrac{1}{1 + \alpha N_0}\right)\left(\dfrac{\alpha N_0}{1 + \alpha N_0}\right)^k$	αN_0
$M > 1$ (Negative Binomial)	$\dbinom{M + k}{k}\left(\dfrac{1}{1 + \alpha N_0}\right)^{M+1}\left(\dfrac{\alpha N_0}{1 + \alpha N_0}\right)^k$	$M\alpha N_0$
	approximation: $\dfrac{(\alpha N_0 M)^k}{k!}\, e^{-(\alpha N_0 M)}$	$M\alpha N_0$

time-bandwidth product $(B_o T)$ of the received field, is always a high number for optical systems. Under the conditions $\alpha N_0 \ll 1$ and $M \gg 1$, the terms in Eq. (3.5.16) have the following limiting forms.

$$\left(\frac{1}{1 + \alpha N_0}\right)^M \to e^{-\alpha N_0 M}$$

$$\binom{M + k}{k} \to \frac{M^k}{k!}$$

$$\left(\frac{\alpha N_0}{1 + \alpha N_0}\right)^k \to (\alpha N_0)^k \qquad (3.5.21)$$

$$L_k^M\left(\frac{-\alpha E_s}{\alpha N_0(1 + \alpha N_0)}\right) \to \frac{1}{k!}\left(M + \frac{E_s}{N_0}\right)^k$$

When these are inserted in Eq. (3.5.16), we have the approximation

$$P(k) \cong \frac{[\alpha E_s + M\alpha N_0]^k}{k!} \exp[-(\alpha E_s + M\alpha N_0)] \qquad (3.5.22)$$

Hence the true Laguerre count can be accurately approximated by a Poisson count under the stated conditions. Recall that we earlier associated Poisson counting with the detection of a deterministic (nonrandom) envelope. This is precisely what the condition $M \gg 1$ implies, since the counting is now averaged over all the modes, and the detector smoothes out the intensity variations when collected over so many modes.

Note that the parameter of the Poisson count in the approximation in Eq. (3.5.22) is the combined field energy $(\alpha E_s + M\alpha N_0)$, and the Gaussian noise field effectively adds in a count rate contribution of $\alpha N_0 M$ counts/second.

Expanding this, we see

$$\alpha N_0 M = (\alpha N_0) D_s B_o T$$
$$= \alpha T [D_s (B_o N_0)] \qquad (3.5.23)$$

Because $B_o N_0$ is the noise power per mode, then $D_s B_o N_0$ is the total noise power over all D_s modes, $(\alpha D_s B_o N_0)$ is the average count rate from this power, and multiplication by T produces the total noise count over T. Hence, Eq. (3.5.22) appears as if the noise is contributing a fixed average count $\alpha M N_0$ to the total detector count.

We can extend to the case where the signal field is itself a Gaussian stochastic process received in the presence of a background noise field. (This occurs, for example, if the signal field is passed through a random channel having scattering, turbulence, or time and space dispersion.) Thus we again expand the total received Gaussian signal field as

$$f_r(t, \mathbf{r}) = \sum_{i=1}^{D_s} a_i(t) e^{j\omega_o t} \Phi_i(\mathbf{r}) \qquad (3.5.24)$$

where again $a_i(t) = s_i(t) + b_i(t)$, as in Eq. (3.5.11). Each $a_i(t)$ is still a Gaussian process, except that both $s_i(t)$ and $b_i(t)$ are complex random time envelopes. As long as both the signal processes $\{s_i(t)\}$ and the noise processes $\{b_i(t)\}$ are independent, the $\{a_i(t)\}$ are independent, and the count probability will be generated as in Eq. (3.5.22) when $M \gg 1$, with E_s now corresponding to the average signal energy of the random signal field.

The CP count has moments directly related to the moments of the variable m_v. For example the mean count now becomes

$$\bar{k} = \sum_{k=0}^{\infty} k P(k)$$
$$= \int_0^{\infty} \left[\sum_{k=0}^{\infty} k \left(\frac{m^k}{k!} e^{-m} \right) \right] p_{m_v}(m) \, dm \qquad (3.5.25)$$
$$= \int_0^{\infty} m p_{m_v}(m) \, dm$$

Therefore,

$$\bar{k} = \overline{m_v}$$
$$= \alpha \int_0^T \overline{n(t)} \, dt \qquad (3.5.26)$$

The mean count is now given by the mean of the random energy parameter m_v. This means the mean count rate function is now

$$\overline{n(t)} = \alpha \int_A \overline{|f_r(t, \mathbf{r})|^2} \, d\mathbf{r}$$

$$= \alpha \int_A R_f(t, t, \mathbf{r}, \mathbf{r}) \, d\mathbf{r}$$

(3.5.27)

and is determined solely by the mutual coherence function of the field.

3.6 PHOTOCOUNTING WITH RANDOM PHOTOMULTIPLICATION

When a photomultiplier is used to enhance the electron flow, the randomness of the photomultiplication mechanism will influence the statistics of the output count. If we let k_1 be the electron count over a volume based on primary electron flow (as computed in Section 3.1) then the output count k_2 (number of total output electrons that flow after photomultiplication) will have a probability

$$P(k_2) = \sum_{k_1 = 0}^{\infty} P(k_2|k_1)P(k_1)$$

(3.6.1)

Here $P(k_1)$ is the primary count probability, and $P(k_2|k_1)$ is the conditional output count probability conditioned on a k_1 primary count. This conditional count depends on the type of photomultiplier used and the mechanism that generates the secondary emissions.

With a PMT having mean gain $\bar{g}$, the conditional probability is generally modeled as

$$P(k_2|k_1) = C \exp\left[\frac{-(k_2 - \bar{g}k_1)^2}{2(\xi\bar{g}k_1)^2}\right]$$

(3.6.2)

where C is a normalizing factor. Equation (3.6.2) is a Gaussian-shaped count envelope (recall k_2 only takes on integer values) centered at $\bar{g}k_1$, with a deviation spread of $(\xi\bar{g}k_1)$, with ξ the PMT spreading factor. The resulting PMT output count probability is obtained by evaluating Eq. (3.6.1) with Eq. (3.6.2) inserted. When the primary count is Poisson with a relatively high rate, $n(t) \gg 1$, $P(k_2)$ is approximately

$$P(k_2) = C \exp\left[\frac{-(k_2 - \bar{g}\bar{k}_1)^2}{2(\xi\bar{g}\bar{k}_1)^2}\right]$$

(3.6.3)

where $\bar{k}_1$ is the mean of the primary count k_1. This is now a discrete count centered at a mean count of $\bar{g}\bar{k}_1$, and standard deviation width of $(\xi\bar{g}\bar{k}_1)$. Thus the PMT high-gain secondary emissions effectively spreads the output count from a Poisson primary count to a more symmetrical Gaussian-shaped output count distribution.

For an APD, the output conditional count probability occurring after avalanche multiplication of the primary count was derived by McIntyre [6], and experimentally verified by Conradi [7], to have the form

$$P(k_2|k_1) = \frac{k_1\Gamma\left(\dfrac{k_2}{1-\gamma}+1\right)}{k_2(k_2-k_1)!\,\Gamma\left(\dfrac{\gamma k_2}{1-\gamma}+1+k_1\right)} \left[\frac{1+\gamma(\bar{g}-1)}{\bar{g}}\right]^{(k_1+\gamma k_2)/(1-\gamma)}$$
$$\cdot\left[\frac{(1-\gamma)(\bar{g}-1)}{\bar{g}}\right]^{k_2-k_1} \tag{3.6.4}$$

where $\Gamma(x)$ is a gamma function, γ is the avalanche ionization coefficient, and $\bar{g}$ is the mean gain. Typically γ has a value of 0.01 to 0.1 for silicon detectors and 0.1 to 0.5 for germanium detectors, and $\bar{g}$ is usually in the range of 10 to 200. Statistical analysis associated with the probability function in Eq. (3.6.1) is hampered by the complexity of Eq. (3.6.4), usually requiring numerical calculation [8]. A useful approximation for the APD $P(k_2)$, when $P(k_1)$ is Poisson in Eq. (3.6.1), is given by Webb [9] as

$$P(k_2) = \frac{1}{(2\pi C_1^2)^{1/2}}\left[\frac{1}{1+\dfrac{(k_2-\bar{g}\bar{k}_1)^{3/2}}{C_1 C_2}}\right]\exp\left\{\frac{-(k_2-\bar{g}\bar{k}_1)^2}{2C_1^2\left[1+\dfrac{(k_2-\bar{g}\bar{k}_1)}{C_1 C_2}\right]}\right\} \tag{3.6.5}$$

where

$$C_1^2 = (\bar{g}\bar{k}_1)^2 F - 1$$
$$C_2 = \bar{g}(\bar{k}_1 F)^{1/2}/(F-1) \tag{3.6.6}$$

and

$$F = \gamma\bar{g} + \left(2 - \frac{1}{\bar{g}}\right)(1-\gamma) \tag{3.6.7}$$

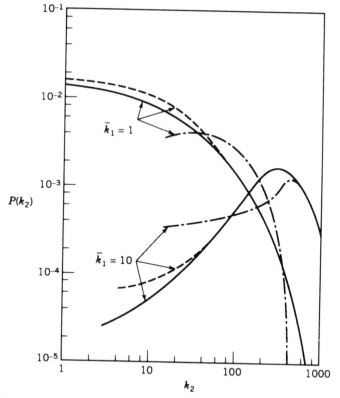

Figure 3.9. Comparison of APD output counting statistics. $\bar{k}_1 =$ average number of primary electrons, $\gamma = 0.028$, $\bar{g} = 50$. —— Eq. (3.6.1); ———— Eq. (3.6.5), – – – – Eq. (3.6.8).

If $\bar{g}(\bar{k}_1/F)^{1/2} \gg 1$, then Eq. (3.6.5) approaches the Gaussian-shaped probability envelope,

$$P(k_2) = \frac{1}{[2\pi(\bar{g}\bar{k}_1)^2(F-1)]^{1/2}} \exp\left[-\frac{(k_2 - \bar{g}\bar{k}_1)^2}{2(\bar{g}\bar{k}_1)^2(F-1)} \right] \qquad (3.6.8)$$

We emphasize that $P(k_2)$ is again a discrete probability on the output counts k_2, and has values only at the nonnegative integers. Figure 3.9 compares the probability envelopes in Eq. (3.6.1) computed by direct numerical calculation with Eqs. (3.6.5) and (3.6.8) at two specific values of $\bar{k}_1$, showing their range of comparison. In particular, we note the departure at the lower values of k_2 and the relatively accurate approximation of Eq. (3.6.8) for k_2 values greater than approximately 50. Because signal counts usually exceed this value, Eq. (3.6.8) indeed serves as a valid approximation to the true $P(k_2)$ envelope, with slight departure from the exact over the tails.

We note that $P(k_2)$ appears with a center value at $\bar{g}\bar{k}_1$ and a variance $(\bar{g}\bar{k}_1)^2(F - 1)$. This latter term identifies the parameter F as the excess noise factor of the APD, as defined in Eq. (3.1.3). Hence F in Eq. (3.6.7) and Table 3.1 indeed defines the noise factor of the APD. We see it depends on the ionization coefficient γ, and, for reasonably large gains, varies as $F \cong 2 + \gamma\bar{g}$, and, therefore, is directly proportional to gain.

3.7 SHOT NOISE PROCESSES

The photodetection output current $i(t)$ is described by the shot noise process in Eq. (3.1.4) as

$$i(t) = \sum_{j=1}^{k(0,t)} g_j h(t - z_j) \tag{3.7.1}$$

where again $k(0, t)$ is the random count process produced over the volume consisting of the detector area and the time interval $(0, t)$. The count probability over any such volume can be determined from the analysis in Sections 3.4 and 3.5. To investigate the statistical properties of this output current, however, we need additional parameter descriptions beyond just the count process. In particular, we need the statistical properties of the gain parameter g_j in Section 3.6 (if photomultiplication is used), and we need the statistics of the occurrence times $\{z_j\}$. These occurrence times of the individual electrons in the shot noise process represent a collection of random locations in time inherently linked to the counting process. In this section, we derive the joint probability density of a given number of such locations in a specified time interval (T_1, T_2) for a Poisson counting process. Specifically, we derive the joint density of the random sequence $\mathbf{z} = (z_1, z_2, \ldots, z_k)$ over (T_1, T_2) given k Poisson counts in that interval. We write this conditional probability density as $p_z(z_1, z_2, \ldots, z_k | k)$, and derive its mathematical form as follows. Consider the interval (T_1, T_2) to contain the infinitesimal slots $(t_1, t_1 + \Delta t)$, $(t_2, t_2 + \Delta t), \ldots, (t_k, t_k + \Delta t)$, as shown in Figure 3.10. The probability of getting a Poisson occurrence during each of the slots above and simultaneously getting exactly k occurrences is simply the probability of getting one occurrence in each slot and not getting any occurrence outside the slots. Because the slots and their exterior regions are disjoint, the independent interval property of Poisson processes means that this joint probability is

$$\text{Prob}[z_1 \in (t_1, t_1 + \Delta t), z_2 \in (t_2, t_2 + \Delta t), \ldots, z_k \in (t_k, t_k + \Delta t)]$$
$$= \prod_{j=1}^{k} \left[\int_{t_j}^{t_j + \Delta t} n(t)\, dt \right] \left\{ \exp\left[-\int_{T_1}^{T_2} n(t)\, dt \right] \right\} \tag{3.7.2}$$

where $n(t)$ is the count intensity of the received field and indicates the rate of

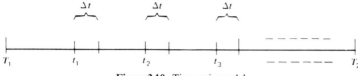

Figure 3.10. Time axis model.

the count occurrences. Now as $\Delta t \to 0$, Eq. (3.7.2) approaches

$$\prod_{j=1}^{k} n(t_j)\Delta t \left\{\exp\left[-\int_{T_1}^{T_2} n(t)\, dt\right]\right\} \tag{3.7.3}$$

which is obtained by again applying Eq. (3.3.10). On the other hand, for small Δt, Eq. (3.7.2) could have been written

$$\text{Prob}[z_1 \in (t_1, t_1 + \Delta t), \ldots, z_k \in (t_k, t_k + \Delta t)]$$
$$= k!\, \text{Prob}[z_1 = t_1, \ldots, z_k = t_k](\Delta t)^k \tag{3.7.4}$$

where the $k!$ term takes into account all the ways in which k events can occur in k slots. Equating Eqs. (3.7.4) and (3.7.3) identifies the joint density as

$$p(z_1, z_2, \ldots, z_k, k) = \frac{1}{k!} \prod_{j=1}^{k} n(z_j) \exp\left[-\int_{T_1}^{T_2} n(t)\, dt\right] \qquad z_j \in (T_1, T_2) \tag{3.7.5}$$

Now the probability of k Poisson occurrences in an arbitrary interval (T_1, T_2), given the count intensity $n(t)$, is

$$P_k(k) = \text{Pos}\left[k, \int_{T_1}^{T_2} n(t)\, dt\right]$$
$$= \frac{[\int_{T_1}^{T_2} n(t)\, dt]^k}{k!} \exp\left[-\int_{T_1}^{T_2} n(t)\, dt\right] \tag{3.7.6}$$

The conditional density of the k occurrence times $\{z_j\}$ is then obtained by dividing Eq. (3.7.5) by Eq. (3.7.6), yielding

$$p[z_1, z_2, \ldots, z_k | k] = \frac{\displaystyle\prod_{j=1}^{k} n(z_j)}{\left[\displaystyle\int_{T_1}^{T_2} n(t)\, dt\right]^k}$$
$$= \prod_{j=1}^{k} \frac{n(z_j)}{(m_v)^k} \tag{3.7.7}$$

where

$$m_v = \int_{T_1}^{T_2} n(t)\, dt \tag{3.7.8}$$

Here m_v is again the mean of the Poisson density in Eq. (3.7.6), or, equivalently, the average number of counts $\bar{k}$ occurring over (T_1, T_2). Note that the desired conditional density for the occurrence times factors into the product of the individual densities

$$p_{z_j}(z) = \frac{n(z)}{\bar{k}} \qquad z \in (T_1, T_2) \tag{3.7.9}$$

for each j. This immediately implies that Poisson occurrence times are independent, each having a location density over t given by the normalized count intensity process $n(t)/\bar{k}$. Note that this latter probability density is nonnegative and integrates to unity. Furthermore, the result is valid for any interval (T_1, T_2). The result also serves as a further interpretation of the meaning of the intensity process $n(t)$; during any time interval, it indicates the probability of releasing an electron at any particular time. Thus, times at which the intensity level is high physically implies times at which electrons are most likely to be released, whereas low intensities can be interpreted as times when electrons are least likely to be released. In particular, times at which $n(t) = 0$ have a zero probability of releasing electrons.

With the statistics known for the count, occurrence times, and gain, the shot noise process $i(t)$ in Eq. (3.7.1) can now be analyzed as a random process in time. As such, this random process will have statistical moments at any t that aid in describing the process itself. These moments can be determined by straightforward analysis. We first calculate conditional moments of $i(t)$ for a given number of counts k in a specified interval with a given intensity function $n(t)$ and then average over k. Using overbars to denote the averaging,

$$\overline{i(t)} = \overline{\sum_{j=0}^{k(0,t)} g_j h(t - z_j)}$$

$$= \int_{-\infty}^{t} \cdots \int_{-\infty}^{t} \overline{\sum_{j=1}^{k} \bar{g} h(t - z_j) p(z_1, z_2, \ldots, z_k \,|\, k)}\, dz_1 \cdots dz_k \tag{3.7.10}$$

where $\bar{g}$ is the mean of g_j. The density in the integrand is given by Eq. (3.7.7), so that Eq. (3.7.10) becomes

$$\overline{i(t)} = \overline{\sum_{j=1}^{k} \int_{-\infty}^{t} \bar{g} h(t - z) p_{z_j}(z)\, dz}$$

$$= k\bar{g} \int_{-\infty}^{t} h(t-z) \overline{\left(\frac{n(z)}{k}\right)} \, dz \qquad (3.7.11)$$

$$= \bar{k}\bar{g} \int_{-\infty}^{t} h(t-z) \left(\frac{\overline{n(z)}}{k}\right) \, dz$$

This reduces to,

$$\overline{i(t)} = \bar{g} \int_{-\infty}^{t} h(t-z)n(z) \, dz \qquad (3.7.12)$$

Thus, the mean of the shot noise appears as a filtering (convolution) of the shot noise count rate process by the electron functions $h(t)$. In particular, we note that the mean depends on t and, therefore, evolves as a function of time. From this point of view, the detector current attempts to "follow" the count rate time process $n(t)$ produced by the impinging field on the photodetector.

The second moment, or mean-square value of the shot noise is related to the power in the process and can be determined similarly,

$$\overline{i^2(t)} = \int_{-\infty}^{t} \cdots \int_{-\infty}^{t} \overline{\left[\sum_{j=1}^{k} g_j h(t-z_j)\right]^2} p(z_1, z_2, \ldots, z_k | k) \, dz_1 \cdots dz_k \qquad (3.7.13)$$

Expanding the square and making use of the independence of the location times and gains, we obtain

$$\overline{i^2(t)} = \left\{ \bar{k} \int_{-\infty}^{t} \overline{g^2} h^2(t-z) p_{z_j}(z) \, dz \right.$$

$$\left. + \overline{(k^2 - k)} \left[\int_{-\infty}^{t} \bar{g} h(t-z) p_{z_j}(z) \, dz \right]^2 \right\} \qquad (3.7.14)$$

where $\overline{g^2}$ is the mean-square gain. To evaluate the averages over the k parameters, we must invoke the Poisson counting moments in Table 3.2. Therefore, Eq. (3.7.14) becomes

$$\overline{i^2(t)} = \bar{k} \int_{-\infty}^{t} \overline{g^2} h^2(t-z) p_{z_j}(z) \, dz + (\bar{g}\bar{k})^2 \left[\int_{-\infty}^{t} h(t-z) p_{z_j}(z) \, dz \right]^2$$

$$= \int_{-\infty}^{t} \overline{g^2} h^2(t-z)n(z) \, dz + (\bar{g})^2 [\overline{i(t)}]^2 \qquad (3.7.15)$$

The variance of the shot noise about the mean is precisely the first term above,

$$\text{var } i(t) = \overline{g^2} \int_{-\infty}^{t} h^2(t-z)n(z) \, dz \qquad (3.7.16)$$

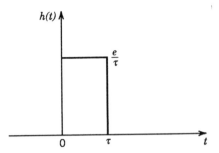

Figure 3.11. Square pulse model for $h(t)$.

The variance of the shot noise process at any t, therefore, also evolves as a filtered version of the count rate process. Note also that the shot noise variance depends on the mean-square value of the photomultiplier gain.

The first two moments of the shot noise are inherently linked to $n(t)$ of the radiation field and we now examine this relation in more detail. Consider the case when the gain is ideal, and the electron functions $h(t)$ are modeled as short rectangular pulses of width τ, as shown in Figure 3.11. The mean and variance of $i(t)$ in Eqs. (3.7.12) and (3.7.16) become

$$\overline{i(t)} = (ge/\tau)\overline{k}(t - \tau, t) \tag{3.17a}$$

$$\text{var } i(t) = (ge/\tau)^2 \overline{k}(t - \tau, t) \tag{3.17b}$$

where

$$\overline{k}(t - \tau, t) = \int_{t-\tau}^{t} n(z)\, dz \tag{3.7.18}$$

is the average count over the interval $(t - \tau, t)$, that is, the count over the last τ seconds at each t. Because the square root of the variance of any random variable is an indication of the average spread about the mean, we can depict the current shot noise process as in Figure 3.12. Here, we have chosen an

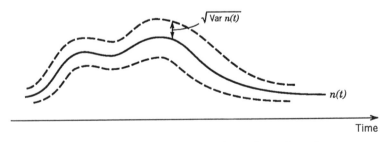

Figure 3.12. Plot of $n(t)$ and spread due to shot noise variance.

arbitrary intensity function $n(t)$ and shown the mean variation and corresponding average spread region of the random process $i(t)$. Because of these results, the spread varies in accordance with the mean. At any time t, we can consider the $i(t)$ to be close, in an average sense, to the count rate function (its mean value) if the spread is small with respect to the mean. Specifically, if we require the squared ratio of the spread to the mean to be small, we would require var $i(t)/(\overline{i(t)})^2 \ll 1$ or equivalently,

$$\overline{k}(t - \tau, t) \gg 1 \tag{3.7.19}$$

Hence, the shot noise random process will closely resemble the count rate process if the average number of electron occurrence in a time interval of its component functions is high. This is equivalent to a condition that the shot noise be generated with high "density" (many electrons overlapping at the same time). As we have seen, this will occur if the count rate is high and the impinging field is collected under high-power conditions. If the overlap count is slight, the shot noise spread is large relative to the mean, and no longer resembles the count rate function, exhibiting the discreteness of the counting. At high count rates, the discreteness is no longer evident, and the photodetector approaches an ideal detector of the count rate function. Because the latter is proportional to the received field power, the photodetector behaves as an instantaneous power detector.

Although the mean and variance of $i(t)$ were readily computed, the actual probability density of $i(t)$ at any t is more difficult to determine. However, if the current rate is high (many counts expected at each t) the shot noise will correspond to a large sum of random variables at any t. By direct application of the Central Limit Theorem [10], it can be inferred that $i(t)$ will approach a Gaussian random process (see also Problem 3.22). This establishes that the first order density of $i(t)$ at any t can be approximated as

$$p_{i(t)}(x) = \frac{1}{[2\pi \text{var } i(t)]^{1/2}} \exp\left[-\frac{(x - \overline{i(t)})^2}{2(\text{var } i(t))}\right] \tag{3.7.20}$$

when $n(t) \gg 1$. This is a Gaussian density at each t, having a time-varying mean and variance. This again substantiates the fact that the individual occurrences of the discrete electron functions "fuses" into a continuous random process, and the discreteness of the photodetection is lost. This Gaussian shot noise model is convenient in communication analysis.

3.8 SPECTRAL DENSITY OF SHOT NOISE

The power spectral density of a random process is important in specifying the power distribution of the process as a function of frequency. In this section, we derive the power spectral density by computing the time averaged power at any radian frequency ω. That is, we consider the current power spectrum of $i(t)$ as

$$S_i(\omega) = \lim_{T \to \infty} \frac{1}{2T} \overline{|X_T(\omega)|^2} \tag{3.8.1}$$

where $X_T(\omega)$ is the Fourier transform of a sample function of the current process $i(t)$ restricted to the interval $(-T, T)$. [Equation (3.8.1) is the classical definition of a power spectrum.] The function $|X_T(\omega)|^2$ is the energy density at each ω for the sample function, and therefore is a random variable at each ω. It must, therefore, be averaged over the statistics of the process. Thus $S_i(\omega)$ is effectively the time average of the ensemble average of the energy density at each ω. Now if we condition upon $\{z_i\}$ and k, we have

$$
\begin{aligned}
X_T(\omega) &= \int_{-T}^{T} \left[\sum_{i=1}^{k} g_i h(t - z_i) \right] e^{-j\omega t}\, dt \\
&= \sum_{i=1}^{k} \int_{-T}^{T} g_i h(t - z_i) e^{-j\omega t}\, dt \\
&= \sum_{i=1}^{k} g_i e^{-j\omega z_i} H_T(\omega)
\end{aligned}
\tag{3.8.2}
$$

where $H_T(\omega)$ is the Fourier transform of $h(t)$, $-T < t < T$. Thus,

$$
\begin{aligned}
|X_T(\omega)|^2 &= X_T(\omega) X_T^*(\omega) \\
&= \sum_{i=1}^{k} g_i e^{-j\omega z_i} H_T(\omega) \sum_{q=1}^{k} g_q e^{j\omega z_q} H_T^*(\omega) \\
&= |H_T(\omega)|^2 \sum_{i=1}^{k} \sum_{q=1}^{k} e^{j\omega(z_q - z_i)} g_i g_q
\end{aligned}
\tag{3.8.3}
$$

The ensemble average over $\{z_j\}$, conditioned upon k, can now be taken. Because the exponent is unity when $q = i$, we have

$$
\begin{aligned}
\overline{|X_T(\omega)|^2} &= |H_T(\omega)|^2 \left[\overline{g^2} k + (k^2 - k) \int_{-T}^{T} g_i e^{-j\omega z_i} p_{z_i}(z_i)\, dz_i \right. \\
&\quad \left. \cdot \int_{-T}^{T} g_q e^{j\omega z_q} p_{z_q}(z_q)\, dz_q \right]
\end{aligned}
\tag{3.8.4}
$$

The subsequent averaging over k, noting that $\overline{k^2} - \overline{k} = (\overline{k})^2$, yields

$$\overline{|X_T(\omega)|^2} = |H_T(\omega)|^2 [\overline{g^2}\,\overline{k}(-T, T) + (\overline{g})^2 N_T(\omega) N_T^*(\omega)] \tag{3.8.5}$$

where

$$\bar{k}(-T, T) = \int_{-T}^{T} n(t) \, dt \tag{3.8.6}$$

$$N_T(\omega) = \int_{-T}^{T} n(u)e^{-j\omega u} \, du \tag{3.8.7}$$

Here $N_T(\omega)$ is the Fourier transform of the count rate function $n(t)$, $-T \leqslant t \leqslant T$. Finally, we have

$$S_i(\omega) = \lim_{T \to \infty} \frac{1}{2T} \overline{|X_T(\omega)|^2}$$

$$= |H(\omega)|^2 [\overline{g^2}\bar{n} + (\bar{g})^2 F_n(\omega)] \tag{3.8.8}$$

where

$$\bar{n} = \lim_{T \to \infty} \bar{k}(-T, T)/2T \tag{3.8.9a}$$

$$F_n(\omega) = \lim_{T \to \infty} |N_T(\omega)|^2/2T \tag{3.8.9b}$$

Here $\bar{n}$ is the time averaged count rate, and $F_n(\omega)$ is the corresponding frequency spectrum of $n(t)$. Note that the spectral density of $i(t)$ obtained in this way now exhibits the frequency content of the received field intensity via the count rate process $n(t)$.

For conditional Poisson (CP) shot noise process, the intensity $n(t)$ is itself a sample function of a random process, and the calculation of the spectrum in Eq. (3.8.8) requires an additional averaging over the ensemble of intensity processes. Hence, we can generalize to

$$S_i(\omega) = |H(\omega)|^2 [\overline{g^2}\bar{n} + (\bar{g})^2 S_n(\omega)] \tag{3.8.10}$$

where $\bar{n}$ is now the statistical mean of the time averaged count rate $n(t)$, and $S_n(\omega)$ is it spectral density. The spectrum in Eq. (3.8.10) is sketched in Figure 3.13 and takes the form of a filtered intensity spectrum $(\bar{g})^2|H(\omega)|^2 S_n(\omega)$ with an additive spectrum $\overline{g^2}\bar{n}|H(\omega)|^2$. For wide bandwidth detectors, $H(\omega) \cong e$ for all ω, and the first term in Eq. (3.8.10) appears as a spectrally white additive process of height $e^2\bar{n}\bar{g}$. Because this appears as an added "noise" to the desired intensity spectrum $S_n(\omega)$ it is called the *shot noise level* of the detector. Note that this noise always appears when photodetecting an optical field, and is due to the discrete nature of the detector model. If the detector has a finite bandwidth, then $|H(\omega)|^2$ effectively "shapes" the signal and shot noise spectrum. As long as the desired frequency content of $S_n(\omega)$ lies within the

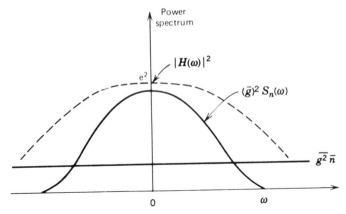

Figure 3.13. Shot noise power spectrum.

frequency band of the $H(\omega)$ function, its spectrum is not distorted. This simply means the detector bandwidth should exceed the desired signal frequency band of $S_n(\omega)$. The intensity spectrum $S_n(\omega)$, in general, contains portions from both the desired source intensity and portions of background noise, if any is present. An examination of these spectra, with specific applications to communication receivers, will be considered in Chapters 4 and 5.

Let us examine the flat shot noise spectrum of the detector output. Specifically, let us compute the area under this portion of the spectrum, which can be considered as the effective added shot noise power. This power becomes

$$\text{shot noise power} = \frac{1}{2\pi} \int_{-\infty}^{\infty} \overline{g^2} |H(\omega)|^2 \bar{n} \, d\omega \qquad (3.8.11)$$

Now by the application of Parseval's theorem, we rewrite this as

$$\text{shot noise power} = \frac{1}{2\pi} \int_{-\infty}^{\infty} \bar{n} |H(\omega)|^2 \, d\omega$$

$$= \overline{g^2} \int_{-\infty}^{\infty} h^2(t) \bar{n} \, dt \qquad (3.8.12)$$

$$= \text{var } i(t)$$

Hence the shot noise power (area under the shot noise portion of the spectrum) is precisely the variance of the detector output process, for stationary intensities. Thus the photodetector output indeed appears as the sum of a signal portion $S_n(\omega)$ and an additive shot noise interference that contributes an additive variance to its observation.

Dark current in photodetectors corresponds to the random emission of electrons at a fixed rate, when no field is being detected. As such, dark current produces its own shot noise process. If the dark current emission rate is n_{dc} counts per second, it will produce an output dc dark current of $I_{dc} = n_{dc}e$ amps. This means the dark current alone will produce a shot noise spectrum given by Eq. (3.8.10), which is

$$
\begin{aligned}
S_{dc}(\omega) &= |H(\omega)|^2 [n_{dc} + 2\pi n_{dc}^2 \delta(\omega)] \\
&= |H(\omega)|^2 \left[\frac{I_{dc}}{e} + 2\pi \left(\frac{I_{dc}}{e} \right)^2 \delta(\omega) \right]
\end{aligned}
\tag{3.8.13}
$$

Thus dark current adds to the shot noise level (first term) and inserts a dc current in the output, both terms dependent on I_{dc} of the detector. The latter is generally a rated parameter of the photodetector, and depends on both its operating temperature and the size of its emitting area, as listed in Table 3.1.

Because the detected spectrum in Eq. (3.8.10) has the form of a signal and noise, there is a tendency to view the photodetected output as itself a signal plus noise waveform. Although such an interpretation can be useful when determining detector output signal-to-noise ratios, the reader is cautioned against a liberal application of this model. The difficulty is that signal and shot noise are not independent, and usual signal plus noise interpretations familiar to communication engineers, can sometimes lead to false conclusions.

PROBLEMS

3.1   Given the Poisson random variable with the discrete probability in Eq. (3.3.19), its jth moment is computed as

$$
\overline{k^j} = \sum_{k=0}^{\infty} k^j \operatorname{Pos}(k, m_v)
$$

Determine the mean value, mean-square value, and variance of the Poisson variable.

3.2   The characteristic function $\Psi(j\omega)$ of any discrete random variable is computed from its probability function $P(k)$ as

$$
\Psi(j\omega) = \sum_{k=0}^{\infty} e^{j\omega k} P(k)
$$

The moment generating function $M(z)$ is obtained from $\Psi(j\omega)$ by replacing $j\omega = \ln(1 - z)$. Show that

$$P(k) = \frac{(-1)^k}{k!} \left[\frac{d^k}{dz^k} M(z) \right]_{z=1}$$

3.3   Use the results of Problem 3.2 to derive the Poisson count probability from its characteristic function.

3.4   Electrons are emitted from a photodetecting surface according to a Poisson probability with mean value m. Suppose, however, the probability that a given emitted electron will be collected at the output is η. Determine the resulting count probability of the electrons collected at the output. *Hint*: Treat the conversion of each emitted electron to an output electron as a binary random variable with probability η.

3.5   Let W_n be the time for n events to occur in a counting process $k(t)$. The variable W_n is called the *waiting time* to the nth event. Show that if the events occur with a Poisson probability with mean rate v events/second then

(a) $$\mathrm{Prob}[W_n > t] = \mathrm{Prob}[k(t) < n] = \sum_{j=0}^{n-1} e^{-vt} \frac{(vt)^j}{j!}$$

(b) $$\bar{W}_n = \frac{n}{v}$$

(c) $$\mathrm{var}\, W_n = \frac{n}{v^2}$$

(d) The time T between events is random with an exponential distribution.

3.6   A laser point source at $\lambda = 1\,\mu m$ produces a power of $10^{-12}\,W$ at the receiver lens. The photodetector has a quantum efficiency of 0.75.

(a) At what rate will the photodetector produce primary electrons?

(b) What will be the average number of electrons emitted over a time period from 2 to 2.5 sec?

(c) What will be the rate in (a) if the receiver lens is reduced by 1/2?

3.7   Consider the laser system in Figure P3.7. The laser operates at $\lambda = 1\,\mu m$, and can deliver $10^{-10}\,W$ to the receiver. The laser is at point A during the first microsecond, and then moves to point B during the second microsecond (still pointed at the receiver). Neglect background light.

(a) Determine the mean count at the photodetector output during the first microsecond.

(b) Repeat for the second microsecond.

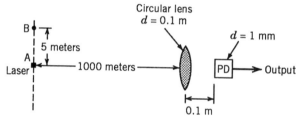

Figure P3.7

3.8 A point light source at $\lambda = 1\,\mu m$ is located at point B in Figure P3.7. At time $t = 0$, it moves with constant velocity of $10^8\,m/sec$ toward point A. The source transmits at all times a plane wave with intensity $10^{-6}\,W/m$ to the same receiver. Sketch a plot of the count rate process of the detector. Label the curve.

3.9 A 10-km fiber is connected directly to a photodetector. The fiber has an attenuation of $2\,dB/km$. A constant power of $1\,mW$ at $\lambda = 1\,\mu m$ is inserted into the fiber with an input coupling loss of $10\,dB$. Assume the photodetecting area is larger than the fiber core. What is the average number of fiber-induced electrons that will be emitted in the photodetector over a 1 nsec (10^{-9} sec) time period?

3.10 A photomultiplier produces primary electrons with a Poisson count having a mean value of 10. The detector has an ideal gain of 100.

 (a) What s the probability that a total of 1000 electrons will occur at the output?

 (b) That a total of 1050 output electrons will occur?

 (c) That a total of 1100 will occur?

 (d) State a gneral expression for the probability that n total output electrons will occur.

3.11 Write the first three moments of a CP count variable k in terms of the moments of the random energy variable m_v.

3.12 Given two conditional Poisson counting processes produced over an area A and time T. Show that the correlation between the random counts from each process is related to the integral of the correlation of the field intensities that generated them.

3.13 If x is a complex Gaussian variable with mean s and variance $\gamma/2$ per component, then $|x|^2$ has the Rician density

$$p_{|x|^2}(z) = \frac{1}{\gamma}\exp\left[\frac{-(|s|^2 + z)}{\gamma}\right]I_0\left(\frac{2|s|z^{1/2}}{\gamma^{1/2}}\right)$$

Determine the characteristic function of $|x|^2$ using the following steps:

(a) Write the characteristic function as an integral.

(b) Modify the integrand with the substitution $y = z(1 - \gamma j\omega)$.

(c) Modify again with the substitution $s' = s/(1 - j\gamma\omega)^{1/2}$.

(d) Rewrite the integral so its value is one.

(e) Show that the remaining factor yields

$$\Psi_{|x|^2}(j\omega) = \left(\frac{1}{1 - j\gamma\omega}\right) \exp\left[\frac{|s|^2 j\omega}{(1 - j\gamma\omega)}\right]$$

3.14 Show that the inverse Fourier transform of Eq. (3.5.14) is Eq. (3.5.16), using the following Laguerre function identity:

$$(1 - z)^{-(\alpha + 1)} \exp\left[\frac{xz}{z - 1}\right] = \sum_{n=0}^{\infty} L_n^\alpha(x) z^n$$

3.15 A Bose-Einstein count probability is given by

$$P(k) = \left(\frac{1}{1 + m_v}\right)\left(\frac{m_v}{1 + m_v}\right)^k$$

(a) Determine its mean value (see Problem 3.1).

(b) Show that its characteristic function is

$$\Psi_k(j\omega) = \frac{1}{1 + m_v(1 - e^{j\omega})}$$

3.16 **(a)** Show that the photodetected count over an interval T due to a single mode of a bandlimited white (N_0, B_o) Gaussian noise field is a Bose-Einstein count if $B_o T = 1$.

(b) Show that the count due to the noise field in Part (a) when $M = B_o T D_s \gg 1$ is given by

$$P(k) = \binom{k + M - 1}{k}\left(\frac{1}{1 + \alpha N_0}\right)^M \left(\frac{\alpha N_0}{1 + \alpha N_0}\right)^k$$

Hint: Start with Laguerre count probability and let the signal component be zero.

3.17 Given a Laguerre count probability with $M = 2$ modes, a signal count of 5, and a background average count of 1 per mode. Compute the probability of exactly one count occurring.

3.18 A unit gain photodetector operates with a diffraction-limited field of view of 1 μrad and a optical bandwidth of 10^{10} Hz. A deterministic signal field produces a signal count rate of 100 counts/sec. A Gaussian noise field has produced the parameter $\alpha N_0 = 10$. The detector output is integrated for T sec. Using the Poisson count approximation, estimate the probability of getting exactly 10 output-detected photoelectrons for the following cases:

(a) $T = 10^{-12}$ sec (1 psec).

(b) $T = 10^{-6}$ sec (1 μsec).

(c) $T = 10^{-12}$ sec, with the field of view increased to 10 μrad and the signal removed.

3.19 A photomultiplier tube produces output counts k from emitted primary counts according to the Gaussian shaped distribution

$$P(k \mid n) = C \exp\left[\frac{-(k - \bar{g}n)^2}{2\sigma^2}\right]$$

Write an integral expression for the probability of getting k output electrons when photodetecting a random field of integrated intensity m having probability density $p(m)$.

3.20 A photodetector receives a plane wave field of power P_r over its detecting area. The detector has the electron response function $h(t) = (ep) \exp[-pt], t > 0$.

(a) Determine the mean of the output current $i(t)$.

(b) Determine the variance of $i(t)$ at any t.

(c) Determine the power spectrum of the output current, and sketch its shape.

3.21 A photodetector has the transfer function $|H(\omega)|^2 = e^2[1 + (\omega/\omega_b)^2]^{-1}$ and receives a plane wave of power P_r. Neglect background light and dark current. Its output current has added to it an independent random white noise current process with spectral level $N_{oC}/2$. Sketch the combined output (photodetector plus noise) spectrum.

3.22 Let $x(t)$ be a random process with semiinvariants χ_i. Define the normalized $x(t)$ as

$$\hat{x}(t) = \frac{x(t) - \chi_1}{\sqrt{\chi_2}}$$

with semiinvariants $\hat{\chi}_i$. If $x(t)$ is a Gaussian process then $\hat{\chi}_1 = 0$, $\hat{\chi}_2 = 1$, and $\hat{\chi}_q = 0$ for all $q > 2$. For a Poisson shot noise process $i(t)$

$$\chi_q = \int_{-\infty}^{\infty} h^q(t - z)n(z) \, dz$$

Determine the conditions on $n(t)$ for $i(t)$ to approach a Gaussian shot noise process.

3.23 A circuit indicates sunrise by switching on when the average (dc) current at the output of a photodetector reaches one microamp. If the photodetector is pointed at the sky with the parameters below, at what sky radiance will the circuit switch on? Field of view = 10^{-2} steradians, gain = 100, receiving area = $10^{-12}\,m^2$, Optical bandwidth = $10^{12}\,Hz$, $\alpha = 10^{18}/1.6$.

3.24 An unmodulated optical plane wave with intensity I watts/area impinges on two photodetectors placed side by side. The detectors have areas A_1 and A_2 and gains g_1 and g_2. The detector output currents are then subtracted. Assuming the $h(t)$ response functions in each detector are delta functions, determine the spectral density of the subtractor output current. Neglect background noise and dark current.

3.25 A plane wave field with intensity I W/area impinges on the detector systems shown in Figure P3.25. Assume all detectors have ideal gain g and efficiency η. Neglect background noise and dark current. Determine the difference (if any) in the spectral densities at the outputs of the two systems.

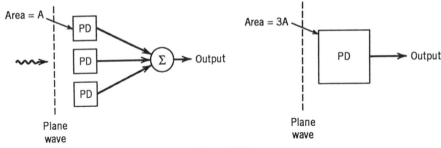

Figure P3.25.

REFERENCES

1. E. O'Niell, *Introduction to Statistical Optics*, Addison Wesley, Reading, MA. 1963.
2. R. Smith, "Photodetectors for fiber systems", *Proc. IEEE*, 68(10), October 1980.
3. T. Lee and T. Li, "Photodetectors", in *Optical Fiber Communications*, edited by S. Miller, Academic Press, New York, 1979.
4. A. Verdeyan, *Laser Electronics*, Prentice Hall, Englewood Cliffs, NJ, 1981.
5. J. Palais, *Fiber Optic Communications*, Prentice Hall Englewood Cliffs, NJ, 1988, Chapter 7.
6. R. McIntyre, Distribution of gains in uniformly multiplying avalanche photodiodes, *IEEE Trans. Electron Devices*, 6, 703–712, June 1972.

7. J. Conradi, Distribution of gains in avalanche photodiodes—experimental results, *IEEE Trans on Electron Devices* 6, 713–718, June 1972.

8. N. Sorensen and R. Gagliardi, Performance of optical receivers with APD, *IEEE Trans. Comm.* 27, pp 1315–1321 November (1979).

9. D. Webb, Properties of avalanche photodiodes, *RCA Rev.* 35, pp 234–278 September (1974).

10. H. Cramer, *Mathematical Methods In Statistics*, Princeton Press, Princeton, NJ, 1946.

4

NONCOHERENT (DIRECT)
DETECTION

In Chapter 3, we concentrated on developing a statistical model for an optical receiver. We are now interested in applying this model to the analysis and synthesis of optical communication systems. We begin in this chapter by devoting attention to a communication system employing noncoherent or direct detection. Noncoherent detection occurs when no use is made of the spatial coherence of the optical field, and the detector responds only to the power in the received field. We consider as a prime objective the accurate transmission of a desired data waveform from the transmitter to the receiver output using this type of system.

4.1 THE NONCOHERENT COMMUNICATION SYSTEM MODEL

The typical model of a noncoherent (direct detecting) communication system is shown in Figure 4.1. The desired information is intensity modulated onto an optical source and transmitted to the receiver over a space or fiber link. The optical receiver collects the field with the receiver area. After preprocessing in the receiver front end (focusing and filtering), the collected field is imaged through the receiver lens system onto the photodetecting surface in the focal plane. In a space system, background radiation is also collected by the lens and focused along with the transmitted field. The photodetector, being basically a power detecting device, responds to the instantaneous field count rate process produced from the receiver area. Its output appears as a shot noise process whose count rate is proportional to the instantaneous receiver power. This output shot noise represents the demodulated optical signal. For the receiver to recover the desired signal, it is necessary that the transmitted information be associated with the intensity variation of the transmitted field. The information waveform may be directly modulated onto the intensity of the transmitted field, or it may be subcarrier-modulated prior to optical intensity modulation. In the latter case, further receiver processing is necessary, following optical

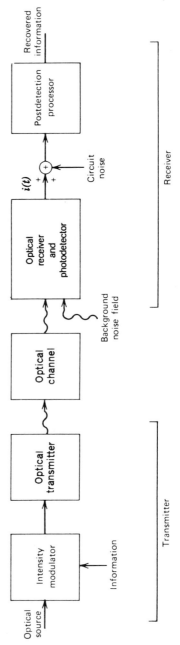

Figure 4.1. The direct detecton optical system.

detection, to recover the information waveform. To design the proper post-detection processing it is first necessary to derive the characteristics of the photodetected output.

The field over aperture area A can be written as the sum of the source signal field $f_s(t, \mathbf{r})$ and any background or input noise field $f_b(t, \mathbf{r})$, if present, as

$$f_r(t, \mathbf{r}) = f_s(t, \mathbf{r}) + f_b(t, \mathbf{r}) \qquad \mathbf{r} \in A \tag{4.1.1}$$

For an intensity modulated laser point source field, $f_s(t, \mathbf{r})$ is a plane wave whose intensity is modulated at the transmitter to produce

$$|f_s(t, r)|^2 = I_s[1 + \beta m(t)] \qquad \mathbf{r} \in A \qquad \text{W/area} \tag{4.1.2}$$

where I_s is the average received intensity, $m(t)$ is the modulating signal, and β is a scale factor. The factor β is the fraction of I_s that represents the linear range of the output power characteristic of the transmitter laser (Figure 4.2), and directly scales the modulation waveform. Since the field intensity in Eq. (4.1.2) is a nonnegative time function, $[1 + \beta m(t)] \geq 0$ for all t, so that

$$\beta|m(t)| \leq 1 \tag{4.1.3}$$

to maintain linear modulation. Since $\beta \leq 1$, the modulation waveform $m(t)$ must be scaled so as to satisfy Eq. (4.1.3).

The input noise field is that due to background noise in a space system, or due to laser or spontaneous emission noise in fiber links, and represents the light collected by the receiver over the optical band B_o. Each noise mode is

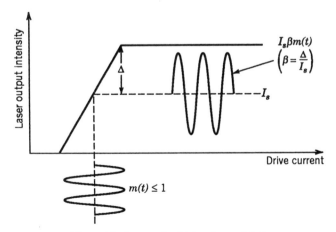

Figure 4.2. Laser output intensity modulation.

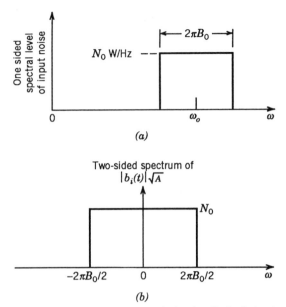

Figure 4.3. Noise spectra. (*a*) Input optical noise. (*b*) Optical noise envelope.

assumed to have the flat spectral level N_0 as shown in Figure 4.3*a*. For blackbody background, the level N_0 is that in Eq. (2.6.10). For spontaneous noise, N_0 is given by the N_{sp} level in Eq. (1.8.2). The input noise power per mode follows as

$$P_{bo} = N_0 B_o \tag{4.1.4}$$

In general we can expand the received field into D_s modes, and Eq. (4.1.1) becomes

$$f_r(t, \mathbf{r}) = \sum_{i=1}^{D_s} a_i(t) e^{j\omega_o t} \Phi_i(\mathbf{r}) \tag{4.1.5}$$

where, as in Eq. (3.5.11),

$$a_i(t) = s_i(t) + b_i(t) \tag{4.1.6}$$

Here $s_i(t)$ and $b_i(t)$ are the signal and noise complex envelope components. The noise envelope process $b_i(t)$ has the magnitude process $|b_i(t)|$ with the normalized spectrum in Figure 4.3*b*, following our discussion in Section 3.4. The resulting photodetector count rate process for this received field can be derived using Eq. (3.3.25) as

$$n(t) = \alpha \int_A |f_r(t, \mathbf{r})|^2 \, d\mathbf{r}$$

$$= \alpha A \sum_{i=1}^{D_s} |a_i(t)|^2$$

(4.1.7)

Thus, the direct detection receiver responds to the incident field in Eq. (4.1.1) by producing a shot noise current process driven by the random count rate in Eq. (4.1.7). This latter process leads directly to the description of the actual current waveform that is optically detected, via our photodetection analysis in Chapter 3. We consider several cases.

4.1.1 Single-Mode Detection

Consider first the case where $D_s = 1$ in Eq. (4.1.5). This means reception is by a diffraction-limited receiver, and only a single plane wave mode occurs. Equation (4.1.7) simplifies to

$$n(t) = \alpha A |s(t) + b(t)|^2$$

(4.1.8)

with $|s(t)|^2$ having the form in Eq. (4.1.2) and $|b(t)|$ having the spectrum in Figure 4.3b. Because this single mode noise envelope process $b(t)$ has the time-average single-mode power in Eq. (4.1.4), we can then write

$$\overline{|b_i(t)|^2} A = N_0 B_o = P_{bo}$$

(4.1.9)

We now wish to determine the spectral density of the photodetected output current for this case. From Eq. (3.8.9) we know

$$S_i(\omega) = |H(\omega)|^2 \, [(\overline{g^2})\bar{n} + (\bar{g})^2 S_n(\omega)]$$

(4.1.10)

where the g terms are the photodetector gain moments, and $H(\omega)$ is the detector transfer function. The parameter $\bar{n}$ is the time average of the statistical average of $n(t)$, and is, therefore,

$$\bar{n} = \alpha A \overline{|s(t) + b(t)|^2}$$

$$= \alpha A \overline{[s(t) + b(t)][s^*(t) + b^*(t)]}$$

(4.1.11)

$$= \alpha A \overline{[|s(t)|^2 + |b(t)|^2 + 2 \, \text{Real}\{s(t)b^*(t)\}]}$$

With zero mean noise, the average in the last term is zero, and

$$\bar{n} = \alpha A[I_s + \overline{|b(t)|^2}]$$

$$= \alpha[P_s + P_{bo}] \tag{4.1.12}$$

Thus the mean detected count rate is directly proportional to the sum of the average powers of the signal and noise fields.

The term $S_n(\omega)$ in Eq. (4.1.10) is the power spectrum of the process $n(t)$, which we expand as

$$n(t) = \alpha A \left[|s(t)|^2 + |b(t)|^2 + n_{sb}(t) \right] \tag{4.1.13}$$

where

$$n_{sb}(t) = 2 \operatorname{Real}\{s(t)b^*(t)\} \tag{4.1.14}$$

To determine the power spectrum of this $n(t)$ we first compute its correlation function,

$$R_n(\tau) = \overline{n(t)n(t + \tau)} \tag{4.1.15}$$

This requires expanding out Eq. (4.1.13) and calculating the average, term by term. Using the fact that the background is a stationary Gaussian, zero-mean complex noise field and is uncorrelated with the signal field, Eq. (4.1.15) becomes

$$R_n(\tau) = \alpha^2 [2P_s P_{bo} + A^2 R_{|s|^2}(\tau) + A^2 R_{|b|^2}(\tau) + A^2 R_{sb}(\tau)] \tag{4.1.16}$$

where

$$A^2 \overline{|s(t)|^2} \; \overline{|b(t)|^2} = P_s P_{bo} \tag{4.1.17}$$

and the individual correlation functions are defined as

$$R_{|s|^2}(\tau) = \overline{|s(t)|^2 |s(t + \tau)|^2} = I_s[1 + \beta^2 R_m(\tau))] \tag{4.1.18a}$$

$$R_m(\tau) = \overline{m(t)m(t + \tau)} \tag{4.1.18b}$$

$$R_{|b|^2}(\tau) = \overline{|b(t)|^2 |b(t + \tau)|^2} \tag{4.1.18c}$$

$$R_{sb}(\tau) = \overline{n_{sb}(t)n_{sb}(t + \tau)} \tag{4.1.18d}$$

These are the correlation functions of the signal field intensity, the background field intensity, and the cross-correlation between these fields. We can now take the Fourier transform (denoted $\mathscr{F}[\cdot]$) of Eq. (4.1.16) to obtain the power

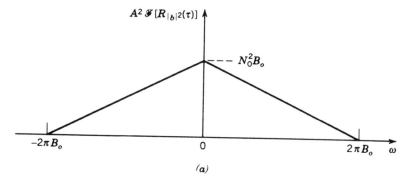

(a)

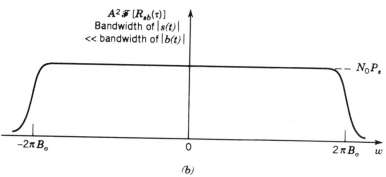

(b)

Figure 4.4. Photodetection noise spectra. (a) Noise intensity spectrum. (b) Signal-to-noise cross spectrum.

spectrum of $n(t)$. The Fourier transform of the signal intensity is

$$A^2\mathscr{F}[R_{|s|^2}(\tau)] = P_s^2[2\pi\delta(\omega)] + P_s^2\beta^2 S_m(\omega) \qquad (4.1.19)$$

with

$$\mathscr{F}[R_m(\tau)] = S_m(\omega) \qquad (4.1.20)$$

Following the steps in Problems 4.3 and 4.4, the Fourier transform of the noise intensity and cross-correlation become

$$A^2\mathscr{F}[R_{|b|^2}(\tau)] = P_{bo}^2(2\pi\delta(\omega)) + \begin{bmatrix} \text{convolution of spectrum of} \\ |b(t)|A^{1/2} \text{ with itself} \end{bmatrix} \qquad (4.1.21a)$$

and

$$A^2\mathscr{F}[R_{sb}(\tau)] = \begin{bmatrix} \text{convolution of spectrum of } |b(t)|\, A^{1/2} \\ \text{with spectrum of } |s(t)|A^{1/2} \end{bmatrix} \qquad (4.1.21b)$$

The convolved spectra in Eqs. (4.1.21) are sketched in Figure 4.4. Since they are each spread over a relatively wide optical bandwidth B_o (hundreds of gigahertz), they will appear flat to typical narrowband modulation bandwidths (hundreds of megahertz). Hence the spectrum of $n(t)$ in Eq. (4.1.13) is approximately

$$S_n(\omega) \cong \alpha^2 (P_s + P_{bo})^2 2\pi\delta(\omega) + (\alpha P_s \beta)^2 S_m(\omega) + \alpha^2 [N_o^2 B_o + 2P_s N_0] \quad (4.1.22)$$

Here the last terms approximate the convolved spectra in Figure 4.4 by a flat spectrum in the vicinity of the origin. Substitution of Eqs. (4.1.12) and (4.1.22) into Eq. (4.1.10) defines the total single mode output current spectrum of the direct detection receiver. We point out that if no signal was present in this mode, $S_n(\omega)$ and $\bar{n}$ are the same as above with P_s set equal to zero.

4.1.2 Multiple-Mode Detection, Single-Mode Signal

We now extend to the case of D_s modes ($D_s > 1$), in which one contains the intensity-modulated plane wave from the source and all the others contain only independent background noise. Again, $n(t)$ is given by Eq. (4.1.7), except now

$$\bar{n} = \alpha(P_s + D_s P_{bo}) \quad (4.1.23)$$

With the $\{a_i(t)\}$ in Eq. (4.1.5) uncorrelated from mode to mode, we now have

$$S_n(\omega) = (\alpha A)^2 \sum_{i=1}^{D_s} [\text{spectrum of each } a_i(t)] \quad (4.1.24)$$

One mode (the signal mode) has the spectrum in Eq. (4.1.22), all the others have noise only. Hence for the multimode case,

$$S_n(\omega) = \alpha^2 (P_s + D_s P_{bo})^2 2\pi\delta(\omega) + (\alpha P_s \beta)^2 S_m(\omega)$$
$$+ \alpha^2 [D_s N_0^2 B_o + 2P_s N_0] \quad (4.1.25)$$

Equation (4.1.25) is the extension of the single-mode case to the multimode case. The resulting photodetector output current spectral density in Eq. (4.1.10) is then generalized to

$$S_i(\omega) = |H(\omega)|^2 [(\overline{g^2})\bar{n} + (\bar{g})^2 S_n(\omega)]$$
$$= |H(\omega)|^2 [(\overline{g^2})\alpha(P_s + P_b) + (\alpha\bar{g})^2 (P_s + P_b)^2 2\pi\delta(\omega) \quad (4.1.26)$$
$$+ (\alpha\bar{g})^2 (P_s \beta)^2 S_m(\omega) + (\alpha\bar{g})^2 (P_b N_0 + 2P_s N_0)]$$

This shows a delta function at $\omega = 0$, a scaled source modulation spectrum, and a flat noise spectrum, all filtered by the $|H(\omega)|^2$ of the photodetector. This

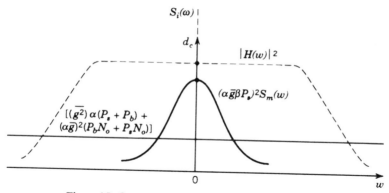

Figure 4.5. Output current spectrum from photodetection.

combined spectrum is shown in Figure 4.5. We see that the current spectrum always contains a constant direct current (dc) component, and a component directly proportional to the spectrum of the signal intensity modulation $m(t)$. Hence the direct detection receiver always demodulates any intensity modulation imposed on the transmitted laser field. Note that the dc component is always present in the current output even if $m(t)$ itself contained no dc term.

In Figure 4.5, we have inserted the parameter $P_b = D_s N_0 B_o$, which is the total input noise power over all modes. For a space system,

$$P_b = D_s P_{bo} = D_s N_0 B_o$$

$$= \left(\frac{\Omega_{fv}}{\Omega_{dL}}\right) W(f) \lambda^2 B_o \qquad (4.1.27)$$

$$= W(f) A \Omega_{fv} B_o$$

Here Ω_{fv} and Ω_{dL} are the receiver field of view and diffraction limited field of view, respectively, and $W(f)$ is the background noise radiance at frequency f. Thus P_b in Eq. (4.1.27) is the total background noise power collected by the receiver, just as we discussed in Section 2.6. Space systems typically have $\alpha N_0 \ll 1$; therefore, the last two terms in Eq. (4.1.26) are negligible compared to the first two terms. Hence, we further approximate the photodetected current spectrum as

$$S_i(\omega) = |H(\omega)|^2 [(\overline{g^2}) \alpha (P_s + P_b) + (\alpha \bar{g})^2 (P_s + P_b)^2 2\pi \delta(\omega)$$

$$+ (\alpha \bar{g} P_s \beta)^2 S_m(\omega)] \qquad (4.1.28)$$

Here the shot noise spectral level is primarily set by the total optical input average power $(P_s + P_b)$ collected by the receiver aperture. In a fiber link with optical amplifiers, the spontaneous noise level $N_0 = N_{sp}$, which may not be negligible, and the last terms in Eq. (4.1.26) may act to increase the shot noise level. This effect will be considered in more detail in Section 7.6.

4.2 DIRECT DETECTION RECEIVER MODEL

The photodetector output current is processed by the general block diagram shown in Figure 4.6. The detector is loaded by the output impedance R_L, producing a voltage signal that can be filtered or amplified for signal processing. The impedance R_L can be a separate load resistance or can represent the input impedance of the postdetection circuitry. Often the resistance R_L is in parallel with any shunt capacitance that also may be present in the detector loading. The linear filter–amplifier is assumed to have a filter function with bandwidth wide enough to cover the modulation process $m(t)$ and an arbitrary voltage gain G. The effect of any postdetection filtering on this modulation can be determined by straightforward circuit analysis.

The photocurrent output will have added to it the detector dark current, the latter having the shot noise spectrum in Eq. (3.8.13),

$$S_{dc}(\omega) = |H(\omega)|^2 \left[\left(\frac{I_{dc}}{e} \right)^2 2\pi\delta(\omega) + \frac{I_{dc}}{e} \right] \tag{4.2.1}$$

Note that the dark current also contributes a constant dc current (I_{dc}), with random, white shot noise added to it. Hence the dark current shot noise level will add directly to the detector noise level in Figure 4.5.

To the detector output, we add in the thermal noise current caused by the local resistance R_L that combines with the detector current. Thermal noise has the two-sided spectral level

$$S_C(\omega) \triangleq N_{oc} = \frac{2\kappa T^\circ}{R_L} \tag{4.2.2}$$

in amps^2/hertz, where T° is the load-resistor temperature in Kelvin, and κ is again Boltzmann's constant. This represents the standard thermal noise associated with any electronic circuit [1, 2].

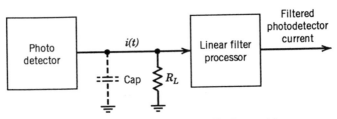

Figure 4.6. Photodetection output filtering model.

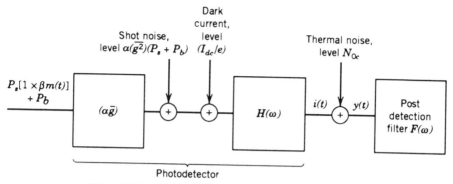

Figure 4.7. Equivalent direct detection optical receiver model.

The combined detector current, dark current, and thermal noise current produce an overall output current $y(t)$. This current is equivalent to that generated by the receiver model depicted in Figure 4.7. The spectral density of this $y(t)$ is

$$S_y(\omega) = |H(\omega)|^2 \left\{ \left[(\alpha\bar{g})^2(P_s + P_b)^2 + \left(\frac{I_{dc}}{e}\right)^2 \right] 2\pi\delta(\omega) \right.$$

$$\left. + (\alpha\bar{g})^2(P_s\beta)^2 S_m(\omega) + \alpha\overline{g^2}(P_s + P_b) + \frac{I_{dc}}{e} \right\} + N_{0c} \tag{4.2.3}$$

This combines the spectra of the individual components at the detector output. Note that, if this output current is amplified by an ideal current amplifier with gain G, each term in Eq. (4.2.3) is multiplied by G^2. In particular, the terms involving the photomultiplier gain $\bar{g}$ are multiplied by this filter gain. If dark current and thermal noise are negligible, postdetection amplification is, therefore, equivalent to an effective increase in the photomultiplication gain. This equivalence is important in system fabrication, because it is generally easier to insert a high-gain electronic amplifier after the photodetector than to design high-gain photomultipliers. We emphasize that this equivalence is not true if thermal noise is significant at the detector output.

Note that Eq. (4.2.2) implies that large load resistors will reduce the thermal noise spectral level. However, it must be remembered also that the combination of this resistor and any shunt capacitance will combine to produce an effective low-pass filtering on the current $y(t)$. This filter will usually have a bandwidth inversely proportional to the resistance-capacitance (RC) time constant, and will therefore decrease as R_L is increased. The overall effect is to insert additive postdetection filtering in Figure 4.7, which can distort the signal modulation. Hence the selection of the R_L loading involves a careful tradeoff between minimizing circuit noise and injecting signal filtering at the detector output [3–5].

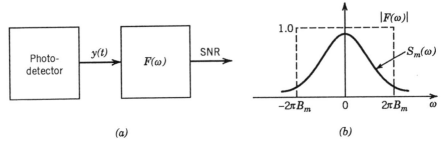

Figure 4.8. Filtering of photodetector output current.

4.3 SIGNAL-TO-NOISE RATIO IN DIRECT DETECTION RECEIVERS

In this section, we compute the electronic signal-to-noise ratio (SNR) following photodetection in a direct detection receiver. We assume an ideal filter $F(\omega)$ is used as in Figure 4.8a to extract the source modulation $m(t)$, the filter having a unit level transfer function over the bandwidth B_m of the signal spectrum $S_m(\omega)$, as shown in Figure 4.8b. The filtered output SNR can now be determined by calculating the output signal and noise powers from the filtered spectrum of $y(t)$ in Eq. (4.2.3). (We neglect the dc contribution to SNR by contending that this dc current can be easily removed and does not contribute to either the signal or noise powers.) The output signal power will be

$$P_{so} = (e\alpha\bar{g}P_s\beta)^2 P_m \tag{4.3.1}$$

where P_m is the total power in the spectrum $S_m(\omega)$. We assume $\beta^2 P_m \leqslant 1$, because $\beta|m(t)| \leqslant 1$ due to the restricted intensity modulation. The output filtered noise can be determined from the contributions from the shot noise and thermal noise. Neglecting the dc terms, this noise power is

$$P_{no} = \left[e^2(\overline{g^2})\alpha(P_s + P_b) + e^2\left(\frac{I_{dc}}{e}\right) + N_{0c} \right] 2B_m \tag{4.3.2}$$

Hence the photodetected SNR in the modulation bandwidth is

$$\begin{aligned}
\text{SNR} &= \frac{P_{so}}{P_{no}} \\
&= \frac{(e\alpha\bar{g}P_s)^2(\beta P_m)^2}{[(\overline{g^2})e^2\alpha(P_s + P_b) + eI_{dc} + N_{oc}]2B_m}
\end{aligned} \tag{4.3.3}$$

The result holds for any modulation signal $m(t)$ as long as the filter $F(\omega)$ is ideally tuned to it (passes it without distortion), and the spectrum $S_m(\omega)$ is within the bandwidth of $H(\omega)$. The SNR in Eq. (4.3.3) is often called the *electronic* or *postdetection* SNR, as opposed to the receiver input optical SNR (P_s/P_b) occurring at the input aperture. We can substitute the detector excess noise factor F in Eq. (3.1.2).

$$F = \overline{g^2}/(\bar{g})^2 \tag{4.3.4}$$

and rewrite

$$\begin{aligned} \text{SNR} &= \frac{(e\alpha\bar{g}P_s)^2(\beta^2 P_m)}{[F(\bar{g}e)^2\alpha(P_s + P_b) + eI_{dc} + N_{oc}]2B_m} \\ &= \frac{(\alpha P_s)(\beta^2 P_m)}{\left[F\left(1 + \dfrac{P_b}{P_s}\right) + \dfrac{eI_{dc}}{(e\bar{g})^2\alpha P_s} + \dfrac{N_{0c}}{(e\bar{g})^2\alpha P_s}\right]2B_m} \end{aligned} \tag{4.3.5}$$

Note that a high gain ($\bar{g}$) photodetector acts to reduce the effect of both the detector dark current (I_{dc}) and the receiver thermal noise (N_{oc}) in computing SNR. Hence high-gain photomultipliers tend to have less detected noise and, therefore, tend to be more sensitive receivers.

When the dark current and thermal noise terms are negligible in Eq. (4.3.5), we say the detector is *shot noise limited*. For this case,

$$\begin{aligned} \text{SNR} &= \frac{\alpha P_s(\beta^2 P_m)}{F\left(1 + \dfrac{P_b}{P_s}\right)2B_m} \\ &= \frac{(\alpha P_s)^2(\beta^2 P_m)}{F\alpha(P_s + P_b)2B_m} \qquad \text{(shot noise limited)} \end{aligned} \tag{4.3.6}$$

Thus shot-noise-limited SNR depends only on the input optical power. Increasing photomultiplier gain $\bar{g}$ beyond the point at which shot-noise-limited operations is achieved does not further improve the SNR. (In fact, F usually increases as $\bar{g}$ is increased, as we have seen.)

If the background power P_b is strong relative to the receiver laser signal power (low optical input SNR), Eq. (4.3.6) behaves as

$$\begin{aligned} \text{SNR} &\cong \frac{(\alpha P_s)^2(\beta^2 P_m)}{(F\alpha P_b)2B_m} \\ &= \left[\frac{\alpha P_s(\beta^2 P_m)}{F\alpha 2B_m}\right]\left(\frac{P_s}{P_b}\right) \end{aligned} \tag{4.3.7}$$

The detected SNR improves as the square of the optical signal power. In particular, it should be noted that a low value of optical P_s/P_b does not necessarily mean a low photodetected SNR.

If $P_s \gg P_b$, and we assume $\beta^2 P_m = 1$ and $F = 1$, then Eq. (4.3.6) becomes

$$\text{SNR}_{\text{QL}} \triangleq \frac{\alpha P_s}{2B_m} \qquad (4.3.8)$$

The above is called the *quantum-limited* SNR of the detector, and denotes the maximum SNR value obtainable during photodetection. In particular, we note that the maximum SNR does not increase without bound as the background and circuit noises are weakened, but rather approaches the quantum-limited SNR. This is because of the inherent properties of the shot noise detection process, and it is precisely this fact that distinguishes the optical communication system from its microwave counterpart. This can be further examined by substituting $\alpha = \eta/hf$ in Eq. (4.3.8) and rewriting

$$\text{SNR}_{\text{QL}} = \frac{\eta P_s}{(hf)2B_m} \qquad (4.3.9)$$

The above now has the appearance of a signal-to-noise ratio in which the signal power is given by ηP_s, whereas the noise appears as the accumulated power in the modulation bandwidth B_m due to a two-sided noise level of hf. This latter term is often called the *quantum spectral level*. From the point of view of SNR, the optical system appears to have an additive quantum noise added during detection, even with no background or circuit noise. This quantum noise has a spectral level that is proportional to the optical frequency f, and places an ultimate limit on detector performance. Note that the detector efficiency factor η directly multiplies the SNR_{QL}, and low-efficiency detectors will produce reduced quantum-limited values. This is why detector efficiency is important to the overall link performance.

The quantum limited noise level in Eq. (4.3.9) can be directly evaluated, following the same substitutions as in Eq. (3.3.22). If we set $F = 1$, $\eta = 1$, and assume an optical wavelength of $\lambda = 1\,\mu$m, the effective quantum noise level becomes

$$hf \cong (6.6 \times 10^{-34})(3 \times 10^{14})$$
$$\cong 20 \times 10^{-20} \qquad (4.3.10)$$
$$\approx -187 \quad \text{dBW/Hz}$$

An RF receiver at a noise temperature of 1000 K produces a thermal noise level of approximately $-199\,$dBW/Hz [6]. Thus the optical quantum-noise level is usually higher than typical RF noise levels. To achieve an $\text{SNR}_{\text{QL}} = 10\,$dB in a bandwidth of $B_m = 10\,$MHz, we would need a received optical signal power of $P_s \cong -104\,$dBW, or approximately $4 \times 10^{-11}\,$W. Obviously, very little optical power is needed to establish a reasonable SNR_{QL}.

The quantum-limited SNR_{QL} can also be interpreted by rewriting Eq. (4.3.8) instead as

$$SNR_{QL} = (\alpha P_s)\left(\frac{1}{2B_m}\right) \tag{4.3.11}$$

The first term is recognized as the rate of detector photoelectron flow due to the average field power P_s. Hence SNR_{QL} is equivalently the number of photoelectrons produced in a time period $1/(2B_m)$ seconds. This means that the number SNR_{QL} can be interpreted either as an effective photodetection SNR or as a desired average electron count generated in a time interval related to the inverse of the modulation bandwidth.

The general direct detection SNR in Eq. (4.3.5) can be rewritten in terms of the quantum-limited SNR in Eq. (4.3.9). After substituting and expanding,

$$SNR = (SNR_{QL})(\beta^2 P_m)\left[F\left(1 + \frac{P_b}{P_s}\right) + \frac{I_{dc}}{(\bar{g})^2 e\alpha P_s F} + \frac{N_{0c}}{(\bar{g}e)^2 \alpha P_s F}\right]^{-1} \tag{4.3.12}$$

Because $\beta^2 P_m \leqslant 1$, the terms in the brackets act as loss factors that reduce the SNR below the maximum SNR_{QL}. The $\beta^2 P_m$ term is an effective modulation loss when the intensity modulation cannot use its full available power, whereas the second bracket is due to the additional receiver noise sources.

Lastly, we can again rewrite the SNR solely in terms of photodetection count rates by introducing:

$$n_s \triangleq \alpha P_s = \text{average number of signal electrons per second}$$

$$n_b \triangleq \alpha P_b = \text{average number of background electrons per second}$$

Then Eq. (4.3.5) can be simplified to

$$SNR = (\beta^2 P_m)\left[\frac{n_s^2}{[F(n_s + n_b) + n_{dc} + n_c]2B_m}\right] \tag{4.3.13}$$

where we have introduced the parameters

$$n_{dc} \triangleq \frac{I_{dc}}{(\bar{g})^2 e} = \text{average dark current count rate in counts per second}$$

$$n_c \triangleq \frac{N_{0c}}{(\bar{g}e)^2} = \text{average circuit noise count rate in counts per second} \tag{4.3.14}$$

Hence the photodetected SNR appears in a more compact form when written

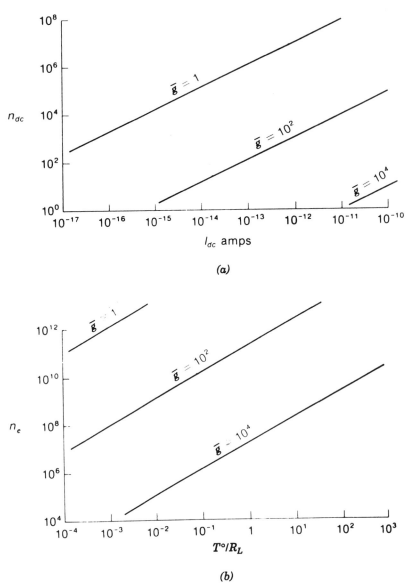

Figure 4.9. Equivalent noise count rate (counts per second). (*a*) Dark current. (*b*) Thermal noise.

in terms of the equivalent count rates instead of power values. This is one reason why photocounting, and count rates, are commonly used in optical receiver analysis. We emphasize that n_c plays the role of a ficticious photodetector count rate that sets its shot noise level equivalent to the added circuit noise level. Figure 4.9 plots typical values of the count rates n_{dc} and n_c versus the receiver dark current I_{dc} and $T°/R_L$ factor for several gain values.

4.4   OPTIMAL PHOTOMULTIPLICATION GAIN

The SNR values in Eq. (4.3.5) or Eq. (4.3.13) for a direct-detection optical receiver shows that performance depends on the photomultiplier through its first two gain moments. The mean gain increased the detected signal power, and the mean-square gain sets the noise factor F, which increases the shot noise level. Because the two are related, and noise factor F invariably increases with gain, it is not evident whether higher gain detectors always improve performance. This suggests the possibility of a gain value that would balance the two effects and produce the best SNR value. To determine this, we rewrite Eq. (4.3.13) as

$$\text{SNR} = \frac{(\beta^2 P_m)n_s^2/2B_m}{Fc_1 + [c_2/(\bar{g})^2]} \qquad (4.4.1)$$

where

$$c_1 = n_s + n_b$$

$$c_2 = \left(\frac{I_{dc}}{e}\right) + \left(\frac{N_{0c}}{e^2}\right) \qquad (4.4.2)$$

Using the fact that F depends on the mean gain, our objective is to determine the value of $\bar{g}$ that maximizes SNR (minimizes the denominator) in Eq. (4.4.1). We consider two cases.

4.4.1   Avalanche Photodetector Receiver

Here we use $F \cong 2 + \gamma\bar{g}$, where γ is the ionization coefficient of the APD material. In this case, the denominator behaves as $[(2 + \gamma\bar{g})(c_1) + c_2/\bar{g}^2]$, which has a minimum at

$$\gamma c_1 - (2c_2/\bar{g}^3) = 0 \qquad (4.4.3)$$

with the solution

$$\bar{g} = \left[\frac{2[(I_{dc}/e) + (N_{0c}/e^2)]}{\alpha(P_s + P_b)\gamma}\right]^{1/3} \qquad (4.4.4)$$

Hence the photomultiplier gain should be set at (or close to) this value to maximize the detected SNR. Lower gains produce reduced signal power, whereas higher gains generate more noise. Note the optimal gain value depends on the received power levels, and therefore must be adjusted to the operating conditions. Figure 4.10 plots $\bar{g}$ in Eq. (4.4.4) as a function of received signal count rate $n_s = \alpha P_s$ for several values of the T°/R_L parameter associated with N_{0c}. We see that only moderate gain values in the detector are required.

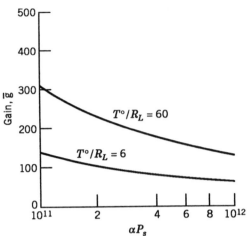

Figure 4.10. Optimal photodetection mean gain $\bar{g}$ vs. signal count rate. $I_{dc} = 0$, $\gamma = 0.028$, $\alpha P_b = 10^9$.

4.4.2   General Photomultiplier Tube

Here we use $F = 1 + (\bar{g})^q$, $0 < q \leqslant 2$, allowing for a slightly faster or slower increase with $\bar{g}$ than the linear APD behavior, as is sometimes measured with various types of PMT devices. We now require

$$(c_1 q)(\bar{g})^{q-1} - \left(\frac{2c_2}{\bar{g}^3}\right) = 0 \tag{4.4.5}$$

or

$$\bar{g} = \left\{\frac{2[(I_{dc}/e) + (N_{0c}/e^2)]}{\alpha(P_s + P_b)q}\right\}^{2+q} \tag{4.4.6}$$

This shows a modified gain value depending directly on the parameter q. Note that for $q > 1$, less gain can be used than in the linear APD case.

It is evident from the above results that using extremely high gain detectors may not always produce the best direct detection performance. It should also be noted that the ability to adjust any photomultiplier gain to a specific value may be a device design problem.

4.5   INTENSITY MODULATED SUBCARRIER SYSTEMS

The previously derived SNR equations pertain to an optical system in which the desired information waveform $m(t)$ is intensity modulated directly onto the optical carrier. An alternative procedure is to make use of an auxiliary subcarrier to carry the information signal. The overall system would appear as

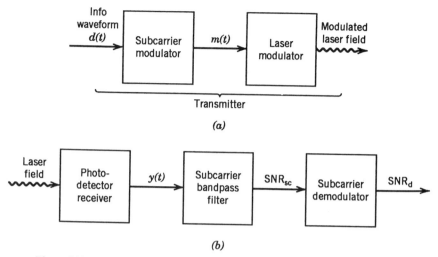

Figure 4.11. Intensity modulated subcarrier systems. (*a*) Transmitter. (*b*) Receiver.

in Figure 4.11. At the transmitter the information waveform is modulated onto a radio-frequency (RF) or an intermediate frequency (IF) subcarrier using standard subcarrier modulation formats such as AM, FM, and so on. The subcarrier $m(t)$ is then intensity modulated onto the main optical carrier. At the receiver, the modulated subcarrier signal is first recovered by photodetection of the optical beam. Note that the entire subcarrier spectrum must be within the detection bandwidth of the photodetector, which limits the upper frequency that may be used for the subcarrier. The recovered subcarrier, along with the detector noise, is then fed into a subcarrier demodulation channel to demodulate the information signal. The advantages of such an operation are (1) possible signal-processing improvement obtained during the subcarrier demodulation, and (2) the use of subcarrier multiplexing, in which a group of subcarriers, each carrying separate source information, can be simultaneously sent over a single optical carrier channel.

In the following we examine some common types of subcarrier modulation formats.

4.5.1 Amplitude Modulation/Intensity Modulation Systems

Consider a system employing amplitude modulation (AM) of the information onto the subcarrier, prior to intensity modulation (IM) of the subcarrier on the optical carrier. Such systems are denoted AM/IM systems. The baseband signal $m(t)$ that we previously defined now has the form

$$m(t) = C[1 + d(t)] \sin(\omega_{sc} t) \qquad (4.5.1)$$

where ω_{sc} is the subcarrier frequency. Here $d(t)$ is the information signal waveform, having a bandwidth B_d in hertz, and normalized such that $|d(t)| \leqslant 1$, with power $P_d \leqslant 1$. The subcarrier signal above occupies a bandwidth $B_m = 2B_d$ in hertz. To satisfy the intensity overmodulation requirement $[|m(t)| < 1]$, we see that in Eq. (4.5.1) we require $C < 1/2$ so that $P_m = C^2(1 + P_d)/2 \leqslant (1 + P_d)/8$. The shot noise limited SNR that occurs in the subcarrier bandwidth at the input to the AM demodulator is obtained directly from Eq. (4.3.6) as

$$(\text{SNR})_{\text{sc}} = \beta^2 \left[\frac{\alpha P_s(1 + P_d)/8}{F(1 + (P_b/P_s))2B_m} \right] \tag{4.5.2}$$

where we have simply used the fact that $m(t)$ now corresponds to the subcarrier signal. An ideal AM demodulator would then yield an output demodulated SNR_d of

$$\begin{aligned}
(\text{SNR})_d &= \frac{\beta^2 C^2(\alpha P_s)P_d}{F[1 + (P_b/P_s)]2B_d} \\
&= \frac{1}{4}\left[\frac{\beta^2 \alpha P_s P_d}{F[1 + (P_b/P_s)]2B_d} \right]
\end{aligned} \tag{4.5.3}$$

It is interesting to compare Eq. (4.5.3) to the result one would obtain if the information $d(t)$ were intensity modulated directly onto the optical carrier. In this case the SNR would be given by Eq. (4.3.6) with $B_m = B_d$ and $P_m = P_d$, which is identical to the bracketed term in Eq. (4.5.3). Thus the SNR is degraded by a factor of 4 (6 dB) if an auxiliary AM subcarrier is used. This poorer performance is simply due to the reduction in the demodulation power caused by the power limitations of the laser. For this reason, AM/IM systems are generally not considered efficient for optical system, and more interest is devoted to FM/IM systems.

4.5.2 Frequency Modulation/Intensity Modulation Systems

In this communication format, the subcarrier is frequency modulated (FM) with the information signal instead of amplitude modulated. The baseband signal that intensity modulates the optical carrier is now

$$m(t) = C \sin\left[\omega_{\text{sc}} t + 2\pi\Delta_f \int d(t)\, dt \right] \tag{4.5.4}$$

where Δ_f is now the subcarrier frequency deviation in hertz from the data signal $d(t)$. We now restrict $C = 1$ to prevent intensity overmodulation, and the subcarrier now occupies an FM bandwidth of

$$B_m = 2\left(\frac{\Delta_f}{B_d} + 1\right) B_d \tag{4.5.5}$$

in hertz, where (Δ_f/B_d) is the FM modulation index. The resulting subcarrier SNR is then

$$SNR_{sc} = \frac{\alpha P_s(C^2/2)}{F[1 + (P_b/P_s)]2B_m}$$

$$= \frac{\alpha P_s/2}{F[1 + (P_b/P_s)]2B_m} \tag{4.5.6}$$

The FM subcarrier demodulating channel, operating above threshold, yields a subcarrier demodulated output SNR_d of

$$SNR_d = 3\left(\frac{\Delta_f}{B_d}\right)^2 \left[\frac{\alpha P_s C^2}{F[1 + (P_b/P_s)]2B_d}\right]$$

$$= 6\left(\frac{\Delta_f}{B_d}\right)^2 \left(\frac{B_m}{B_d}\right)[SNR_{sc}] \tag{4.5.7}$$

The bracketed term indicates the improvement possible during the FM subcarrier demodulation and depends explicitly on the transmitted subcarrier deviation Δ_f. Thus, by using wideband FM subcarriers, an FM/IM system can produce larger output SNR than one using direct intensity modulation.

4.5.3 Multiplexed FM/IM Systems

Here we assume a group of N FM subcarriers, spaced contiguously in frequency, are summed and used to intensity modulate the laser. This is a popular form of multiplexing in optical cable systems. Thus, $m(t)$ now has the form

$$m(t) = \sum_{i=1}^{N} C \sin\left[\omega_{si} t + 2\pi\Delta_f \int d_i(t) \, dt\right] \tag{4.5.8}$$

The photodetector must have an optical bandwidth that will pass the total intensity spectrum (sum of all the subcarrier bandwidths). To overcome the source saturation during modulation, it will again be necessary to require the condition $|m(t)| \leqslant 1$, which means

$$\sum_{i=1}^{N} C = NC \leqslant 1 \tag{4.5.9}$$

so that $C \leqslant 1/N$. The resulting demodulated single-channel SNR_d is now $1/N^2$ times the result in Eq. (4.5.7). Hence multiplexing of the N FM channels may produce a $1/N^2$ power loss in each channel from the amplitude restriction in Eq. (4.5.9). With many channels, this can be a significant reduction that would not occur if the modulation was ideally linear. For this reason, linearization of external optical modulators is an important element in subcarrier multiplexing with direct detection systems.

4.6 POSTDETECTION INTEGRATION

An important processing operation following optical photodetection is finite-time integration. This particular type of processor is common in digital decoding and acquisition systems, where the integration intervals are clocked to specific bit times and acquisition periods.

The receiver processing will appear as in Figure 4.12 with the integrator replacing the receiver filter in Figure 4.8. The system can be approximated by replacing the integrator by an appropriate lowpass filter and by using the filter analysis in Section 4.3. However, since the processor produces a direct integration of the photodetector output, an exact analysis of the integrated variable can be developed. If we again let $y(t)$ represent the detector current, then the integrated output in Figure 4.12 produces, at time t, the integration of $y(t)$ over the last T seconds. Hence,

$$
\begin{aligned}
v(t) &= \int_{t-T}^{t} y(\rho) \, d\rho \\
&= \int_{t-T}^{t} i(\rho) \, d\rho + \int_{t-T}^{t} i_n(\rho) \, d\rho
\end{aligned}
\tag{4.6.1}
$$

where $i_n(t)$ is the thermal noise current in Figure 4.7. This second integral is the integral of a Gaussian noise process and, therefore, evolves at each t as a Gaussian random variable, with zero mean and variance $(N_{0c} T/2)$, with N_{0c} given in Eq. (4.2.2). The first integral is simply the short-term integration of a

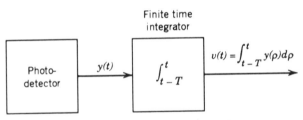

Figure 4.12. Photodetector output integrator.

shot noise process. Assuming an integration time T much larger than the transit time of the photoelectrons, the first integral becomes

$$\int_{t-T}^{t} i(\rho)\,d\rho = \int_{t-T}^{t} \sum_{j=1}^{k(0,\rho)} gh(\rho - t_j)\,d\rho$$

$$= egk(t - T, t) \tag{4.6.2}$$

$$= ek_2(t - T, t)$$

where $k_2(t - T, t)$ is the random output count over the integration interval. This first term is therefore directly proportional to the shot noise output counting process, with statistics that depend on the photomultiplier gain variable g and on the primary count. If g is not random (ideal photomultiplier), then the output count is a Poisson variable. If g is random, the output count k_2 depends on the photomultiplier conversion probabilities, as discussed in Section 3.6. In either case, $v(t)$ in Eq. (4.6.1) is the sum of a discrete count variable plus a continuous Gaussian noise variable. As a result, its probability density at any t is a mixture density and, because of the discreteness of the count variable, this density has the form

$$p_{v_t}(v) = \sum_{k=0}^{\infty} P(k) \left[\frac{\exp[-(v - ek)^2/2(N_{0c}T/2)]}{\sqrt{2\pi(N_{0c}T/2)}} \right] \tag{4.6.3}$$

with $P(k)$ the discrete count probability of the output photomultiplier counts k_2. Note that Eq. (4.6.3) is itself a continuous density in v due to the additive thermal noise. Only if this noise spectral level is extremely weak will the discreteness of the counting be apparent, and we say the integrator output is "counting photons." When the noise is present, however, the discrete count observable is lost among the noise.

In Chapter 6 we will see how these various integrator statistics will lead to different computations for the decoding error probabilities in digital systems. It is precisely for this reason that the actual photodetection counting statistics had to be developed in detail to completely characterize the receiver.

4.7 DIRECT DETECTION WITH MULTIMODE SIGNALS

The SNR in Section 4.3 was based on a modulated source field occupying a single-field mode. That is, the source was considered a point source transmitting to the receiver with a single mode. When a source is spread over many modes (as with an extended source space field or multimode fiber), the arriving field now requires a multimode description. Because noise may be present in

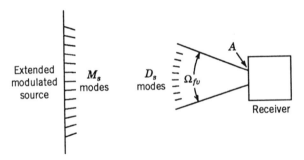

Figure 4.13. Extended source model.

all modes as well, the resulting computation of the photodetected SNR at the receiver must be modified. In this section, we examine photodetection of these multimode signals.

Consider the space system in Figure 4.13. A direct detection receiver with an area A and field of view Ω_{fv} encompasses D_s spatial modes. An extended source transmits signal energy in M_s signal modes. If $M_s < D_s$ the receiver observes the entire source in its field of view. If $M_s > D_s$, only a portion of the source is observed. In the ith mode, the signal field intensity is assumed

$$|s_i(t)|^2 = I_{si}[1 + m(t)] \tag{4.7.1}$$

where $m(t)$ is again the modulating signal imposed at the transmitter, I_{si} is the signal intensity in the ith mode, and we use $\beta = 1$. Equation (4.7.1) implies a signal field with the source intensity modulation $m(t)$ distributed over all modes. The average signal power in the ith mode is then

$$P_{si} = I_{si}A \tag{4.7.2}$$

We now have a total combined received count rate over the D_s receiver modes of

$$n(t) = \alpha A \sum_{i=1}^{D_s} |s_i(t) + b_i(t)|^2 \tag{4.7.3}$$

with $b_i(t)$ the complex noise field of the ith mode. Following the expansion as in Eq. (4.1.13), the count process is

$$n(t) = \alpha A \sum_{i=1}^{D_s} [|s_i(t)|^2 + |b_i(t)|^2 + 2\,\text{Real}\{s_i(t)b_i^*(t)\}] \tag{4.7.4}$$

Its mean value is

$$\bar{n} = \alpha A \sum_{i=1}^{D_s} [\overline{|s_i(t)|^2} + \overline{|b_i(t)|^2}]$$

$$= \alpha \sum_{i=1}^{D_s} (P_{si} + P_{bo}) \tag{4.7.5}$$

The signal terms in $n(t)$ are obtained from

$$\alpha A \sum_{i=1}^{D_s} |s_i(t)|^2 = [1 + m(t)]\alpha \sum_{i=1}^{D_s} P_{si} \tag{4.7.6}$$

Note that both Eq. (4.7.5) and Eq. (4.7.6) are directly proportional to the available signal power collected over the D_s spatial modes. To examine this further, assume equal signal power $P_{si} = P_{so}$ in each mode. This means the total signal power collected is $M_s P_{so}$ if $M_s \leqslant D_s$ and is $D_s P_{so}$ if $M_s > D_s$. The shot-noise-limited receiver SNR corresponding to Eq. (4.3.6) is now one of two cases. If $M_s > D_s$,

$$\text{SNR} = \frac{(\alpha D_s P_{so})^2}{\alpha D_s (P_{so} + P_{bo}) 2FB_m}$$

$$= D_s \left[\frac{(\alpha P_{so})^2}{\alpha (P_{so} + P_{bo}) 2FB_m} \right] \tag{4.7.7}$$

where B_m is the bandwidth of $m(t)$. Because the bracketed term is the single mode SNR, Eq. (4.7.7) indicates that SNR performance is always improved by increasing D_s (opening the receiver field of view) as long as $M_s \geqslant D_s$. When $M_s < D_s$,

$$\text{SNR} = \frac{(\alpha M_s P_{so})^2}{\alpha (M_s P_{so} + D_s P_{bo}) 2FB_m}$$

$$= \frac{(\alpha P_s)^2}{\alpha (P_s + D_s P_{bo}) 2FB_m} \tag{4.7.8}$$

where $P_s = M_s P_{so}$ is the total available source power. We now decrease the SNR if we continue to increase D_s. Because all the signal power has been collected, increasing the number of receiver modes only adds to the detector noise. Thus, for this example, we should always open the field of view until the entire source field is exactly encompassed (i.e., $D_s = M_s$) and not increase it any further. That is, we should only look at a field of view that encompasses the source.

It should be noted that Eqs. (4.7.7) and (4.7.8) assumed that the mode modulations $m(t)$ were in exact time phase when summed in Eq. (4.7.6). This means that there is no time delay between the signals of the various modes and the modulation added coherently. Time delays between the modes may in fact cause a noncoherent addition of the modulation waveforms that may instead produce a power loss from the value used in Eq. (4.7.6). For example, let $m(t)$ be a random modulation process with correlation function $R_m(\tau)$, with $R_m(0) = 1$, and consider two spatial modes with differential time delay t_d. The power in the signal term in Eq. (4.7.6) from those two modes is now

$$\overline{[\alpha P_{so} m(t) + \alpha P_{so} m(t - t_d)]^2} = (\alpha P_{so})^2 [2R_m(0) + \overline{2m(t)m(t - t_d)}]$$
$$= (\alpha P_{so})^2 [2 + 2R_m(t_d)]$$

(4.7.9)

Because $R_m(t_d) \leqslant 1$, this is always less than the signal power value in the numerator of Eq. (4.7.7), which would be $4(\alpha P_{so})^2$. As the delay t_d approaches the correlation time of the modulation process so that $R_m(t_d) \approx 0$, the power no longer adds coherently as predicted in Eq. (4.7.6) but instead is reduced to $2(\alpha P_{so})^2$. Extending to M_s modes shows that the postdetection SNR in Eq. (4.7.8) with noncoherent mode combining and $D_s = M_s$ is now

$$\text{SNR} = \frac{M_s(\alpha P_{so})^2}{\alpha(M_s P_{so} + M_s P_{bo})2FB_m}$$
$$= \frac{(\alpha P_{so})^2}{\alpha(P_{so} + P_{bo})2FB_m}$$

(4.7.10)

The SNR above is identical to the single mode SNR. Thus the noncoherent combining of the M_s field modes has not improved the SNR over that of a single mode, and the performance is in fact independent of the actual number of modes collected.

4.8 OPTIMAL COLLECTION OF MULTIMODE SIGNAL POWER

The analysis in Section 4.7 assumed equal signal power in all spatial modes of a multimode source. The receiver with coherent combining maximized SNR by opening the field of view to collect all modes. Suppose now that the source mode powers P_{si} were not equal in all modes but instead had a spatial distribution of its power over the set of receiver modes. Opening the receiver view may no longer produce optimal power collection.

Let us first assume that the receiver had the capability of separately detecting each mode and applying arbitrary photodetection gain to each mode. Let g_i be the gain (nonrandom) applied to the ith mode during photodetection.

Because the gain multiplies the detected field power, the combined outputs of the individual detectors would have the SNR given by

$$\text{SNR} = \frac{\left(\alpha \sum_i g_i P_{si}\right)^2}{\alpha \sum_i g_i^2 (P_{si} + P_{bo}) 2FB_m} \tag{4.8.1}$$

If we rewrite the numerator sum as

$$\sum_i g_i P_{si} = \sum_i (g_i \sqrt{P_{si} + P_{bo}}) \left(\frac{P_{si}}{\sqrt{P_{si} + P_{bo}}}\right) \tag{4.8.2}$$

we can apply the Schwarz inequality,

$$\left|\sum_i g_i P_{si}\right|^2 \leqslant \sum_i (g_i \sqrt{P_{si} + P_{bo}})^2 \sum_i \left(\frac{P_{si}}{\sqrt{P_{si} + P_{bo}}}\right)^2 \tag{4.8.3}$$

This establishes that Eq. (4.8.1) has the bound

$$\text{SNR} \leqslant \sum_i \frac{(\alpha P_{si})^2}{\alpha(P_{si} + P_{bo}) 2FB_m} \tag{4.8.4}$$

Furthermore, the bound occurs only when the equality in the Schwarz inequality holds,

$$g_i \sqrt{P_{si} + P_{bo}} = C\left[\frac{P_{si}}{\sqrt{P_{si} + P_{bo}}}\right]$$

or when

$$g_i = C\left[\frac{P_{si}}{P_{si} + P_{bo}}\right] \tag{4.8.5}$$

where C is any scale factor. This represents the optimal photomultiplication gain for collecting the ith signal mode. Note that it depends on the relative signal and noise power in that mode. Modes with higher signal powers should use more gain that those with strong noise [since the gains in Eq. (4.8.5) are all scaled, it is only the relative gains that are important]. Equation (4.8.5) implies that all modes having signal power should be collected, although some may require lower gain values, and all modes with the same signal power should have the same gain applied.

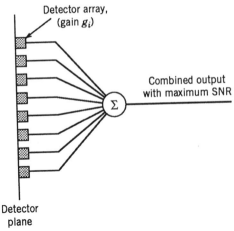

Figure 4.14. Focal-plane array processor.

Individual spatial modes can only be detected by focal plane array processing, as shown in Figure 4.14. Theoretically, each detector is aligned with a single Airy pattern, using a gain given by Eq. (4.8.5), and the outputs of all detectors are summed to produce a single output current with the SNR in Eq. (4.8.4). Arrays of charge-coupled devices (CCD) with micron detecting areas have been proposed for this purpose. However the use of small, low-gain detectors within the array make it difficult to achieve the shot-noise-limited operation assumed in the derivation of Eq. (4.8.5). Use of postdetection amplifiers to achieve the gain distribution over the array can also be considered, but only if the thermal noise is negligible.

Note that a set of individual mode detectors requiring the same detector gain can be combined into a single larger detector with combined collecting area of the set. This means the optimal array can be approximated by partitions of larger detectors encompassing focal-plane areas that have roughly the same signal power. For example, with circular symmetric signal fields, concentric circular detectors of various detecting widths and properly selected gain values can be used to collect the focused field and serve as an approximation to an optimal focal-plane processor. The signal intensity distribution in the detector plane, needed to design the detectors shapes, can often be estimated from knowledge of the receiver coherence functions, using the transform analysis in Section 2.4.

PROBLEMS

4.1   (a) Using Figure 4.2, show that β in Eq. (4.1.2) is equal to the ratio $\Delta/I_s m_p$, where Δ is the range of output power that defines the linear range of the modulator and m_p is the peak value of $m(t)$.

(b) Assuming maximum linear range around I_s, write β in terms of the maximum and minimum values of the output intensity that define the linear range of the modulator.

4.2   (a) Given the circuit loading in Figure P4.2 at the output of a photo-detector, show that the processor input current is a low-pass filtered version of the detector output current.

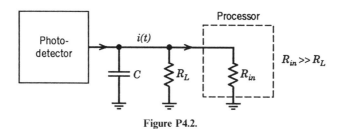

Figure P4.2.

(b) Determine the 3-dB bandwidth of this filter.

(c) Show that if the bandwidth of this filter is to be set at B Hz for any capacitance value C, the noise current spectral level due to R_L in Eq. (4.2.2) increases with B.

4.3   Consider two independent, zero mean, complex random processes

$$s(t) = s_r(t) + js_I(t)$$
$$b(t) = b_r(t) + jb_I(t)$$

each with uncorrelated real and imaginary parts, with the individual spectral densities

$$S_{s_r}(\omega) = S_{s_I}(\omega) = (1/2)S(\omega)$$
$$S_{b_r}(\omega) = S_{b_I}(\omega) = (1/2)B(\omega)$$

Let $S(\omega)$ and $B(\omega)$ have transforms $R_s(\tau)$ and $R_B(\tau)$. Define a new process

$$y(t) = 2\,\mathrm{Real}\{s(t)b^*(t)\}$$

(a) Compute the correlation of $y(t)$ in terms of $R_s(\tau)$ and $R_B(\tau)$.

(b) Write an expression for the spectrum $S_y(\omega)$.

(c) Evaluate $S_y(0)$ for the special case where $B(\omega) = N_0$ (white noise) and $|s(t)|$ has power $P_s = (1/2)R_s(0)$.

4.4   For a Gaussian random process $x(t)$, the process $z(t) = x^2(t)$ has the correlation function

$$R_z(\tau) = R_x^2(0) + 2R_x^2(\tau)$$

Use this fact to write the correlation function of the complex envelope process $|b(t)|^2$ in Problem 4.3, assuming its real and imaginary parts are Gaussian processes.

4.5   An optical system operates at a 0.6-μm wavelength with a photodetector of efficiency 50 percent, a load resistance of 100 ohms, and a receiver temperature of 300 K. The receiver has an optical bandwidth $B_o = 10^{10}$ Hz, an aperture area of 0.5 m^2, and a diffraction limited field of view.

   (a) Determine the approximate signal count rate needed to achieve shot-noise-limited operation. (Assume the latter is achieved if the shot noise level is 10 times the thermal level).

   (b) Convert (a) to watts.

   (c) If a blackbody background at 1000 K is present, determine the signal power needed to achieve quantum-limited performance.

   (d) Determine the quantum-limited SNR in a 1-MHz bandwidth.

4.6   A shot-noise-limited direct detection receiver operates with a background noise count rate of 10^8 cps. How much signal power is needed to achieve a detected SNR of 10 in a bandwidth of 5 MHz. Use $\alpha = 10^{18}$ and $F = 1$.

4.7   An optical link uses a receiver designed for a $SNR_{QL} = 10$, using a laser source producing a detected signal count of 10^9 cps. What will the SNR be when a background with radiance 10^4 μW/cm-sr-micron is present? Assume $\lambda = 1$ μm, an optical bandwidth of 10 Å, a receiver area of 1 m^2, and a receiver field of view with 10^4 spatial modes. Use $\alpha = 10^{18}$.

4.8   A laser transmits a field that is intensity modulated with $\sin(2\pi f_s t)$. The field arrives at a direct detection receiver in two modes, with one having a time delay of t_d seconds relative to the other, and each having a mode power of $P_s/2$. Neglect background noise.

   (a) Determine the photodetected signal current time variation.

   (b) For a delay of $t_d = 10^{-9}$ sec, find the maximum frequency f_s such that the amplitude of the signal current will not be degraded by more than 1/2.

4.9   The noise equivalent power (NEP) of a direct-detection receiver is defined as the value of signal power P_s needed to produce a $\sqrt{SNR} = 1$ in a bandwidth of $B_m = 1$ Hz.

(a) Derive an expression for NEP in terms of the optical parameters P_b, N_{0c}, and I_{dc}.

(b) Repeat (a) assuming $P_b \gg P_s$.

(c) Given the NEP of a receiver, write an expression for the SNR that will occur with an arbitrary power P_s in a bandwidth B_m Hz.

4.10 A single optical source simultaneously sends two plane waves, one at wavelength λ_1 and one at λ_2. Each wave has intensity I_s watts/area. The direct detection receiver has collecting area A. Neglect the background, dark current, and thermal noise, and assume $\alpha = 10^{18}$.

(a) What is the *dc* value of the output current?

(b) Sketch the spectrum at the output.

4.11 A baseband signal with unit power and bandwidth of $B_d = 1$ MHz is to be modulated on to a subcarrier, with the latter intensity modulated (IM) on to an optical carrier at 10^{14} Hz. The optical receiver is operated quantum limited, and a subcarrier SNR of 20 dB is required for subcarrier demodulation.

(a) How much optical receiver power is needed to operate the system if an AM/IM system is used.

(b) What is the demodulated AM SNR?

(c) What is the subcarrier bandwidth needed if FM/IM is used with a subcarrier FM deviation of 10 MHz?

(d) What is the required optical power and demodulated SNR for the FM/IM system in (c)?

4.12 A source is effectively spread by turbulence to a solid angle of 2.5×10^{-4} sr when viewed from a receiver. The receiver lens has diameter 0.01 m at the optical frequency $f = 6 \times 10^{14}$ Hz.

(a) Determine the number of modes into which the source has been extended.

(b) Assuming a constant radiance background, compute the decibel improvement in SNR over single-mode reception, obtained by opening the receiver field of view to that of the extended source.

REFERENCES

1. A. Van der Ziel, *Noise in Solid-State Devices and Circuits*, Wiley, New York, 1986.

2. M. Buckingham, *Noise in Electronic Devices and Systems*, Wiley, New York, 1983.

3. S. Personick, Receiver design for optical fiber systems, *Proc. IEEE*, 65, 1670–1678, December 1977.

4. M. Brian and T. Lee, Optical receivers for lightwave communication systems, *J. Lightwave Technol* 3, 1281–1300, December 1985.

5. B. Kaspar, "Receiver design," in *Optical Fiber Telecommunications*, Vol. 2, S. E. Miller, ed., Academic Press, New York, 1988.

6. R. Gagliardi, *Introduction to Communication Engineering*, Wiley, New York, 1988, Chapter 4.

5

COHERENT(HETERODYNE) DETECTION

In the preceding chapter, we studied direct, or incoherent, detection of the received optical field. In such a system we found that the receiver may have to combat the background radiation and the internal receiver thermal noise while attempting to recover the intensity modulation of the transmitted beam. The receiver noise is overcome by achieving shot-noise-limited conditions, and the resulting performance then depends only on an ability to transmit signals whose power level dominates that of any received background. Shot-noise-limited operation is achieved with exceptionally strong received signal fields, or by the use of high-quality photomultipliers. When high-performance photomultipliers are not available, other means must be used to improve detectability. An alternative method is the use of heterodyning.

In heterodyne detection, the receiver operates by optically adding a locally generated field to the received field prior to photodetection. The prime objective is to use the added local field to improve the detection of the weaker received field in the presence of the receiver thermal noise. The combined field is then photodetected, as if it were a single received optical field. Since the addition of two electromagnetic fields requires spatial alignment of the fields, the use of heterodyne detection is often called (spatial) *coherent* detection.

5.1 THE HETERODYNE RECEIVER

A typical heterodyning receiver is shown in Figure 5.1. The received optical field is projected onto the photodetector surface by the receiver lens and front-end system. A local optical field, generated by a receiver source, is diffracted by a receiver lens and aligned by means of a mirror with the received field in the photodetector. The detector responds to the combined field of the received and local sources by producing a detector shot noise process in the usual way. The mixing of the two fields can be described in terms of diffraction patterns in the focal plane.

151

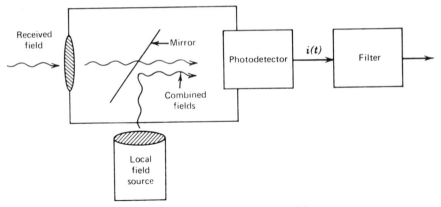

Figure 5.1. Heterodyne receiver model.

Let the received field be the sum of a laser source field and a input noise field

$$f_r(t, \mathbf{r}) = f_s(t, \mathbf{r}) + f_b(t, \mathbf{r}) \qquad (5.1.1)$$

This field is focused by the receiver lens through the mirror onto the detector plane as $f_d(t, \mathbf{q})$, where $\mathbf{q} = (u, v)$ represents vector points in the focal plane. The local field is reflected by the mirror and imaged on the detector plane as $f_L(t, \mathbf{q})$. The combined focal plane field is then

$$\text{focal plane field} = f_d(t, \mathbf{q}) + f_L(t, \mathbf{q}) \qquad (5.1.2)$$

The photodetector of area A_d collects the focal plane field and produces the photodetected count rate process

$$
\begin{aligned}
n(t) &= \alpha \int_{A_d} |f_d(t, \mathbf{q}) + f_L(t, \mathbf{q})|^2 \, d\mathbf{q} \\
&= \alpha \int_{A_d} |f_d(t, \mathbf{q})|^2 \, d\mathbf{q} + \alpha \int_{A_d} |f_L(t, \mathbf{q})|^2 \, d\mathbf{q} \qquad (5.1.3) \\
&\quad + 2\alpha \operatorname{Real}\left\{ \int_{A_d} f_d(t, \mathbf{q}) f_L^*(t, \mathbf{q}) \, d\mathbf{q} \right\}
\end{aligned}
$$

The first two terms account for the intensity of the individual fields, and they each can be analyzed similarly to the direct detection analysis of the previous sections. The third term in Eq. (5.1.3) is the crossterm, or beat term, between the received and local fields. In heterodyne detection, this beat term is of prime importance for signal recovery.

Consider first the case where the local field is generated as a pure unmodulated plane wavefield that is focused to a single Airy pattern,

$$f_L(t, \mathbf{q}) = a_L e^{j(\omega_L t + \theta_L)} \phi_L(\mathbf{q}) \tag{5.1.4}$$

when $\phi_L(\mathbf{q})$ is the diffraction pattern of the local field lens. Let the received field $f_r(t, \mathbf{r})$, when focused into the detector plane, have the general focal-plane expansion,

$$f_d(t, \mathbf{q}) = \sum_{i=1}^{D_s} [s_i(t) + b_i(t)] e^{j\omega_o t} \phi_i(\mathbf{q}) \tag{5.1.5}$$

where $s_i(t)$ and $b_i(t)$ are again the complex signal and noise envelopes, ω_o is the received optical frequency, and $\phi_i(\mathbf{q})$ denote the D_s receiver Airy patterns distributed in the focal plane.

We concentrate first on the beat term in Eq. (5.1.3). Using Eqs. (5.1.4) and (5.1.5), this becomes

$$n_{RL}(t) = 2\alpha \, \text{Real} \left\{ \sum_{i=1}^{D_s} [s_i(t) + b_i(t)] \, a_L \exp\{j[\omega_o - \omega_L)t - \theta_L]\} \right.$$
$$\left. \times \int_{A_d} \phi_i(\mathbf{q}) \phi_L^*(\mathbf{q}) \, d\mathbf{q} \right\} \tag{5.1.6}$$

Note that Eq. (5.1.6) corresponds to a sum of time varying modulated exponentials, each weighted by a particular integral involving the focused spatial patterns of the received and local fields. Thus the beat term $n_{RL}(t)$ involves products of both time and integrated spatial terms.

When the received signal field is a modulated plane wave from a normal direction of arrival, it will be focused down to a single (say the first) Airy pattern, so that

$$s_i(t) = a_s(t) e^{j\theta_s(t)} \qquad i = 1$$
$$= 0 \qquad i \neq 1 \tag{5.1.7}$$

Here $a_s(t)$ and $\theta_s(t)$ are the signal amplitude and/or phase modulation imposed on the signal field. Again let each $b_i(t)$ be the complex noise envelope of a bandpass optical noise field collected over the aperture area A, so that the power spectrum of each $|b_i(t)| A^{1/2}$ process is N_0 over a bandwidth $(-B_o/2, B_o/2)$, as shown in Figure 4.3. The local-field Airy pattern is assumed to be identical to the focused received normal plane wave, so that

$$\phi_i(\mathbf{q}) = \phi_1(\mathbf{q}) \tag{5.1.8}$$

That is, $\phi_L(\mathbf{q})$ is exactly matched to the first Airy pattern $\phi_1(\mathbf{q})$, and spatially orthogonal to all other $\phi_i(\mathbf{q})$. Equation (5.1.6) now becomes

$$
\begin{aligned}
n_{RL}(t) &= 2\alpha\,\mathrm{Real}\{a_L[a_s(t)e^{j\theta_s(t)} + b_1(t)]\exp\{j[(\omega_o - \omega_L)t - \theta_L]\}\int_{A_d}|\phi_1(\mathbf{q})|^2\,d\mathbf{q} \\
&= 2\alpha a_L a_s(t)A\cos[(\omega_o - \omega_L)t + \theta_s(t) - \theta_L] \\
&\quad + 2\alpha a_L A\,\mathrm{Real}\{b_1(t)\exp\{j[(\omega_o - \omega_L)t - \theta_L]\}\}
\end{aligned}
\tag{5.1.9}
$$

where we have used the fact that the Airy pattern magnitude integrates to the value A of the aperture area, as was shown in Eq. (1.6.10). The first term in Eq. (5.1.9) is the signal term, which now corresponds to a modulated carrier having the same amplitude and phase modulation as the optical carrier, except the latter is now shifted to the difference frequency $(\omega_o - \omega_L)$. By carefully selecting the local frequency ω_L relative to the laser frequency, this difference frequency can be set to a desired RF carrier (at megahertz or gigahertz). Thus the optical mixing with the local field at the photodetector has resulted in the generation of an RF carrier after photodetection containing the same amplitude and phase modulation of the transmitted optical carrier. Note that this optical mixing to a precise RF carrier requires setting the local field wavelength to within a small fraction (≈ 0.01 percent) of the received field wavelength.

When the difference frequency is set to an appropriate RF, the system is referred to as a heterodyne system. When the local field frequency is set identical to the received field frequency, $\omega_L = \omega_o$, the difference frequency is zero, and the system is referred to as a *homodyne* system. In homodyning, the received field is effectively beaten down to zero frequency and the signal term appears as a baseband waveform (in which the spectrum is centered at zero frequency). Note that any phase modulation is lost (or distorted) during homodyning and only amplitude modulation is preserved.

The second term in Eq. (5.1.9) is the effect of the input noise field being shifted to the difference frequency $\omega_H \triangleq (\omega_o - \omega_L)$ with the same random envelope modulation as the optical noise. Thus, the mixed noise complex process has the two-sided spectral level $N_0/2$ over the optical bandwidth B_o. The mixed noise term in Eq. (5.1.9) involves only the real part of this complex noise and, therefore, has the two-sided spectral level $N_0/4$ over the same bandwidth B_o shifted to the frequency ω_H. Note that even though the received noise field was collected by the lens over D_s spatial modes of the receiver field of view, only one mode (that of the local field) appears in Eq. (5.1.9). Thus, the focusing of the local field has limited the input noise to only a single mode. In effect, the local field has set the effective receiver field of view. Recall that for

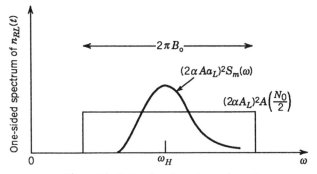

Figure 5.2. Heterodyned spectrum of $n_{RL}(t)$.

direct detection the receiver field of view was determined by the detector size. Here the focused Airy pattern sets the field of view. Note also that both the noise term and the signal term are scaled by the local field amplitude a_L.

The overall power spectrum of the beat term $n_{RL}(t)$ is sketched in Figure 5.2. It corresponds to a scaled version of the carrier intensity spectrum $S_m(\omega)$, the latter being the spectrum of the modulated heterodyned carrier

$$m(t) = a_s(t) \cos(\omega_H t + \theta_s(t) - \theta_L) \tag{5.1.10}$$

Note that both the amplitude and phase of the optical carrier and the local laser field are superimposed on this heterodyne carrier.

It should be emphasized that Eq. (5.1.9) was based on the fact that the local-field pattern at the photodetector matched the complex received-field pattern, $\phi_L(\mathbf{q}) = \phi_1(\mathbf{q})$. This means it must match exactly in shape, phase, polarization, and location on the detector surface. The effect of field mismatches are considered in Section 5.4.

Using Eq. (5.1.9) in Eq. (5.1.3) we can now combine and write the total photodetected counting process as

$$n(t) = \alpha A \sum_{i=1}^{D_s} |s_i(t) + b_i(t)|^2 + \alpha a_L^2 A + n_{RL}(t) \tag{5.1.11}$$

The first and second terms are now the individual intensities of the received and local field on the detector and are added directly to the beat terms. From Eq. (5.1.11), we can compute the contributions to the resulting photodetector output spectrum. The time-averaged count rate is now

$$\bar{n} = \alpha A \left[\overline{|s_1(t)|^2} + \sum_{i=1}^{D_s} \overline{|b_i(t)|^2} \right] + \alpha a_L^2 A \tag{5.1.12}$$

because the time average of the heterodyne carrier in Eq. (5.1.10) is zero. Letting

$$P_L = a_L^2 A \quad = \text{average power of the local field}$$

$$P_s = \overline{|a_s(t)|^2} A = \text{time average power of the received optical signal field} \tag{5.1.13}$$

$$P_b = D_s N_0 B_o = \text{total average input noise power in the } D_s \text{ noise modes}$$

we then have

$$\bar{n} = \alpha[P_s + P_b + P_L] \tag{5.1.14}$$

The spectral components of $n(t)$ in Eq. (5.1.11) are derived from the intensity variations of the first two terms, plus the spectrum of the beat term $n_{RL}(t)$, shown in Figure 5.2. These spectral terms must be combined with the filtering, dark current, and photomultiplication effects of the photodetector itself. The overall heterodyned output current spectrum is shown in Figure 5.3.

From the spectral plot, we can make the following observations. The detected output always contains the heterodyne carrier with the amplitude, frequency, or phase modulation of the transmitted optical carrier. Recall that the direct detection system could only detect intensity modulations on the carrier. Furthermore we see that the entire modulated carrier spectrum must lie within the photodetector bandwidth $H(\omega)$ to prevent carrier distortion. In addition, the photodetected spectrum contains a dc term plus spectral components due to the intensity variations of the input received fields. As discussed in Chapter 4, these are usually concentrated at low frequencies, and are caused by the time variations of the intensity $|a_s(t)|^2$, the envelope of $b_i(t)$, and their beat terms. The latter are spread over a wide bandwidth and are usually much

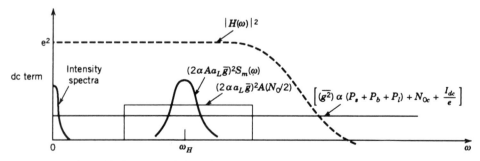

Figure 5.3. Spectrum of photoelectronic output spectrum with heterodyning.

weaker in level than the shot noise levels. The shot noise level is set by the total power on the detector surface, as given in Eq. (5.1.14).

In deriving Figure 5.3 we also assumed an idealized local field (constant amplitude–constant phase). Variations in the local field amplitude a_L or phase θ_L in Eq. (5.1.4) will be directly impressed on the heterodyne term in Eq. (5.1.9) as an effective distortion of the heterodyned carrier. Thus, local optical sources should be fairly stable in both amplitude and phase for distortionless heterodyning.

In a homodyne system, the beat frequency is at zero, and the heterodyne carrier term $m(t)$ is now

$$m(t) = a_s(t) \cos[\theta_s(t) - \theta_L] \qquad (5.1.15)$$

The spectrum of this mixed term is now superimposed on the spectral terms associated with the intensity variations of the signal, noise, and local fields. The latter intensity variations will, therefore, fall in-band to a homodyne receiver.

In typical heterodyne or homodyne systems, the power P_L of the local field will be much stronger than the received field power (typically milliwatts of local laser power versus microwatts of received power). This means that $\bar{n}$ in Eq. (5.1.14) is dominated by αP_L, and the local power alone sets the shot noise level in Figure 5.3. This also means that the low-frequency intensity noise is primarily that of the local field and will only be produced from time variations in the amplitude a_L. This strong local field condition assures a high count rate at the detector output (even for low power levels of the received fields). Because this is the required condition for the photodetector shot noise to appear Gaussian in nature, heterodyned detector outputs are almost always assumed as Gaussian processes, with the signal term corresponding to the modulated carrier, and all other spectral components (beat noise, shot noise, detector noise, and intensity noise) considered as additive Gaussian noise with the spectra given.

5.2 HETERODYNE SIGNAL-TO-NOISE RATIOS

The modulated heterodyne carrier generated from photodetection can now be filtered out to recover the modulations. The overall system following the photodetector is shown in Figure 5.4a. A bandpass filter $F(\omega)$ is used to recover the modulated carrier, which is then RF demodulated to produce the information signal. The filtered output spectrum then appears as

$$S_F(\omega) = |F(\omega)|^2 [S_i(\omega) + N_{0c}] \qquad (5.2.1)$$

where $S_i(\omega)$ is the photodetected output current spectrum, and N_{0c} is the

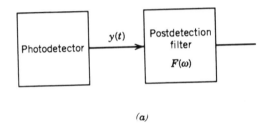

(a)

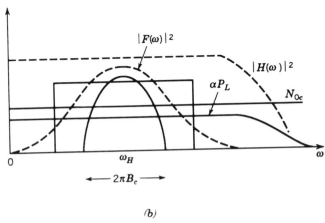

(b)

Figure 5.4. (a) Postdetection filtering. (b) Output filtered spectrum (strong local field).

spectral level of the output thermal noise current. The filter is generally tuned to the bandpass characteristic of the desired signal spectrum. If we assume a wideband optical detector $[H(\omega) = e]$ and a flat filter function $[F(\omega) = 1]$ over the bandwidth of the heterodyned signal, the filtered signal power after heterodyning is

$$P_{so} = (2e\alpha\bar{g})^2 P_L A I_m \qquad (5.2.2)$$

where $I_m = \overline{|a_s(t)|^2}/2$ is the time-average signal intensity of the heterodyned carrier in Eq. (5.1.10). Since $I_m A$ is the heterodyned carrier power, we can use Eq. (5.1.13) to write this in terms of the received laser power P_s in Eq. (5.1.13), and Eq. (5.2.2) becomes

$$P_{so} = 2(e\alpha\bar{g})^2 P_L P_s \qquad (5.2.3)$$

The bandpass noise in the heterodyned signal bandwidth can be determined by collecting the noise terms. Using the spectral levels in Figure 5.4b, the total detected noise power in the signal carrier bandwidth B_c of the filter $F(\omega)$ is then

$$P_{no} = [e^2\overline{g^2}\alpha P_L + (2\bar{g}e\alpha)^2 P_L N_0/4 + N_{0c}]2B_c \qquad (5.2.4)$$

The terms in brackets represent the two-sided noise levels due to shot noise, background, and circuit noise respectively. (We neglect dark current here.) The heterodyned SNR is then

$$\begin{aligned}
\text{SNR} &= \frac{2(e\alpha\bar{g})^2 P_L P_s}{[e^2\overline{g^2}\alpha P_L + (2\bar{g}e\alpha)^2 P_L(N_0/4) + N_{0c}]2B_c} \\
&= \frac{2\alpha P_s}{[F + \alpha N_0 + (N_{0c}/(\bar{g}e)^2\alpha P_L)]2B_c}
\end{aligned} \qquad (5.2.5)$$

This represents the detected carrier SNR developed at the output of a heterodyning receiver over a bandwidth B_c about frequency $\omega_H = \omega_o - \omega_L$. Note that this SNR depends directly on the laser power P_s collected over the receiver aperture area A. Clearly, this area should be as large as possible, provided the signal always corresponds to a single spatial mode. The premise is inherent in the single-mode model used in deriving this SNR. We shall find in Section 5.5, however, that for certain operating conditions, this receiver collecting area A sometimes cannot be made arbitrarily large.

The heterodyned SNR is also interesting when compared with the corresponding result for direct detection (Eq. 4.3.5). We first note that in both cases P_s represents the received averaged laser field power. The direct detected power, however, is reduced by the factor $\beta^2 P_m (\beta^2 P_m < 1)$, because the baseband signal must be intensity modulated onto the linear portion of the laser modulator. The bandwidths B_m and B_c each represent signal bandwidths, except B_c refers to the optical carrier modulated bandwidth, and B_m is the information bandwidth prior to intensity modulation. Hence, Eqs. (4.3.5) and (5.2.5) can be compared directly in terms of power and bandwidth. Second, we note that the circuit noise can be eliminated, and shot-noise-limited behavior is achieved using a strong local source such that $(\bar{g}e)^2\alpha P_L \gg N_{0c}$. In this sense the local source power P_L is playing the role of the photomultiplied signal power $\bar{g}P_s$ in Eq. (4.3.5). Thus, effective signal amplification, as far as achieving shot-noise-limited operation, can be provided by a strong local source. Third, we notice, however, that this local source cannot eliminate the effect of the input noise. In fact, we notice that even with a strong source and no photomultiplication ($\bar{g} = 1, F = 1$), Eq. (5.2.5) becomes

$$\text{SNR} = \frac{2\alpha P_s/2B_c}{1 + \alpha N_0} \qquad (5.2.6)$$

and we never actually reach the quantum-limited condition of Eq. (4.3.8). In space systems, with background noise as input noise, $\alpha N_0 \ll 1$, and it is usually argued that Eq. (5.2.6) is approximately a quantum-limited result. We often write

$$\text{SNR} \approx \frac{2\alpha P_s}{2B_c}$$

$$= \frac{2\eta P_s}{(hf)2B_c} \tag{5.2.7}$$

as the heterodyning quantum-limited bound. On the other hand, we should be aware that for conditions such that $\alpha N_0 \geq 1$ (input noise due to amplifier or laser noise, or high-temperature background sources such that $\eta N_0 \geq hf$), (5.2.6) is instead

$$\text{SNR} \cong \frac{2\alpha P_s}{(\alpha N_0)2B_c}$$

$$= \frac{P_s}{N_0 B_c} \tag{5.2.8}$$

Therefore, the receiver is noise limited, and not quantum limited, even though a strong local source was used. The "hot" noise sources prevent attainment of the quantum-limited bound of heterodyne detection predicted by Eq. (5.2.7). Note that Eq. (5.2.8) is the SNR that would be computed for an RF front-end system undergoing RF or microwave mixing. Hence optical heterodyning is analogous to RF mixing in terms of SNR values. This also explains why quantum-limited operation cannot be achieved by electronic mixing at microwave frequencies. The RF background noise level N_0 is always much larger than the quantum noise level (hf) at these frequencies, and Eq. (5.2.8) defines the resulting RF mixing SNR rather than a quantum noise result.

5.3   DEMODULATED SIGNAL-TO-NOISE RATIO FOLLOWING OPTICAL HETERODYNING

The SNR derived in Eqs. (5.2.4), (5.2.6), and (5.2.8), for optical heterodyning refers to signals at the output of the photodetector filter $F(\omega)$ (see Fig. 5.4). Of ultimate interest, however, is the SNR of the recovered information waveform. To determine this, we must take into account the postdetection demodulation and its effect on the previously derived heterodyned SNR, as shown in Figure

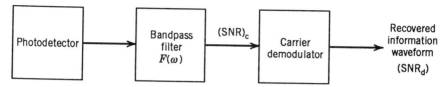

Figure 5.5. Heterodyne postdetection filtering and demodulation.

5.5. This processing depends on the manner in which the information waveform $d(t)$ is demodulated from the heterodyned carrier.

5.3.1 Amplitude Modulation with Heterodyne Detection

If amplitude modulation (AM) is used to modulate an information waveform $d(t)$ onto the optical carrier, we can write $a_s(t)$ in Eq. (5.1.7) as

$$a_s(t) = I_s^{1/2}[1 + d(t)] \tag{5.3.1}$$

where we constrain $|d(t)| < 1$ to prevent overmodulation, and let I_s be the average intensity of the optical signal field. Note that the carrier amplitude is nonnegative and its intensity spectrum is

$$S_a(\omega) = I_s[2\pi\delta(\omega) + S_d(\omega)] \tag{5.3.2}$$

where $S_d(\omega)$ is the spectrum of $d(t)$. [This is valid whether $d(t)$ is deterministic or random, so long as $S_d(\omega)$ is properly defined.] The total power in the received optical signal field is then

$$P_s = AI_s(1 + P_d) \tag{5.3.3}$$

where P_d is the power in $d(t)$. The spectrum of the heterodyned carrier $m(t)$ in Eq. (5.1.10), with $\theta_s(t) = 0$, is then

$$S_m(\omega) = \frac{1}{4} S_a(\omega + \omega_H) + \frac{1}{4} S_a(\omega - \omega_H) \tag{5.3.4}$$

corresponding to the shift of the spectrum in Eq. (5.3.2) to ω_H. The bandwidth occupied by the AM optical carrier term at frequency ω_o, is $B_c = 2B_d$, where B_d is the bandwidth of $d(t)$. After photodetecting and filtering, the heterodyned signal, appears as an AM signal carrier at frequency ω_H, and with bandwidth $2B_d$. Ideal AM demodulation now produces an improvement factor of two from the AM carrier SNR_c to the demodulated SNR_d [1]. From Eq. (5.2.7), the AM quantum-limited SNR_c [for the shot-noise-limited case, the denominator must be modified according to Eq. (5.2.6)] is then

$$SNR_c = \frac{2\alpha P_s P_d}{2B_c}$$
$$= \frac{2\alpha P_s P_d}{4B_d}$$

(5.3.5)

where $P_s = I_s A$. For the case $P_d = 1$, the demodulated SNR is then

$$SNR_d = 2SNR_c$$
$$= \frac{\alpha P_s}{B_d}$$

(5.3.6)

which represents the achievable (upperbound) demodulated SNR_d for the information waveform following heterodyning and ideal AM demodulation. Note that when Eq. (5.3.6) is compared to the equivalent result for AM/IM operation with direct detection (Eq. 4.5.3) we see that the AM SNR_d is improved by a factor of four with perfect heterodyning. This can be directly attributed to the fact that the information is amplitude modulated directly onto the optical carrier rather than onto an intensity-modulated subcarrier.

An alternative method is to amplitude modulate directly onto the optical carrier and use optical homodyning at the receiver, obviating the need for AM demodulation following photodetection. In this case, the filtered output of the photodetector yields directly the spectrum of Eq. (5.3.2), in the absence of input noise. The recovered quantum limited SNR_d is then

$$SNR_d = \frac{2\alpha a_L^2 (I_s A) P_d}{a_L^2 (2B_d)}$$
$$= \frac{\alpha P_s}{B_d}$$

(5.3.7)

which is identical to Eq. (5.3.6) for $P_d = 1$. Thus, we see that there is no difference in demodulated SNR_d between homodyning and heterodyning with AM demodulation. Some confusion often occurs on the point, primarily because there is a tendency to compare the homodyne SNR_d in Eq. (5.3.7) with the heterodyned carrier SNR_c in Eq. (5.3.5). For the case $P_d = 1$, $P_s = I_s A$, and we see that the former is larger than the latter by a factor of two. The difference, of course, is that in the homodyne case this factor of two is already included, whereas in heterodyning this improvement is yet to be recovered (by the ideal AM detector).

5.3.2 Frequency Modulation with Heterodyne Detection

When the information $d(t)$ is frequency modulated (FM) onto the optical carrier, the complex envelope now has the form

$$a(t) = I_s^{1/2} \exp\left[j\, 2\pi \Delta_f \int d(t)\, dt \right] \qquad (5.3.8)$$

where Δ_f is the transmitted frequency deviation imposed on the optical carrier at the transmitter, and $P_s = I_s A$ is the average signal power in the received signal mode. The bandwidth occupied by the heterodyned FM carrier $m(t)$ [2] in (5.1.10) is approximately

$$B_c \cong 2(b + 1)B_d \qquad (5.3.9)$$

where $b = \Delta_f / B_d$ is the optical FM modulation index, and B_d is the information bandwidth. After heterodyning and filtering, the quantum limited SNR_c is then

$$SNR_c = \frac{2\alpha P_s}{2B_c} \qquad (5.3.10)$$

Subsequent FM demodulation of the FM carrier term at frequency ω_H generates the well-known FM improvement [2] and yields an output demodulated SNR_d of

$$\begin{aligned} SNR_d &= 3b^2 \left(\frac{B_c}{B_d} \right) SNR_c \\ &= 6b^2(b + 1) \left(\frac{\alpha P_s}{B_c} \right) \end{aligned} \qquad (5.3.11)$$

provided Eq. (5.3.10) is above the required FM demodulation threshold. Therefore, the output SNR_d can be increased by using large values of b, that is, wideband optical FM. The result is larger by a factor of two over that possible with FM/IM and direct detection.

5.4 THE ALIGNMENT AND FIELD-MATCHING PROBLEM

In the preceding sections, we considered an idealized heterodyning operation in which both the received field and the local field are perfectly aligned and each generates identical and overlapping diffraction patterns in the focal plane. In practice, it is somewhat difficult to obtain this ideal heterodyning condition, and effects of misalignment and local-field distortion must be considered.

Minimizing these anomalies, therefore, becomes an important part of system design.

Let us reconsider the basic heterodyning equation (5.1.3), when the local and received fields are not identical. Recognizing that the spatial integral is simply a complex number, having a given magnitude and vector angle, the signal term in Eq. (5.1.6) generalizes to

$$n_{RL}(t) = \left[2\alpha a_L \int_{A_d} \phi_1(\mathbf{q})\phi_L^*(\mathbf{q}) \, d\mathbf{q} \right] m(t) \tag{5.4.1}$$

where $m(t)$ is again given by Eq. (5.1.10). Note that the integral magnitude term now becomes a multiplying factor on the signal amplitude. It is convenient to rewrite Eq. (5.4.1) as

$$n_{RL}(t) = (2\alpha a_L A)L_H m(t) \tag{5.4.2}$$

where we have introduced the heterodyning amplitude loss factor

$$L_H = \frac{1}{A} \left| \int_{A_d} \phi_1(\mathbf{q})\phi_L^*(\mathbf{q}) \, d\mathbf{q} \right| \tag{5.4.3}$$

By a simple application of the Schwartz inequality, it can be easily established that $L_H \leqslant 1$, and therefore acts as a true amplitude loss on the signal term. Thus the effect of field mismatching in heterodyning is to cause an amplitude loss in the heterodyned signal. Furthermore, this loss depends directly on the integrated product of the two focused fields (spatial correlation). Hence, field mismatch losses can be determined by directly evaluating this integral.

We can recompute the resulting SNR after mismatched heterodyning. The power in the signal term is now

$$\begin{aligned} P_{so} &= (2e\alpha a_L A)^2 L_H^2 (I_s/2) \\ &= 2(e\alpha)^2 P_L P_s L_H^2 \end{aligned} \tag{5.4.4}$$

The additive shot noise in the heterodyned carrier bandwidth with a strong local field is then

$$\begin{aligned} P_{no} &= \left[e^2 \alpha a_L^2 \int_{A_d} |\phi_L(\mathbf{q})|^2 \, d\mathbf{q} \right] 2B_c \\ &= [e^2 \alpha P_L] 2B_c \end{aligned} \tag{5.4.5}$$

where we have used the fact that the local field integral is always normalized as

$$\int_{A_d} |\phi_L(\mathbf{q})|^2 \, d\mathbf{q} = A \tag{5.4.6}$$

The resulting SNR in Eq. (5.2.5) is now modified to

$$\text{SNR} = \frac{2\alpha P_s L_H^2}{[F + \alpha N_0 + (N_{0c}/(\bar{g}e)^2 \alpha P_L)]2B_c}$$ (5.4.7)

The quantum-limited result (with $\alpha N_0 \ll 1$) is then

$$\text{SNR} = \left[\frac{2\alpha P_s}{2B_c}\right] L_H^2$$ (5.4.8)

and the squared heterodyne loss factor (L_H^2) directly suppresses the SNR.

Let us consider first the effect of only field misalignment. We assume the diffraction patterns are identical in shape, phase, and polarization, but the local field is not perfectly aligned with the focused received field. As a result, the identical diffraction patterns on the detector surface do not overlap, as shown in Figure 5.6. The resulting heterodyne loss term is then

$$L_H = \frac{1}{A}\left|\int_{A_d} \phi_1(u, v)\phi_L^*(u - u_o, v - v_o)\,du\,dv\right|$$ (5.4.9)

where (u_o, v_o) are the offset distances from the normal in the focal plane due to the misalignment. Recall that for an arriving plane wave of wavelength λ arriving normal to a square aperture of width d, the Airy pattern is given by

$$\phi_1(u, v) = \Gamma(u, v)\left(\frac{A}{\lambda f_c}\right)\left[\frac{\sin(\pi du/\lambda f_c)}{(\pi du/\lambda f_c)} \cdot \frac{\sin(\pi dv/\lambda f_c)}{(\pi dv/\lambda f_c)}\right]$$ (5.4.10)

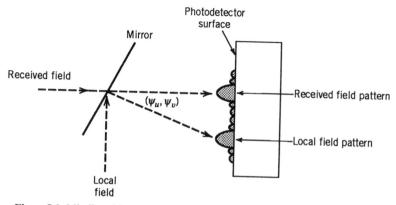

Figure 5.6. Misaligned heterodyne receiver. (ψ_u, ψ_v) = misaligned offset subangles.

Using this pattern for both fields in Eq. (5.4.9) the L_H term evaluates to

$$L_H = \left| \left(\frac{\sin \Psi_u}{\Psi_u} \right) \left(\frac{\sin \Psi_v}{\Psi_v} \right) \right| \tag{5.4.11}$$

where

$$\Psi_u = \left(\frac{\pi d}{\lambda f_c} \right) u_o \approx \left(\frac{\pi d}{\lambda} \right) \psi_u$$

$$\Psi_v = \left(\frac{\pi d}{\lambda f_c} \right) v_o \approx \left(\frac{\pi d}{\lambda} \right) \psi_v \tag{5.4.12}$$

Here (ψ_u, ψ_v) are the misalignment subangles (azimuth and elevation) defined by $\psi_u \cong u_o/f_c$, $\psi_v \cong v_o/f_c$. Thus, the misalignment causes the heterodyne power loss to fall off as $|\sin x/x|^2$ as the offset angle increases. To maintain a small loss factor, it is necessary to carefully control these angles. For example, if no more than a 3-dB loss can be accepted, both angles must be maintained within the range $\psi \leqslant 1.39$. ($|\sin x/x| = 1/2^{1/2}$ at $x = 1.39$). Thus, we require, for either angle, ψ_u or ψ_v, that

$$\left(\frac{\pi d}{\lambda} \right) \psi \leqslant 1.39$$

or

$$\psi \leqslant 0.44 \left(\frac{\lambda}{d} \right) \tag{5.4.13}$$

At $\lambda = 1 \, \mu m$, and say $d = 0.1$ m, an alignment to within 4 μrad is required. That is, the local and received fields must be focused to within a few microradians to satisfy this heterodyning operation. Note that as the aperture width d increases, the alignment requirement becomes tighter. Thus increasing aperture size to achieve a larger collecting area A (and higher SNR) requires a correspondingly more accurate field alignment.

The angle misalignment may be due to either offsets in the local field (as shown in Fig. 5.6) or a misalignment of the received field relative to the normal. The latter impacts the receiver pointing problem—the ability to accurately point the receiver optics directly at the laser source. Hence the receiver pointing in space optical systems becomes particularly important in heterodyne systems. This pointing will be addressed in Chapter 10.

A heterodyning loss occurs even if the field polarizations of the received and local fields are not aligned. [Recall our discussion in Section 1.3 associated with Eq. (1.3.5) in which the latter assumed a linearly polarized field.] If both fields are linearly polarized at different polarization angles, the integral in Eq. (5.4.9) becomes a dot products of these vector fields. Thus, even if the fields are matched in shape and alignment, the loss factor now becomes

$$L_H = \frac{|A\cos(\theta_p)|}{A} = |\cos\theta_p| \tag{5.4.14}$$

where θ_p is the difference angle between the two linear polarizations. Hence the heterodyned signal power degrades as the squared cosine of the polarization mismatch. This particular type of loss may become important if the received field polarization shifts in time during propagation. Unless the local field polarization adaptively corrects, the polarization loss can produce time-varying signal fadings during communications.

Perfect heterodyning requires focusing to the same diffraction patterns. This, in turn, requires accurate lens and mirror design. Often, simpler systems are used in which the local field is purposely not focused. For example, consider the case where the received field is focused by the input lens to its Airy pattern, as in Eq. (5.4.10), but the local field is supplied by simply illuminating the detector with an unfocused plane wave. This would occur if a smaller local lens is used to intentionally defocus the local field or, in the limit, no lens is used at all and the local field is projected directly on the detector. The normalized local field pattern at the detector is then $\phi_L(u, v) = (A/A_d)^{1/2}$. The heterodyne loss term now becomes

$$L_H = \frac{1}{A}\left|\int\int_{A_d} \phi_1(u, v)\, du\, dv\right| \tag{5.4.15}$$

For the square input aperture of width d, this is

$$L_H = \frac{d}{\lambda f_c(A_d)^{1/2}}\left|\int\int_{A_d}\left(\frac{\sin(\pi\,du/\lambda f_c)}{(\pi\,du/\lambda f_c)}\right)\left(\frac{\sin(\pi dv/\lambda f_c)}{(\pi dv/\lambda f_c)}\right) du\, dv\right| \tag{5.4.16}$$

Changing variables, and assuming a square photodetector area of width b, this becomes

$$
\begin{aligned}
L_H &= \left(\frac{2\lambda f_c}{bd\pi^2}\right)\left|\int\int_0^{(\pi db/2\lambda f_c)}\left(\frac{\sin x}{x}\right)\left(\frac{\sin y}{y}\right) dx\, dy\right| \\
&= \left(\frac{2\lambda f_c}{bd\pi^2}\right)\left|S_i\left(\frac{\pi db}{2\lambda f_c}\right)\right| \\
&= \left|\frac{S_i^2(b/d_o)}{\pi(b/d_o)}\right|
\end{aligned}
\tag{5.4.17}
$$

where we have used

$$S_i(a) = \int_0^a \left(\frac{\sin x}{x} \right) dx \qquad (5.4.18)$$

and defined $d_o = 2\lambda f_c/\pi d$ for convenience. The SNR suppression is therefore a function of the normalized detector size, b/d_o, when heterodyning with a local plane wave. The maximum value of the braces in Eq. (5.4.17) occurs when $b/d_o = 2.2$, for which the associated suppression factor L_H^2 is $(1.29)^2/\pi^2 \cong 0.168$. Thus the heterodyned SNR is reduced by approximately 7.7 dB when plane-wave heterodyning is used and the detector area is properly selected. Note that the required detector size is $b = 2.2d_o = 1.4(\lambda f_c/d)$, where we recognize $(2\lambda f_c/d)$ as the distance between zeros in the received field diffraction pattern. The

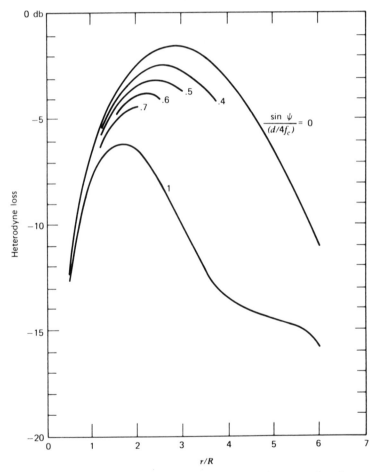

Figure 5.7. Heterodyne loss in plane wave heterodyning r = detector radius, d = aperture diameter, $R = 2\lambda_L f_c/\pi d$, ψ = offset angle.

suppression loss is directly attributable to the fact that the local field is spread out over a wider area of the detector surface than that actually necessary for the received field. This excess field power is therefore wasted power in terms of the heterodyne operation; that is, it should have been concentrated over the received field pattern.

For the circular aperture and detector, the misalignment suppression loss in Eq. (5.4.15) is instead

$$L_H = \left(\frac{2}{r/R}\right)\left|\int_0^{r/R} J_1(\rho)J_0\left(\frac{4f_c\rho\sin\psi_L}{d}\right)d\rho\right| \qquad (5.4.19)$$

where r is the detector radius, d the aperture diameter, $R \triangleq 2\lambda f_c/\pi d$ and ψ_L is the local oscillator misallignment angle (see Problem 5.10). A plot of Eq. (5.4.19) is shown in Figure 5.7. Note again that alignment losses can be reduced by proper selections of detector size, the latter dependent on the degree of misalignment. It is interesting to note that the losses with misalignment angles out to about $\psi_L < \sin^{-1}(d/4f_c)$ are only several decibels greater than that for a perfectly aligned plane wave ($\psi_L = 0$), with detectors having $r \cong 3R$. This means that except for the loss factor of using a mismatched plane wave, Eq. (5.4.19) shows an insensitivity to alignment errors for this range of ψ_L. This implies the feasibility of using a single unfocused local plane wave for heterodyning.

5.5 MULTIMODE HETERODYNING

In Eq. (5.1.3), we wrote the general expression for heterodyning with a received field and a local field. We then specialized to the case where the signal was confined to a single diffraction pattern and the local field mode was spatially matched to it. In this section, we examine several generalizations of this case.

Let the received field produce the general expansion in Eq. (5.1.5), and assume the local field is itself expanded into the same set of orthogonal diffraction functions,

$$f_L(t, \mathbf{q}) = \sum_{i=1}^{D_s} a_L \phi_i(\mathbf{q}) \qquad (5.5.1)$$

This corresponds to a multimode version of the local field, with each mode having power $P_L = a_L^2 A$. The heterodyned waveform is again given by Eq. (5.1.3), with the crossterm now being

$$n_{RL}(t) = (2\alpha a_L A)\,\text{Real}\left\{\sum_{i=1}^{D_s} [s_i(t) + b_i(t)]e^{j(\omega_H t - \theta_L)}\right\} \qquad (5.5.2)$$

This represents the generalized crossterm produced from multimode heterodyning. When expanded in this way, we see that $n_{RL}(t)$ is now the sum of envelope time functions, one from each mode. If the local field existed in only one mode, only the crossterm from that one mode would appear. Hence multimode heterodyning adds in contributions from all other heterodyned signal modes.

We now examine some special cases. Consider a strong local field occupying D_L spatial modes, heterodyned with a single mode signal field in additive multimode input noise. We assume each local field mode has equal amplitude a_L, and the signal is confined to the first mode. As before, the signal component in the first mode generates a power term of

$$P_{so} = (4e^2\alpha^2 A P_L)(I_s/2) \tag{5.5.3}$$

For each of the local field modes, we generate a noise term

$$P_{no} = e^2[\alpha P_L + \alpha^2 P_L N_0]2B_c \tag{5.5.4}$$

over a bandwidth B_c. The total noise is that contributed from all heterodyned modes on the receiver surface, plus that added by the circuit noise to the detector output. The heterodyned SNR becomes

$$\begin{aligned} \text{SNR} &= \frac{2(\alpha^2 P_L)A I_s}{[D_L P_{no} + (N_{0c}/e^2)]2B_c} \\ &= \left(\frac{1}{D_L}\right)\left[\frac{2\alpha P_s/2B_c}{[1 + \alpha N_0 + (N_{0c}/e^2\alpha P_L D_L)]}\right] \end{aligned} \tag{5.5.5}$$

Clearly, we have reduced the resulting SNR from the idealized case in Eq. (5.1.22) by the factor D_L, due to the number of unnecessary local field modes. These excess modes are heterodyning input noise, while producing additive shot noise. We are, however, using the excess mode power to aid in overcoming the circuit noise; that is, to produce shot-noise-limited operation. Note that the division by D_L can be considered a form of signal power suppression, similar to the suppression effect in Eq. (5.3.3). In the latter case, the cause is field mismatch, whereas in the present case the fields are matched, but unnecessary modes are used. In essence, the local field source is overspread for the received field.

Now extend to the case where both the local and received fields occupy more than one mode. Let the local field have D_L modes and the received field D_s modes. We again describe the transmitted signal field as $a_s(t)\exp(j\omega_0 t)$ with average intensity I_s referred to the receiver. We assume, for simplicity, the received signal field power is distributed equally over all D_s modes, and we describe the received field in the ith mode as $(1/D^{1/2})a_s(t-\tau_i)\exp[j\omega_i(t-\tau_L)]$, where τ_i is the mode delay and ω_i is the mode optical frequency. The differences in delays $\{\tau_i\}$ over the modes is due to the channel delay dispersion, and the differences in mode frequencies $\{\omega_i\}$ may be due to possible doppler spreading over the modes. Each mode has signal intensity I_s/D_s, so that the total available intensity I_s is spread equally over all modes.

We consider the local field to be $a_L \exp[j\omega_L t]$ in each mode. Hence the heterodyned signal term is

$$
n_{RL}(t) = (2\alpha a_L A) \, \text{Real} \left\{ \sum_{i=1}^{D_L} \left(\frac{1}{\sqrt{D_s}} \right) a_s(t - \tau_L) \exp j \left[\omega_i (t - \tau_i) - \omega_L t \right] \right.
$$

$$
= (2\alpha a_L A) \sum_{i=1}^{D_L} \left(\frac{1}{\sqrt{D_s}} \right) a_s(t - \tau_i) \cos \left[(\omega_i t - \omega_L t) - \omega_i \tau_i \right]
\tag{5.5.6}
$$

Equation (5.5.6) corresponds to the heterodyned collection of delayed field modes, and therefore represents the heterodyned counterpart to our multimode direct detection result in Eq. (4.7.7). If the dispersion in frequency (referred to the heterodyne frequency ω_H) and delay is excessive, Eq. (5.5.6) corresponds to a sum of interfering waveforms, and the structure of the waveform $n_{RL}(t)$ is destroyed.

On the other hand, if we assume the frequency and delays are not widely dispersed, the terms in Eq. (5.5.6) combine to produce a reinforced signal. For example, if we assume the frequency spread is neglible ($\omega_i \approx \omega_o$ for all i) and the delays are all identical ($\tau_i \approx 0$), then the heterodyne SNR can be computed as in Eq. (5.5.5). For the case $D_s > D_L$,

$$
P_{so} = (2\alpha a_L A)^2 \left[\sum_{i=1}^{D_L} \left(\frac{1}{\sqrt{D_s}} \right) a_s(t) \cos(\omega_H t) \right]^2
$$

$$
= (2\alpha a_L A)^2 D_L^2 (I_s/D_s)
\tag{5.5.7}
$$

Whereas, for $D_s < D_L$,

$$
P_{so} = (2\alpha a_L A)^2 D_s I_s
\tag{5.5.8}
$$

In either case the noise power is

$$
P_{no} = [\alpha D_L P_L + D_L(\alpha^2 P_L) N_0 + N_{0c}] 2B_c
\tag{5.5.9}
$$

and, therefore, increases with D_L. In Eq. (5.5.7) signal power is lost by not using all available signal modes. In Eq. (5.5.9), the performance can clearly be improved by reducing D_L to D_s, which does not effect P_{so} but reduces P_{no}.

When $D_L = D_s$, the resulting SNR is

$$\text{SNR} = \frac{2\alpha A I_s/2B_c}{[1 + \alpha N_0 + (N_{0c}/D_L \alpha P_L e^2)]} \tag{5.5.10}$$

If single mode heterodyning had been used, the SNR would be given by Eq. (5.5.10) with the mode intensity I_s replaced by I_s/D_s. Thus Eq. (5.5.10) is D_s times larger than if single-mode heterodyning were used. The system is, therefore, benefitting from the coherent combining of the signal modes when the frequency and delay dispersion are negligible.

We emphasize that the delay dispersions in Eq. (5.5.6) must be small relative to a single period of the heterodyne frequency ω_H. [Recall that, in direct detection, the delay spread is relative to the time variations of the intensity modulation $a_s(t)$.] For example, if we assume that $\{\tau_i\}$ are independent delays, and each produces uniformly distributed delays over a period of ω_H, then Eq. (5.5.7) becomes

$$P_{so} = (2\alpha a_L A)^2 D_L(I_s/D_s) \tag{5.5.11}$$

and Eq. (5.5.10) is reduced by the factor $1/D_L$. In this case, we are no longer coherently combining the heterodyned modes and the multimode advantage is lost. Thus, an advantage can be attained in heterodyning with many field modes only if the mode delay dispersion is small or can be reduced. This suggests the possible use of delay equalization over the heterodyned modes as a tradeoff for field power and bandwidth in multimode detection. Just as in direct detection, this can theoretically be achieved in a space channel by spatially separating the heterodyned modes (heterodyne and detect each mode separately with a detector array) and attempting to phase correct (perhaps from phase measurements obtain by previous channel probing or by estimating each mode phase) before combining. The nonstationarity of the space channel may seriously hinder the practical implementation of this type of equalizer.

The local field still has the requirement of matching and aligning to each field mode, but, in addition, it must now have enough strength in each mode to overcome the circuit noise of each individual detector. If the field is not properly matched, a power suppression factor must be included for each detector, which may negate the coherent combining advantage if too severe. Again, delay equalization may have its most practical application to the fiberoptic channel, where fibers tend to produce many signal modes. Heterodyning over a single spatial mode produces power loss because of the unusable modes. Therefore, one must improve the fiber (reduce the number of modes) or attempt to regain lost power by multimode heterodyning with delay equalization. Because it is difficult to separate modes spatially in an optical guide, equalization must be achieved directly with the heterodyned signal in Eq. (5.5.6). This operation is hindered by the fact that when the fiber modes are in "equilibrium," each mode itself has a delay spread that must undergo a comparable equalization.

5.6 HETERODYNING WITH RANDOM SIGNAL FIELDS

We pointed out that if an optical signal field is transmitted over a turbulent path, the effect is to break up the optical beam spatially. In discussing noncoherent detection in Chapter 4, we considered this effect as an apparent extension of the optical source size. In dealing with coherent detection it is more convenient to consider the turbulent effect as one of converting a coherent optical field to a random field. This randomness is over the receiver spatial variable $\mathbf{r}$ and corresponds to a point-to-point loss in coherence over the receiver area. We are here interested in assessing the spatial effects of turbulence on heterodyning. Let us first examine the detected field at a fixed time t. Subsequent time averaging will allow comparison to our earlier SNR results.

Consider the received signal field to be represented as $a(t, \mathbf{r}) \exp(j\omega_o t)$ at the receiver aperture, where $a(t, \mathbf{r})$ is again the complex signal envelope. A local heterodyning field source is assumed to produce a single mode diffraction pattern at the photodetector surface. We again describe this local field function over the detector area as $a_L \exp(j\omega_L t)$. We can again describe the field mixing in terms of the spatial integral over the detector surface. Instead, let us rewrite this heterodyne term in terms of an equivalent integral over the receiver aperture. (Recall that Parcevals theorem allows us to equate an integral in one domain in terms of integrals involving their Fourier transforms.) A focused local field, with an Airy pattern matching the aperture lens, can therefore be referred back to the aperture as a plane wave field. Hence the heterodyned term is equivalent to

$$n_{RL}(t) = (2\alpha) \, \text{Real} \left\{ e^{j\omega_n t} \int_A a_L \, a(t, \mathbf{r}) \, d\mathbf{r} \right\} \tag{5.6.1}$$

where the integral now involves the receiver aperture area A. If the received field and local field are aligned deterministic plane waves in the same mode $[a(t, \mathbf{r}) = a(t), \mathbf{r} \in A]$, then Eq. (5.6.1) reduces to Eq. (5.1.9), with I_s the average intensity of $a(t)$. The resulting SNR is then identical to Eq. (5.2.6) when operating under shot-noise-limited conditions.

When the received signal field is random, however, the field in the integral in Eq. (5.6.1) must be considered a spatially random field. The integral, therefore, evolves as a random variable at each t, and $n_{RL}(t)$ is now a random process in time. Its mean-square value (total spatial power at time t) is then

$$P_s = \overline{n_{RL}^2(t)} = (2\alpha a_L)^2 \left| \int_A a(t, \mathbf{r}) \, d\mathbf{r} \right|^2 \tag{5.6.2}$$

where the overbar denotes random field averaging. Expanding the integral term, we can write

$$\left| \int_A a(t, \mathbf{r}) \, d\mathbf{r} \right|^2 = \int_A \int_A \overline{a(t, \mathbf{r}_1) a^*(t, \mathbf{r}_2)} \, d\mathbf{r}_1 \, d\mathbf{r}_2 \tag{5.6.3}$$

The integrand is the mutual coherence function of the signal field at the receiver aperture. For stationary, coherence-separable fields,

$$\overline{a(t, \mathbf{r}_1) a^*(t, \mathbf{r}_2)} = \overline{I}_s R_s(\mathbf{r}_1, \mathbf{r}_2) \tag{5.6.4}$$

where $\overline{I}_s = R_t(0)$ is the mean field intensity at any t, and $R_s(\mathbf{r}_1, \mathbf{r}_2)$ is the normalized spatial coherence. Thus the time-averaged heterodyned signal power is then

$$P_s = 2(\alpha a_L)^2 \overline{I}_s \int_A \int_A R_s(\mathbf{r}_1, \mathbf{r}_2) \, d\mathbf{r}_1 \, d\mathbf{r}_2 \tag{5.6.5}$$

The shot noise power collected over the receiver area A can be computed as before. For a strong local source the shot noise contributes a power of P_L. The effect of additive input noise appears only in the portion of the noise heterodyned by the local source. Hence, the total noise power collected at the detector (shot noise plus heterodyned-input noise) is identical to that of single-mode detection. Therefore, the random field SNR is

$$\text{SNR} = \frac{2\alpha \overline{I}_s (a_L)^2}{[(1 + \alpha N_0) a_L^2 A] 2B_c} \left[\int_A \int_A R_s(\mathbf{r}_1, \mathbf{r}_2) \, d\mathbf{r}_1 \, d\mathbf{r}_2 \right] \tag{5.6.6}$$

If we compare this to our earlier result in Eq. (5.2.6), we see that Eq. (5.6.6) can be rewritten as

$$\text{SNR} = \frac{2\alpha \overline{I}_s A_r}{(1 + \alpha N_0) 2B_c} \tag{5.6.7}$$

where we have defined an *effective* receiver area

$$A_r \triangleq \frac{1}{A} \int_A \int_A R_s(\mathbf{r}_1, \mathbf{r}_2) \, d\mathbf{r}_1 \, d\mathbf{r}_2 \tag{5.6.8}$$

If we interpret $P_s = \overline{I}_s A_r$ as the received signal power, we see that the parameter A_r plays the role of an effective signal-collecting aperture area. We therefore conclude that, as far as SNR is concerned, the effect of randomized input fields on heterodyning can be accounted for by replacing the true receiver aperture area A by the effective area A_r in determining signal power. The size of this collecting area depends on the mutual coherence function of the random input signal field relative to the true receiver area.

We note two immediate results. If the received field is spatially coherent over the receiver area A [$R_s(\mathbf{r}_1, \mathbf{r}_2) = 1$ for all $\mathbf{r}_1$, and $\mathbf{r}_2$ in A], then

$$\int_A \int_A R_s(\mathbf{r}_1, \mathbf{r}_2)\, d\mathbf{r}_1\, d\mathbf{r}_2 = A^2 \qquad (5.6.9)$$

and the effective area is $A_r = A^2/A = A$. Thus the effective signal collecting area is the true receiver area A as long as the received field is entirely coherent over A.

Now assume the received field is coherent over an area A_c less than A. That is,

$$R_s(\mathbf{r}_1, \mathbf{r}_2) = 1 \quad \text{for all } \mathbf{r}_2 \text{ within an area } A_c \text{ around each } \mathbf{r}_1$$
$$= 0 \quad \text{elsewhere} \qquad (5.6.10)$$

Then

$$\int_A \int_A R_s(\mathbf{r}_1, \mathbf{r}_2)\, d\mathbf{r}_1\, d\mathbf{r}_2 = \int_A \left[\int_{\mathbf{r}_1 + A_c} d\mathbf{r}_2 \right] d\mathbf{r}_1 = A_c A \qquad (5.6.11)$$

and the signal-collecting area is now $A_r = AA_c/A = A_c$. We see that the collecting area is no larger than the field coherence area when the latter is less

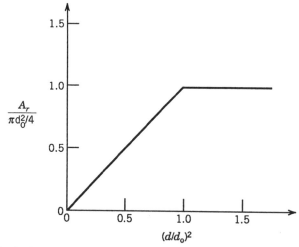

Figure 5.8. Effective heterodyne collecting area A_r versus receiver aperture diameter. $d_o =$ coherence diameter of received field.

than the receiver aperture area. In this case, no additional signal power is collected, even if we further increase the physical receiver aperture area A beyond the area A_c.

The signal collecting area for these two limiting cases is sketched in Figure 5.8, as a function of a circular receiver aperture diameter d, assuming d_o is the circular coherence diameter of the received field. For $d < d_o$, the effective area (and therefore the heterodyned SNR) is directly related to d and increases as the receiver area increases, as it should for single-mode heterodyning. When d exceeds d_o, the effective area A_r no longer depends on d and never exceeds the value at $d = d_0$. Hence, no further improvement is possible by increasing d beyond the coherence distance. Physically, the randomness breaks up the optical beam to such an extent that increasing receiver size no longer collects useful (coherent) signal energy. The principal conclusion here is that when performing single-mode heterodyning there is no advantage in using receiver diameters larger than the coherence distance of the received optical field. Furthermore, the maximum attainable signal power is limited to that collectable over such a coherence area. This effect is studied in more detail in Chapter 9, where actual coherence functions produced by the atmosphere are considered.

PROBLEMS

5.1   A heterodyne system uses an RF bandpass filter following photodetection. The filter is tuned to a frequency of 1 GHz with a bandwidth of 10 MHz. The local oscillator operates at frequency 10^{14} Hz and the transmitter carrier is at an optical frequency such that oscillator mixing produces the proper RF for the filter.

   (a) How much optical Doppler (frequency) shift of the received carrier is acceptable in the system? Neglect carrier modulation.

   (b) Convert (a) to normal velocity between transmitter and receiver.

   (c) How much local oscillator frequency instability can occur?

5.2   A 1-mW optical transmitter uses a 6-in. transmitting lens at 10^{14} Hz. The beam is transmitted over a 500 mile free space path to a heterodyning receiver. The receiver has a collecting area of 4 in.^{2}.

   (a) How much local signal power is needed to achieve a strong oscillator condition, assuming a 20-dB local-to-received field power ratio is needed?

   (b) How much is needed to overcome circuit noise with the same ratio as (a), when the detector load resistance is 100 ohms and operates at room temperature?

   (c) If the detector had gain g, determine how the answers in (a) and (b) are affected.

5.3   A coherent optical system receives an unmodulated laser field with a power of 100 mW. The local laser power is 1 mW, with a heterodyning frequency set at 1 MHz. Assume $e\alpha = 2$. Neglect input noise, dark current, and thermal noise.

(a) How much photodetector output power is in the band from 0 to 100 Hz?

(b) Repeat if the local field is turned off.

5.4   The input to a heterodyne system is an unmodulated laser field at $f = 10^{14}$ Hz with a power of 100 μW and no input noise. It heterodynes with a local field at $f_L = (10^{14} + 1 \text{ Mhz})$ with a power of 1 mW. A power meter measuring only dc ($f = 0$) current power is placed at the photodetector output. How will the power meter reading react (increase or decrease) if the local laser field is removed? Explain with equations.

5.5   A laser plane wavefield at frequency ω_o is split into two paths. The upper path has its frequency shifted to $(\omega_o + 1 \text{ GHz})$. The fields are then recombined (summed) and illuminate a photodetector. Assume the upper arm wave has an intensity of 20 $W^{1/2}/m$ and the lower arm 30 $W^{1/2}/m$.

(a) If a dc ammeter is placed at the photodetector output, what will it read?

(b) If a power meter with bandwidth 100 MHz centered at 1 GHz is placed at the output, what will it read?

5.6   A heterodyne system is designed in which the local laser field is transmitted directly with the modulated laser field. Let the AM carrier have average intensity I, frequency ω_o and modulated carrier bandwidth B_c. Let the transmitted heterodyning field have amplitude a_L and optical frequency ω_L. Assume both fields are normal plane waves at the receiver aperture (area A) Neglect background noise.

(a) Determine the heterodyned signal–to–shot–noise-power ratio after photodetection.

(b) Compare the quantum-limited SNR of an ideal local heterodyne system (strong local laser condition) to the result in (a) for $a_L \gg I^{1/2}$ and $a_L \ll I^{1/2}$.

5.7   An optical space system operates with a received signal power P_s and input noise power P_b watts.

(a) Determine the detected shot-noise-limited SNR in a bandwidth B_n Hz for a direct detection optical receiver with ideal detector gain g.

(b) Determine the shot-noise-limited SNR of an ideal (strong local laser, matched fields) heterodyne receiver with the same noise bandwidth.

(c) Determine the condition on the ratio P_s/P_b for the direct detection system to have the higher SNR.

5.8 A heterodyne system does not use an aperture lens. Instead it places a square photodetector (area A_d) in the receiver plane. An unmodulated plane wavefield at wavelength λ_1 with intensity I_1 arrives along the normal. A local field is at λ_2 with intensity I_2 is added at angle 10^{-8} rad from the normal. Determine the power in the output signal component at wavelength $(\lambda_1 - \lambda_2)$. Neglect noise. Assume $\lambda_1 \approx \lambda_2 = 10^{-6}$.

5.9 In a heterodyne receiver the local field is perfectly aligned on the center of the detector. The received field operates at $\lambda = 1\,\mu m$ and the receiver aperture is a 6-in. square. What is the minimal offset angle that can be allowed to an arriving plane wave in order to guarantee that the alignment power loss is less than 3 dB?

5.10 Show that if a circular receiver lens of diameter d is used when heterodyning with two matched Airy patterns, the equivalent suppression factor L_H for misaligned angles is proportional to

$$\int_{A_d} \left(\frac{J_1 |\mathbf{u}|}{|\mathbf{u}|}\right)\left(\frac{J_1|(|\mathbf{u} - \rho_o|)|}{|\mathbf{u} - \rho_o|}\right) d\mathbf{u}$$

5.11 A receiver with a circular lens of diameter d is focused onto a detector of radius r. Heterodyning is done with an unfocused plane wave on the detector.

  (a) Show that the heterodyne loss factor is given by the loss factor L_H given in Eq. (5.4.19).

  (b) Set $2f_c \sin \psi/d = 1$, and evaluate the integral by using the first two terms of an expansion of J_o as a power series.

5.12 Given the heterodyne signal term in (5.5.6) and assuming that the random delays are small so that $s(t - \tau_i) \cong s(t)$. Define $x_i = \omega_i \tau_i$.

  (a) Write a general expression for the multimode signal power for an arbitrary joint distribution $p(x_1, x_2, x_3 \cdots x_{D_L})$.

  (b) Show that if the $\{x_i\}$ are independent and uniformly distributed over $(0, 2\pi)$, then Eq. (5.5.7) follows.

REFERENCES

1. A. Carlson, *Communication Systems*, 3rd ed., McGraw Hill, New York, 1986, Chapter 6.

2. R. Gagliardi, *Introduction to Communication Engineering*, 2nd ed., Wiley, New York, 1988, Chapter 5.

6

OPTICAL DIGITAL
COMMUNICATIONS

In the analog transmission of information, the primary objective is to transmit a source waveform from the transmitter to the receiver with as little distortion as possible. In Chapters 4 and 5, this distortion was measured in terms of a postdetection signal-to-noise ratio (SNR) following photodetection. Another way of sending information is by digital transmission, in which the desired message is converted to binary symbols (bits) and transmitted as modulated light fields. The receiver now has the task of detecting, or decoding, the transmitted symbols from the received optical field. The performance criterion is based on the accuracy of this symbol decoding rather than on the SNR considered previously. The optical digital system is the objective of study in this chapter.

The transmission of the digital bits over the optical link can be done on a bit-by-bit basis (binary encoding) or on a bit word basis (block encoding). We first consider binary encoding, then extend to block coding later in Section 6.6.

6.1 BINARY DIGITAL OPTICAL SYSTEMS

The block diagram of a general optical system transmitting digital bits is shown in Figure 6.1. Each bit (herein denoted by the symbols 0 and 1) are sent individually by transmitting one of two optical fields to represent each bit. Each bit field is of finite time length, say T_b seconds long, so that one data bit is sent every T_b seconds. Hence the system transmits at the bit rate

$$R_b = 1/T_b \quad \text{bits/sec} \tag{6.1.1}$$

The optical field representing the bit is transmitted to the optical receiver as a modulated light field. In a space system, this light field is transmitted as an optical beam to the optical receiving lens. In a fiber system, the light field is

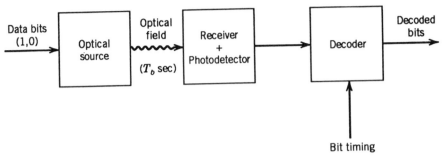

Figure 6.1. Digital optical communication system model.

inserted into the fiber and detected at the output receiver after fiber propagation.

The overall system may be designed as a direct detection (incoherent) system in which intensity modulation (IM) is used, or can be a coherent system with direct optical carrier modulation. At the receiver, the field is photodetected (by direct detection or by heterodyning) and the output current is processed to determine the transmitted bit during each T_b-second bit interval. Since the encoding operation is known at the receiver, a correct bit decision can be made by deciding which of the two optical fields was received. The resulting sequence of decoded bits then represents the received version of the transmitted bits. Hence, accurate bit transmission is directly related to the ability to identify the photodetected optical field. Background light, input noise fields, and detector noise incurred during photodetection, will cause decoding bit errors, degrading the digital peformance. Thus the probability that any given bit will be decoded in error (bit error probability) becomes an important performance measure in digital systems.

Optical bit modulation can be classified as either pulsed or continuous wave (CW). In a pulsed system, the bits are transmitted by pulsing the light source (laser, LED, etc). In a CW system the bits are modulated so that the laser source is continually emitting. Pulsed modulation is almost always used with direct detection, whereas CW modulation may involve either direct detection or heterodyning systems.

The most common forms of pulsed modulation in binary direct detection receivers are on–off keying (OOK) and Manchester coding. In OOK, each bit is transmitted by either pulsing the light source on or off during each bit time. This represents the most basic type of optical signaling and corresponds to merely blinking the light source for digital encoding. In Manchester coding, a sequence of two on–off pulses is used for each bit. In either case, the light source is effectively intensity modulated by the coded pulse waveform, and the direct detection receiver attempts to decode this intensity pattern.

The common forms of CW encoding use subcarrier intensity modulation of the light source for direct detection and direct laser carrier modulation with heterodyne systems. The bit modulation on the subcarrier or optical carrier can be digital amplitude, frequency, or phase modulation, just as in RF communications.

As noted in Figure 6.1, digital decoding requires the establishment of exact bit timing between the transmitter encoder and the receiver decoder. That is, the decoder must know exactly when each T_b bit interval begins and ends during the decoding. This time synchronization is generally accomplished by a timing subsystem that operates in parallel with the electronic decoder following photodetection. As we see later, the ability to maintain accurate receiver timing can become an important part of the overall system design.

6.2 ON–OFF KEYING

In OOK encoding, the optical light is pulsed on or off during each bit time, as shown in Figure 6.2a. Each "1" bit is encoded into an optical pulse and each "0" bit is encoded into an off (no field) bit. Clearly, the maximum source bit rate is directly related to the rate at which the source can be switched on and off. Since lasers and LEDs can be switched at rates up to hundreds of megahertz, the OOK modulation represents a simple procedure for producing relatively high bit rates. (In fact, gigabit rates can be approached by interlacing OOK bit streams from several different lasers.) The light pulses of the OOK

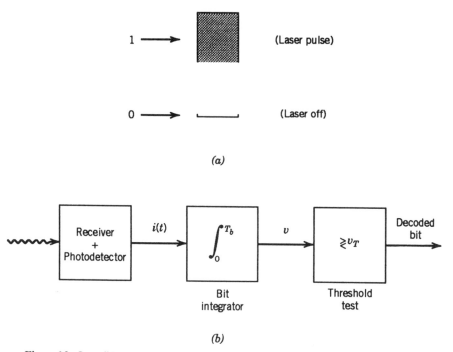

Figure 6.2. On–off keyed (OOK) system. (a) Encoding format. (b) Receiver and decoder.

bit sequence are then transmitted as a pulsed laser beam in a space system, or can be inserted into the fiber core in a fiberoptic link.

The OOK direct detection optical receiver is shown in Figure 6.2b. Decoding is achieved by determining the presence or absence of the received pulse field during each bit time. This is accomplished by integrating over the bit interval at the photodetector output and determining if the integrated current sample is above or below some threshold value. Since the integration of the photodetector output is a measure of the received field energy, OOK decoding is based on whether the received optical field energy is high enough. The threshold value is selected to yield the best decoding performance, that is, yield the lowest probability of making bit errors.

The bit error probability depends on the receiver and on the photodetection characteristics. A decoding error occurs if a zero bit (no optical field) is sent and the detected integration sample is above threshold or if a light pulse is sent and the sample is below threshold. The probability of these events depends on the statistical characteristics of the receiver, and the exact probability densities of the photodetector output are needed. This is why the analytical detail of the photodetection operation in Chapter 3 is essential for evaluating digital performance.

In Section 4.6, it was shown that the output of a current bit–time integrator following direct photodetection of a received optical field is the random variable

$$v = ek_2(0, T_b) + v_n \tag{6.2.1}$$

where e is the electron charge, $k_2(0, T_b)$ is the output count collected over the bit time $(0, T_b)$, and v_n is the integrated thermal noise Gaussian variate, having zero mean and variance $N_{0c} T_b$. (We neglect the dark current noise here. If present, its noise level can be added to that of N_{0c}). The resulting threshold-detection performance of the OOK decoder, therefore, depends on the statistics associated with the variable v. Several cases can be considered based on the assumptions concerning the photodetector. In the following we examine some specific models.

6.2.1 Shot-Noise-Limited–Ideal-Gain Case

In this case, the receiver thermal noise is negligible relative to the photodetector shot noise, and the integrated current sample is directly proportional to the photodetector count. This model would be typical of space systems with high background noise levels and high gain PMT. Thus in Eq. (6.2.1), for a specific bit interval,

$$v = egm \tag{6.2.2}$$

where $m = k(0, T_b)$ is the primary count variable collected from the bit integration. For an ideal gain detector, the probability that $m = k$ is the Poisson probability,

$$\text{Prob}[m = k] = \frac{(K_s + K_b)^k e^{-(K_s + K_b)}}{k!} \tag{6.2.3}$$

where K_s and K_b are the contributions to the average count from the source field and background or input noise, respectively. If the received optical pulse power is P_p and if P_b is the average background power, then

$$K_s = \alpha P_p T_b$$
$$K_b = \alpha P_b T_b \tag{6.2.4}$$

with $\alpha = \eta/hf$. If a zero bit (no pulse) occurred during the bit interval, then $K_s = 0$ in Eq. (6.2.3). Note that K_s depends on the total received optical pulse power collected at the photodetector from the source. Note also that the pulse power P_p differs from the average received optical power P_r, since the optical source is on only half the time (for equal likely bits). Thus $P_p = 2P_r$, and the OOK system takes advantage of the peak power, rather than the average power.

A threshold comparison of v in Eq. (6.2.2) to a threshold v_T is equivalent to a comparison of the count variable m to a threshold $m_T = v_T/eg$. That is, $v \gtreqless v_T$ if $m \gtreqless m_T$. A decoding error will occur if $m < m_T$ when a "1" bit is sent, or if $m > m_T$ when a "0" bit is sent. Hence the decoding bit probability of error (PE) for equally probable bits is

$$\text{PE} = \tfrac{1}{2}\text{Prob}[m > m_T|0] + \tfrac{1}{2}\text{Prob}[m < m_T|1] \tag{6.2.5}$$

When the Poisson count probabilities in Eq. (6.2.3) are inserted,

$$\text{PE} = \frac{1}{2}\sum_{k=0}^{m_T}\frac{(K_s + K_b)^k e^{-(K_s + K_b)}}{k!} + \frac{1}{2}\sum_{k=m_T}^{\infty}\frac{K_b^k e^{-K_b}}{k!} \tag{6.2.6}$$

The above represents the decoding PE for the shot-noise-limited OOK direct detection receiver and can be easily evaluated from readily available Poisson summation tables [1]. Note that PE depends on the choice of the threshold m_T (or v_T) selected. This threshold can be selected to minimize PE. This occurs at the value of m_T, where $d\text{PE}/dm_T = 0$ in Eq. (6.2.6), or where (Problem 6.2)

$$\text{Pos}(m_T, K_s + K_b) = \text{Pos}(m_T, K_b) \tag{6.2.7}$$

This occurs at

$$m_T = \frac{K_s}{\log\left(1 + \dfrac{K_s}{K_b}\right)} \tag{6.2.8}$$

Some plots of Eq. (6.2.6) are shown in Figure 6.3 as a function of K_s for several values of K_b, using the threshold in Eq. (6.2.8). The curves permit one to determine the amount of received optical pulse power in Eq. (6.2.4) needed to achieve a desired PE with a given amount of background light present. The curve for $K_b = 0$ would be used if no background light power was present, as would occur in a fiber channel with no input noise light. The OOK error probability in Eq. (6.2.6) can also be written in terms of other classical functions (see Problem 6.3), which can also be used for its evaluation.

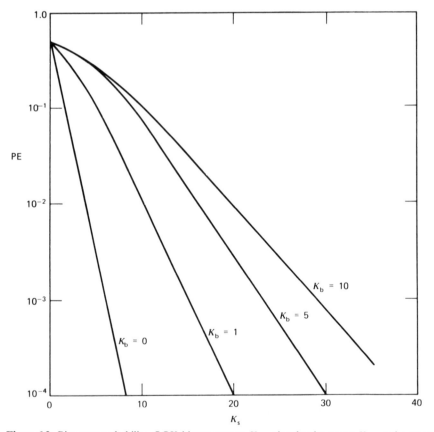

Figure 6.3. Bit error probability, OOK binary system. K_s = signal pulse count. K_b = noise count per bit interval.

The fact that the threshold in Eq. (6.2.8) depends on both the received signal power and the background power points out a basic problem with the OOK system; that is, the signal and noise power must be known exactly to optimally set the threshold.

The shot-noise-limited–ideal-gain case considered here is commonly referred to as the *Poisson optical channel*, for which much has been written concerning communication analysis and design [2, 3]. We emphasize that it represents a special case of the general optical channel that would occur only with ideal photomultiplication. Nevertheless, it represents the upper bound in PE performance achievable in an optical system limited only by the power levels that are delivered to the detector.

6.2.2 Avalanche Photodiode Receivers With Thermal Noise

When the photodetector is an APD and the receiver contains thermal noise, the previous statistics must be modified to account for the random gain and the additive Gaussian noise with spectral level N_{0c}. The integrated detector current in Eq. (6.2.1) now has the continuous probability density in Eq. (4.6.3) due to the mixture of the APD counting and the Gaussian thermal noise, which we rewrite here as

$$p(v \mid \bar{k}) = \sum_{k_2=0}^{\infty} P(k_2 \mid \bar{k}) G(v, ek_2, \sigma_n^2) \tag{6.2.9}$$

where

$$G(v, a, \sigma_n^2) = \frac{1}{\sqrt{2\pi}\sigma_n} e^{-(v-a)^2/2\sigma_n^2}$$

$$\sigma_n^2 = N_{0c} T_b = (2\kappa T^\circ / R_L) T_b \tag{6.2.10}$$

and $P(k_2 \mid \bar{k})$ is the discrete APD count probability from Eq. (3.6.5) when the primary mean count is $\bar{k}$. With a threshold of v_T, the probability of bit error is then

$$\text{PE} = \frac{1}{2} \int_0^{v_T} p(v \mid K_s + K_b) \, dv + \frac{1}{2} \int_{v_T}^{\infty} p(v \mid K_b) \, dv \tag{6.2.11}$$

Note the test is now an integrated current threshold test and no longer a count test as in the Poisson channel. The threshold v_T is the value of v_T where

$$p(v_T \mid K_s + K_b) = p(v_T \mid K_b) \tag{6.2.12}$$

This threshold may have to be found by numerical computation or iteration due to the complexity of the densities. When Eq. (6.2.9) is used in (6.2.11) we obtain

$$
\begin{aligned}
PE = \frac{1}{2} \int_0^{v_T} \sum_{k_2=0}^{\infty} P(k_2 \mid K_s + K_b) G(v, ek_2, \sigma_n^2)\, dv \\
+ \frac{1}{2} \int_{v_T}^{\infty} \sum_{k_2=0}^{\infty} P(k_2 \mid K_b) G(v, ek_2, \sigma_n^2)\, dv
\end{aligned}
\tag{6.2.13}
$$

Interchanging summation and integration simplifies this to

$$
\begin{aligned}
PE = \frac{1}{2} \sum_{k_2=0}^{\infty} \Bigg\{ P(k_2 \mid K_s + K_b) \left[1 - Q\left(\frac{v_T - ek_2}{\sigma_n}\right) \right] \\
+ P(k_2 \mid K_b) Q\left(\frac{v_T - ek_2}{\sigma_n}\right) \Bigg\}
\end{aligned}
\tag{6.2.14}
$$

with $Q(x)$ the Gaussian tail integral,

$$
Q(x) = \int_x^{\infty} G(v, 0, 1)\, dv
\tag{6.2.15}
$$

Furthermore, by combining the summations in Eq. (6.2.14), we can write

$$
PE = \frac{1}{2} + \frac{1}{2} \sum_{k_2=0}^{\infty} [P(k_2 \mid K_b) - P(k_2 \mid K_s + K_b)]\, Q\left(\frac{v_T - ek_2}{\sigma_n}\right)
\tag{6.2.16}
$$

This allows direct computation of the exact PE from the APD count probabilities using an auxilary Q- function subroutine. The system performance will depend on the APD mean gain $\bar{g}$ relative to the receiver thermal noise level, which determines σ_n. Results of this computation for some special cases are shown in Figure 6.4. In Figure 6.4a is the idealized case for $\sigma_n = 0$, corresponding to a shot-noise-limited APD receiver, with various gain values and signal counts K_s. The threshold was set by numerically solving Eq. (6.2.12). The pure Poisson case from Figure 6.3 is also included. The result shows the degradation due to the random APD gain when thermal noise is not a factor. That is, the performance is degraded by the added randomness of the counts more than it is improved by the increase in the mean signal count. Superimposed is the corresponding PE using the Webb approximation in Eq. (3.6.8) for the discrete APD density. Note that the former gives almost exact PE values for $K_s \gtrsim 50$.

Figure 6.4b shows PE for normalized receiver temperatures T°/R_L and two different background noise levels, with the mean gain $\bar{g}$ set at 100. Also included is the APD shot-noise-limited case from Figure 6.4a. It is evident that thermal noise, just like background noise, degrades the decoding performance.

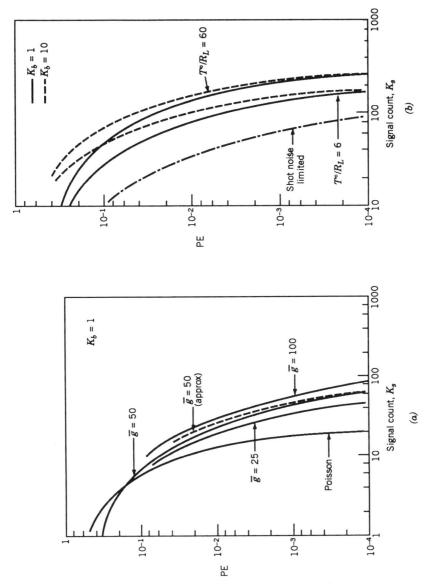

Figure 6.4. OOK PE versus K_s. (a) Shot noise limited. (b) With thermal noise.

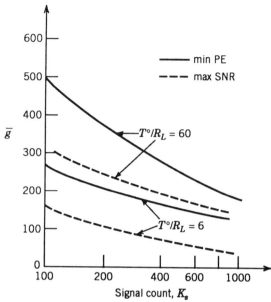

Figure 6.5. Optimal APD gain versus signal pulse count K_s. OOK signaling. $\gamma = 0.028$. $T_b = 10^{-9}$ sec. $K_b = 1$.

Because PE depends on the gain value through both its amplification factor and its excess noise factor, it is possible to optimize APD gain. This was achieved computationally by varying $\bar{g}$ in steps, each time recomputing the required threshold v_T and the corresponding PE, and noting the minimal PE and the gain value at which it occurs. The optimal gain is shown in Figure 6.5. Also included is the optimal gain value for maximizing SNR derived earlier in Section 4.4. This shows that higher gains are needed in the digital system for optimal performance than in the corresponding analog systems.

Figure 6.6 plots the resulting PE when the optimal gain is used at each noise temperature, as a function of the signal count K_s. As a comparison, the fixed gain results of Figure 6.4b are included. Note the rather severe penalty paid in PE when the optimal gain is not used. Unfortunately, optimal gain selection depends on knowledge of the expected power levels, again demonstrating this disadvantage of OOK systems. In addition, it should be pointed out that as the thermal noise level increases, higher gain values are required. This could present a hardware problem, because most APD devices become extremely sensitive when operated at relatively high gain values.

When the average counts K_s and K_b are sufficiently high, the discrete APD counting probabilities can be approximated by continuous Gaussian densities, as was verified in Figure 6.3. The integrated detector variable in Eq. (6.2.1) can be considered Gaussian with the same mean and variances, depending on the OOK bit. Using $(1, 0)$ subscripts to indicate the (on,off) pulse condition, the variable v would then be a Gaussian variate with the means and variances,

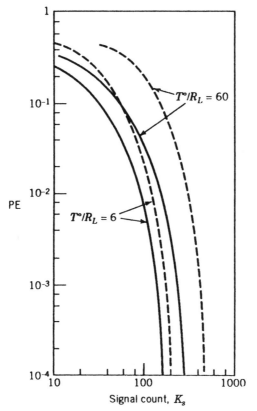

Figure 6.6. OOK bit error probability with thermal noise. — with optimal gain, – – – with $\bar{g} = 100$. $K_b = 1$, $T_b = 10^{-9}$ sec, $\gamma = 0.028$.

conditioned on the "1" and "0" bits, as

$$
\begin{aligned}
m_1 &= \bar{g}e(K_s + K_b) \\
m_0 &= \bar{g}eK_b \\
\sigma_1^2 &= \bar{g}^2 Fe^2(K_s + K_b) + \sigma_n^2 \\
\sigma_0^2 &= (\bar{g}e)^2 FK_b + \sigma_n^2
\end{aligned}
\tag{6.2.17}
$$

with σ_n^2 given in Eq. (6.2.10). With a decision threshold v_T, the OOK bit error probability for the Gaussian model is

$$
\text{PE} = \frac{1}{2}Q\left(\frac{m_1 - v_T}{\sigma_1}\right) + \frac{1}{2}Q\left(\frac{v_T - m_0}{\sigma_0}\right)
\tag{6.2.18}
$$

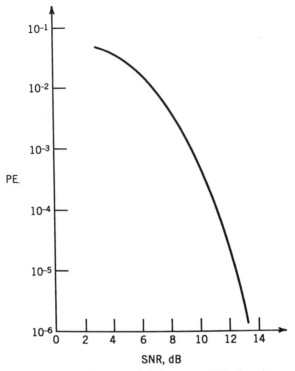

Figure 6.7. OOK bit error probability versus SNR, Gaussian model.

The optimal threshold in Eq. (6.2.12) producing the minimal PE is now

$$v_T = \frac{m_1 \sigma_0 + m_0 \sigma_1}{\sigma_1 + \sigma_0} \tag{6.2.19}$$

When this threshold is used, the resulting PE in Eq. (6.2.18) combines to

$$PE = Q\left(\frac{m_1 - m_0}{\sigma_1 + \sigma_0}\right) \tag{6.2.20}$$

Note that with the Gaussian assumption, PE depends only on the difference of the photodetected mean values. Thus any contribution to both means, such as from dark current or background noise, would not effect the $(m_1 - m_0)$ term. (They will however contribute to the variances.) We often write

$$PE = Q(\sqrt{SNR}) \tag{6.2.21}$$

where

$$\text{SNR} = \frac{(m_1 - m_0)^2}{(\sigma_1 + \sigma_0)^2} = \frac{(\bar{g}eK_s)^2}{(\sigma_1 + \sigma_0)^2} \qquad (6.2.22)$$

We see that this SNR, which determines PE in noncoherent OOK, is slightly different from that derived for the direct detection receiver in Chapter 4, based on postdetection filter bands. Figure 6.7 shows a plot of the PE in Eq. (6.2.21) as a function of the SNR parameter in Eq. (6.2.22).

6.3 MANCHESTER PULSED SIGNALS

The basic disadvantage of OOK signaling is that key receiver parameter values, such as power levels, must be known, or measured, to optimally set the threshold. A pulse format that avoids this difficulty uses pulse-to-pulse comparisons for decoding. A way to do this is to send the optical pulse in one of

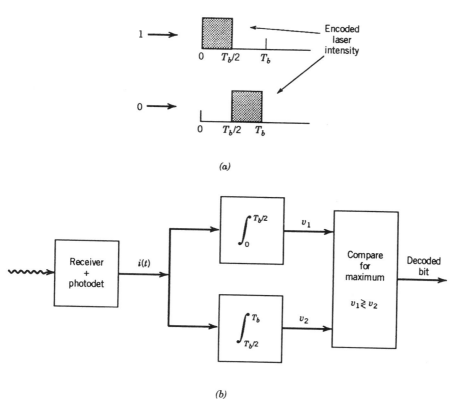

(a)

(b)

Figure 6.8. Manchester signaling. (*a*) Pulse encoding format. (*b*) Receiver and decoder.

two adjacent time intervals and to then compare the integrated output current over each interval. The bit is decoded according to which integration produces the higher value, and no threshold need be selected. This format is referred to as Manchester coding.

The encoding operation is shown in Figure 6.8a. A "1" bit is sent as a pulse in the first half of the bit interal, and a "0" bit as a pulse in the second half. The decoder (Figure 6.8b) separately integrates the detector output over the two half-bit intervals and compares for bit decoding. The system still uses pulse signaling, but the pulse time is one half the bit time. The decoder integrators, therefore, have shorter time intervals (higher bandwidths) than for the OOK system.

The bit error probability is now the probability that the bit half-interval containing the pulse does not produce the higher integrator output. This again depends on the type of optical channel.

6.3.1 Shot-Noise-Limited–Ideal-Gain Poisson Channel

For the Poisson channel, the bit error probability PE of the Manchester coded system is simply the probability that one Poisson count containing the pulse energy does not exceed another Poisson count containing no pulse. Hence,

$$
\begin{aligned}
\text{PE} &= \sum_{k_1=0}^{\infty} \sum_{k_2=k_1+1}^{\infty} \text{Pos}(k_1, K_s + K_b)\, \text{Pos}(k_2, K_b) \\
&+ \frac{1}{2} \sum_{k_1=0}^{\infty} \text{Pos}(k_1, K_s + K_b)\, \text{Pos}(k_1, K_b)
\end{aligned}
\tag{6.3.1}
$$

where K_s and K_b are now the signal and noise counts per pulse interval, which is one half the bit interval. The second term in Eq. (6.3.1) accounts for the possibility of equal counts in each interval, in which case a random choice will be made. These PEs are again in the form of discrete summations, and therefore they are amenable to numerical computation. Figure 6.9 shows some plots of PE in terms of K_s and K_b. Care must be used in making a direct comparison with the OOK Poisson system. The OOK uses pulses twice as long as the Manchester pulses, and therefore has a higher noise count K_b. If the systems are compared at the same K_s, the Manchester coding will show the better PE performance. If the systems are compared at the same average signal power, the OOK has twice the signal energy, and therefore the better performance.

It is important to note that PE in Eq. (6.3.1) depends on both the signal and noise counts K_s and K_b, and not simply on their ratio. This fact is emphasized in Figure 6.10, in which PE is plotted as a function of K_b for two fixed ratios of K_s/K_b. Notice that PE can vary several orders of magnitudes at the same ratio, depending on the value of K_b. This is an important point, since the ratio

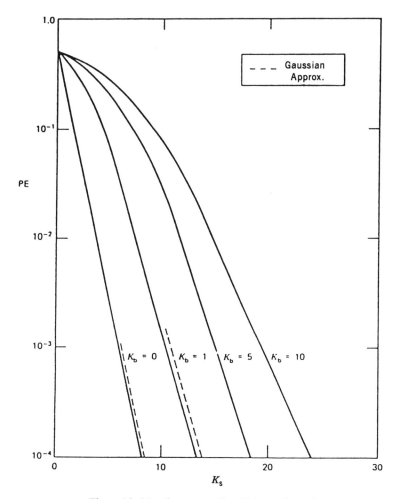

Figure 6.9. Manchester encoding, Poisson channel.

K_s/K_b is often used as an indication of system quality, and, as we can see, can often be misleading in digital systems. It is this dependence on both energies that distinguishes the Poisson optical channel from the analagous Gaussian channel.

6.3.2 Avalanche Photodiode Gain and Thermal Noise

With an APD receiver with additive thermal noise, the Manchester bit error probabilities depend on the continuous integration variables, rather than discrete counts. If v_1 and v_2 are the integrated voltages during the first and second half of the bit intervals, then PE is the probability that the correct v

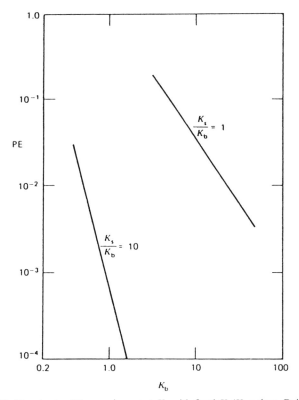

Figure 6.10. Manchester PE vs. noise count K_b with fixed K_s/K_b values, Poisson channel.

(containing the pulse) does not exceed the incorrect v. Hence,

$$\text{PE} = \int_{-\infty}^{\infty} \int_{v_1}^{\infty} p(v_1 \mid K_s + K_b)p(v_2 \mid K_b) \, dv_2 \, dv_1 \tag{6.3.2}$$

where $p(v \mid \bar{k})$ are the mixture densities in Eq. (6.2.9). Again, this PE can be expanded as weighted summations of Gaussian integrals, with the weighting dependent on the APD count probabilities,

$$\text{PE} = \sum_{k_1=0}^{\infty} \sum_{k_2=0}^{\infty} P(k_1 \mid K_s + K_b)P(k_2 \mid K_b)Q\left[\frac{e\bar{g}(k_1 - k_2)}{\sqrt{2}\sigma_n}\right] \tag{6.3.3}$$

With moderately high average counts, we can again model the integrated outputs as Gaussian variates, and the correct and incorrect integrations will have means

$$m_1 = \bar{g}e(K_s + K_b) + I_{dc}T_b/2$$
$$m_0 = \bar{g}eK_b + I_{dc}T_b/2$$

(6.3.4)

where K_s and K_b are now the pulse interval counts, and I_{dc} is the average dark current. The variances are

$$\sigma_1^2 = (\bar{g}e)^2 F(K_s + K_b) + \sigma_n^2$$
$$\sigma_0^2 = (\bar{g}e)^2 FK_b + \sigma_n^2$$

(6.3.5)

with σ_n given in Eq. (6.2.10) with T_b replaced by $T_b/2$. The decoding test decides the larger of v_1 and v_2, or equivalently decides if $(v_1 - v_2) \gtrless 0$. Because $(v_1 - v_2)$ is Gaussian, with mean $(m_1 - m_0)$ and variance $(\sigma_1^2 + \sigma_0^2)$, the bit error probability is then

$$PE = Q\left(\frac{m_1 - m_0}{\sqrt{\sigma_1^2 + \sigma_0^2}}\right)$$

(6.3.6)

This can be rewritten as

$$PE = Q(\sqrt{SNR})$$

(6.3.7)

where now

$$SNR = \frac{(\bar{g}e)^2 K_s^2}{(\bar{g}e)^2 F(K_s + 2K_b) + 2\sigma_n^2}$$
$$= \frac{K_s^2}{F(K_s + 2K_b) + 2K_n}$$

(6.3.8)

Here $K_n = n_c T_b/2$ is the equivalent thermal noise count in $T_b/2$, with the count rate $n_c = \sigma_n^2/(\bar{g}e)^2$, as introduced in Eq. (4.3.13). Note that Eq. (6.3.8) is a modified version of the direct detection SNR in Eq. (4.3.14), corresponding to photodetection followed by filtering at a bandwidth equal to the bit rate. Figure 6.7 can again be used to determine the Manchester PE in Eq. (6.3.6), except the SNR value is defined in Eq. (6.3.8). The receiver may be quantum limited, shot noise limited, or thermal noise limited, depending on the relative values of the terms in the denominator of this SNR.

A common design requirement is to determine the amount of optical signal power needed to achieve a prescribed PE when operating with a particular receiver in either an OOK or Manchester format. This requires reading off the value of SNR at the prescribed PE in Figure 6.7, then solving for the value of K_s needed to obtain this SNR, using either Eq. (6.2.22) or Eq. (6.3.8). The required P_p can then be computed from Eq. (6.2.4) with the proper time interval inserted. Note, however, that solving SNR for K_s in general involves a quadratic equation in K_s (see Problem 6.10). Usually the receiver is either quantum limited or noise limited, for which this solution simplifies. For the quantum limited case, $K_s \gg [2K_b + (2K_n/F)]$, and

$$K_s \cong F(\text{SNR}) \tag{6.3.9}$$

The required value for K_s is therefore directly proportional to the required SNR needed to achieve the desired PE. For the noise-limited case, $(2K_b F + 2K_n) \gg FK_s$ and

$$K_s = [\text{SNR}(2K_n + 2K_b/F)]^{1/2} \tag{6.3.10}$$

These equations lead to simple computations for the required signal count once the desired SNR is determined. From this, we can determine the required signal pulse power that must be delivered to the optical receiver, using Eq. (6.2.4).

6.4 DIGITAL SUBCARRIER INTENSITY-MODULATED SYSTEMS

The previous encoding formats involved optical pulsed digital signaling. A continuous wave direct detection system can be designed by first digitally modulating the bits on to an RF subcarrier and intensity modulating the subcarrier on to an optical carrier, as shown in Figure 6.11a. The latter can be transmitted as space beam or inserted into a fiber. After photodetecting at the receiver, the subcarrier can be decoded by standard RF electronics to recover the data bits. (Figure 6.11b). The subcarrier encoding can use any binary method, the most power efficient being binary phase shift keying (BPSK) [4, 5].

The system is similar to the intensity-modulated subcarrier systems in Section 4.4. Note that the laser source is continuously transmitting instead of being pulsed. After photodetection, the output current is decoded in the RF-BPSK decoder. It is well known that with optimal phase-coherent decoding of the BPSK subcarrier, the decoding bit error probability of the subcarrier is given by

$$\text{PE} = Q(\sqrt{2E_b/N_{0d}}) \tag{6.4.1}$$

where E_b/N_{0d} is the ratio of subcarrier bit energy to the total noise spectral level at the decoder input. This equation assumes that the coherent subcarrier decoder will track out all the phase noise appearing on the RF subcarrier. Using the analysis in Section 4.3, the subcarrier bit energy after photodetection is $E_b = (\beta^2 P_s/2)(\alpha \bar{g} e)^2 T_b$ where β is the modulation index, P_s is the average

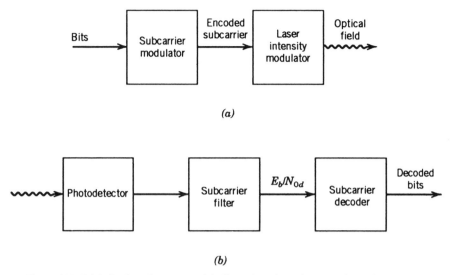

Figure 6.11. Digital subcarrier system with direct detection. (a) Transmitter. (b) Receiver.

received laser power, and T_b is the bit time. The noise level has the contribution from the shot noise, dark current, and thermal noise. The resulting PE in Eq. (6.4.1) is then

$$PE = Q(\sqrt{SNR}) \tag{6.4.2}$$

where

$$SNR = \left(\frac{\beta^2}{2}\right) \left[\frac{(\alpha P_s)^2}{[F\alpha(P_s + P_b) + (N_{0c}/(\bar{g}e)^2)]R_b}\right] \tag{6.4.3}$$

where $R_b = 1/T_b$ is the bit rate. Converting to equivalent signal and noise counts in a bit time, we can rewrite

$$SNR = \left(\frac{\beta^2}{2}\right) \left[\frac{K_s^2}{F(K_s + K_b) + K_n}\right] \tag{6.4.4}$$

Note that the subcarrier SNR is effectively reduced by the $(\beta^2/2)$ factor in decoding from that of an OOK system. The two factor comes from the fact that only half the peak laser power is available from the subcarrier modulation, whereas the β^2 factor is caused by the limited linear modulation range of the laser. Hence subcarrier systems are not as power efficient as optically pulsed systems.

6.5 DIGITAL SIGNALING WITH HETERODYNE DETECTION

Digital bits can also be encoded directly on the phase or frequency of the laser carrier itself if heterodyne detection is used at the receiver. This would represent a coherent digital system, as opposed to the direct detection systems of the previous sections. Heterodyning allows the received modulated laser carrier to be translated to a lower RF frequency, where the digital modulation can then be decoded by standard RF decoding methods.

The most common carrier modulations are BPSK and noncoherent frequency shift keying (FSK) [5, 6]. For the laser link, this would correspond to direct phase or frequency shifting of the laser spectral line itself (Figure 6.12a). This would mean that narrow line width lasers must be used for these systems, to separate out the modulations from the laser phase and frequency noise itself. Phase and frequency modulation of laser sources can be achieved by basic phase-retarding circuitry or by piezoelectric control of the laser cavity. Direct heterodyning at the receiver (Figure 6.12b) via a local offset laser mixes to RF,

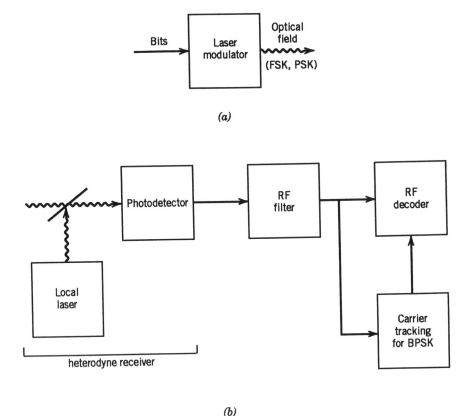

(a)

(b)

Figure 6.12. Digital heterodyne system. (a) Encoder. (b) Receiver and decoder.

producing the RF carrier with the same modulation. This is then decoded by RF BPSK or FSK detecting circuits, generating the decoded bits.

If a strong local laser field is used for the heterodyning, the photodetected mixed field can be modeled as a Gaussian process. (that is, the heterodyned carrier plus additive Gaussian noise). The decoding bit error probability for Gaussian RF channels with BPSK or FSK modulation are given by

$$PE = \begin{cases} Q\sqrt{2E_b/N_{0d}} & \text{BPSK} \\ (\frac{1}{2})\exp(-E_b/2N_{0d}) & \text{FSK} \end{cases} \tag{6.5.1}$$

where E_b/N_{0d} is again the decoder signal bit energy to noise level ratio [5, 6]. From Eq. (5.2.2), the heterodyned carrier energy in a bit time T_b is given by

$$E_b = 2(e\alpha)^2 P_s P_L T_b \tag{6.5.2}$$

The heterodyned noise level from Eq. (5.2.3) is

$$N_{0d} = e^2[\alpha P_L + \alpha^2 P_L N_0 + N_{0c}] \tag{6.5.3}$$

Thus the heterodyned decoding will be based on

$$\frac{E_b}{N_{0d}} = \left[\frac{2\alpha P_s T_b}{1 + \alpha N_0 + (N_{0c}/\alpha P_L e^2)}\right] \tag{6.5.4}$$

It is convenient to again denote $K_s = \alpha P_s T_b$ as the detected average signal count per bit and to again assume $\alpha P_L \gg N_{0c}/e^2$. In this case, Eq. (6.5.4) simplifies to

$$\frac{E_b}{N_{0d}} = \frac{2K_s}{1 + \alpha N_0} \tag{6.5.5}$$

Figure 6.13 plots PE in Eq. (6.5.1) for both BPSK and FSK heterodyne decoding, as a function of the parameter $K_s/1 + \alpha N_0$.

Although BPSK exhibits better performance, it inherently assumes perfect phase referencing in the RF decoder. Phase referencing in BPSK is typically achieved by squaring or Costas loops [7] with specified loop bandwidths B_L. The latter must be wide enough to track all the phase noise on the RF carrier at the decoder input. In the laser heterodyne system, the phase noise (laser line broadening) of both the transmitter laser and the local laser is transferred directly to the RF carrier. Hence the decoder phase-referencing loop must have wide enough bandwidths to track out this total laser phase noise on the RF carrier. On the other hand, the larger the B_L the more thermal noise in the tracking loop, degrading the accuracy of the phase referencing. Hence loop design requires a balance of loop thermal noise and phase noise tracking.

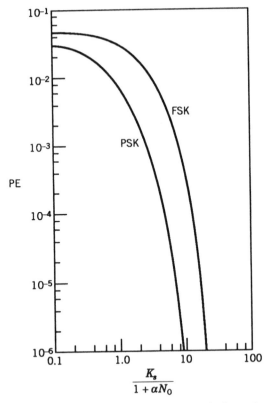

Figure 6.13. Bit error probability vs. normalized signal count for heterodyne digital systems.

Another important effect in BPSK heterodyning is that the phase referencing loop bandwidth must be a small fraction (typically approximately 1/10)
of the carrier modulation bandwidth to prevent bit waveform degradation
during the phase referencing. Because the modulation bandwidth in BPSK is
approximately R_b Hz, and B_L will be approximately equal to the phase noise
bandwidth set by the laser line width, it follows that ideal BPSK bit decoding
can only occur if the laser line halfwidth Δf and the bit rate roughly satisfies

$$R_b > 10(\Delta f) \tag{6.5.6}$$

Thus the modulation bit rate must be about 5 times the laser full linewidths
for a successful BPSK heterodyne system. This means that only high-rate data
or low-linewidth lasers should be used in such a system.

Frequency-shift-keying heterodyne systems, on the other hand, require no
phase referencing and, therefore, completely avoid this condition. However, the
optical signaling frequencies (wavelengths) must be sufficiently separated to

avoid laser phase noise crosstalk from one signaling frequency band to the other, in order to achieve the performance in Figure 6.13. Increasing the wavelength separation, however, expands the required tuning range of the source laser to accomplish the frequency-shifting modulation.

6.6 BLOCK ENCODING AND PULSE POSITION MODULATION

In the previous systems, only binary encoding was considered, in which bits were sent one at a time over the optical link. An alternative is the use of block encoding in which bits are transmitted in blocks instead of one at a time. Optical block encoding is achieved by converting each block of b bits into one of $M = 2^b$ optical fields for transmission. Decoding of each block is achieved at the receiver by determining which of the M fields is being received during each block time.

One of the more popular forms of optical block encoding is by pulse position modulation (PPM). In PPM an optical pulse is placed in one of M adjacent time slots to represent the data block, as shown in Figure 6.14a. The M slots constitute a PPM frame, or word time T_f seconds. The location of the pulse in the frame, therefore, determines the data word. The system uses pulsed optics, and is, therefore, compatible with direct detection receivers. The pulse width is equal to the slot width T_s (if we assume ideal pulses and neglect slot overlaps due to possible pulse spreading). The laser need only produce pulses at the frame rate and, therefore, operates at the pulse repetition frequency (PRF) of $1/T_f = 1/MT_s$. Since the M-slot PPM frame can represent $\log_2 M$ bits, the PPM transmitter sends bits at the rate

$$R_b = \log_2 M/MT_s$$
$$= \log_2 M(\text{PRF}) \tag{6.6.1}$$

given in bits/second. The PPM decoder must decide which of the M slots occuring during a frame time contains the optical pulse. The optical PPM direct detection receiver block diagram is shown in Figure 6.14b. The incoming field is photodetected, and a slot integration is made for each slot time by a synchronized slot clock. The sequence of slot integrations, $(v_1, v_2, \ldots, v_M)$ collected over a frame time are then compared for the maximum, with the largest one identifying the signal slot for that frame. This maximum comparison among the slot values is in fact the decoding test producing the minimum probability of a decoding error (see Problem 6.13).

A decoding word error will be made if an incorrect slot produces a higher integration value than the correct slot. Let $p(v \mid 1)$ be the probability density of v for the signal slot, and $p(v \mid 0)$ for the incorrect slots. Then, because the

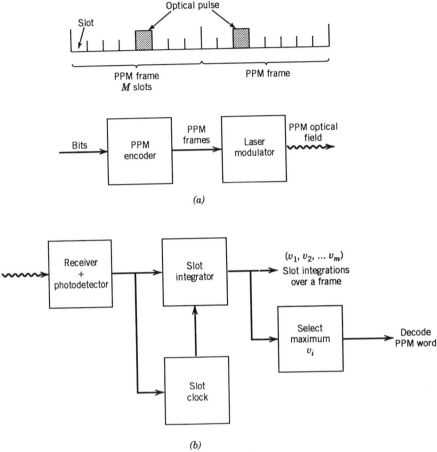

Figure 6.14. Pulse position modulation format. (a) Encoder. (b) Receiver and decoder.

integrations are independent from slot to slot, the probability of a word error (PWE) in PPM is

$$\text{PWE} = 1 - \int_0^\infty p(v_1 \mid 1) \left[\int_0^{v_1} p(v_2 \mid 0) \, dv_2 \right]^{M-1} dv_1 \qquad (6.6.2)$$

The integrator densities depend on the photodetection model. For the Poisson receiver model (ideal gain, shot noise limited), the integrations produce count variables, and the densities in Eq. (6.6.2) become discrete Poisson probabilities. The PWE is then the probability that the signaling slot produces a count that is not larger than the $M - 1$ nonsignaling slot counts. (If a count equality occurs a random choice is made among the competing slots.) Considering all possibilities,

$$
\begin{aligned}
\text{PWE} = 1 &- \frac{\exp[-(K_s + MK_b)]}{M} \\
&- \sum_{k_1=1}^{\infty} \text{Pos}(k_1, K_s + K_b) \left[\sum_{k_2=1}^{k_1-1} \text{Pos}(k_2, K_b) \right]^{M-1} \\
&- \sum_{r=1}^{M-1} \frac{(M-1)!}{r!(M-1-r)!(r+1)} \\
&\times \sum_{k=1}^{\infty} \text{Pos}(k, K_s + K_b) [\text{Pos}(k, K_b)]^r \left[\sum_{j=0}^{k-1} \text{Pos}(j, K_b) \right]^{M-1-r}
\end{aligned}
\tag{6.6.3}
$$

where K_b is the noise count per slot interval, and K_s is the signaling pulse count. The first exponential term accounts for the case when a zero count occurs in all slots, the first summation term accounts for the case where at least one nonsignaling slot count exceeds the correct slot count, and the last summation accounts for all the ways in which a count equality with the correct slot occurs among the remaining $M - 1$ slots.

The terms in Eq. (6.6.3) have been digitally computed, and some plots are shown in Figure 6.15 for various values of M with K_b fixed. Although the curves show that the performance is worsened as higher frame sizes M are used, care must be used in making a direct comparison at fixed K_s and K_b. This is due to the fact that (1) the higher M systems are transmitting higher data rates at fixed frame times, so that the curves are not rate normalized, and (2) the parameter K_s, when written in terms of the average laser power P_s, becomes

$$
\begin{aligned}
K_s &= \alpha P_p T_s \\
&= \alpha(MP_s)T_s
\end{aligned}
\tag{6.6.4}
$$

Hence the PPM systems convert average laser power to peak power directly proportional to M. Thus the higher M systems will operate at higher K_s values, and Figure 6.15 should not be compared at the same values of K_s when the average source power is fixed. This, of course, assumes that the laser can produce an optical pulse at M times its average power level.

For the Gaussian APD detector model, the densities in Eq. (6.6.2) become Gaussian variables, with

$$
\begin{aligned}
p(v \mid 1) &= G[v, \bar{g}e(K_s + K_b), \sigma_1^2] \\
p(v \mid 0) &= G(v, \bar{g}eK_b, \sigma_0^2)
\end{aligned}
\tag{6.6.5}
$$

where σ_1^2 and σ_0^2 are given in Eq. (6.3.4) with counts collected over T_s seconds.

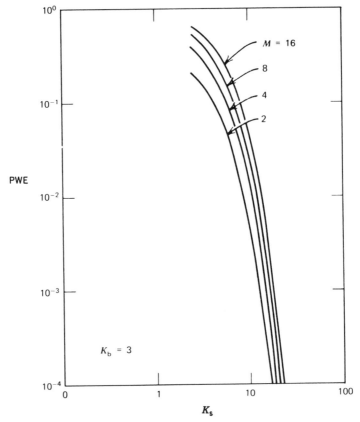

Figure 6.15. Word error probability for M-ary PPM decoding. Poisson channel. $K_b = 3$ counts/ slot. $K_s =$ signal count.

Note that the densities have different means and variances. Figure 6.16 shows the resulting PWE for the Gaussian case with some specific receiver parameters inserted.

The exact PWE in Eq. (6.6.2) can be accurately approximated by the union bound [8], which states

$$\text{PWE} \cong (M - 1)\text{Prob[one incorrect slot } v \text{ exceeds the correct slot } v] \quad (6.6.6)$$

The probability in brackets is the binary probability between a pulsed slot and a nonpulsed slot, and is, therefore, similar to the Manchester Gaussian PE given in Eq. (6.3.2). Thus, we can use

$$\text{PWE} \cong (M - 1)Q(\sqrt{\text{SNR}}) \quad (6.6.7)$$

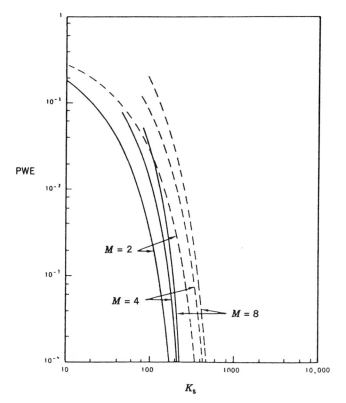

PWE

K_s

Figure 6.16. M-ary PPM, thermal noise, APD receiver (optimal gain). — $T°/R_L = 6$, – – – $T°/R_L = 60$. $K_b = 1$.

with SNR given in Eq. (6.3.8) with K_b and K_n referred to slot integration times, and K_s given by Eq. (6.6.4). Hence, PPM word error probabilities can be approximated by substituting from the Manchester PE curves.

We emphasize that PWE is the word error probability, or the probability of decoding the incorrect PPM pulse position. This will decode an incorrect bit word (block). However, the incorrect decoded word may still produce some correct bits. Hence, the bit error probability PE of the PPM system is different from the PWE. This relation can be obtained by determining the probability that a given bit of the word will be incorrect after incorrect decoding. If the incorrectly decoded word is equally likely to be any of the remaining $M - 1$ words, then a given bit will be decoded as any of the bits in the same position of each word. In M equally bit patterns, a given bit position will be a one or a zero $M/2$ times. The chance of being the incorrect bit from the $M - 1$ incorrect patterns is then $(M/2)/(M - 1)$. This means that the probability of a given bit being in error is this probability times the probability that the word was in error. Hence the bit error probability PE is related to the word error probability PWE in PPM by

$$PE = \left(\frac{M/2}{M-1}\right) PWE \qquad (6.6.8)$$

Thus, the PE can be computed from the previous PWE curves at any M value by modifying according to Eq. (6.6.8).

When the union bound in Eq. (6.6.7) is combined with Eq. (6.6.8), the bit error probability for the PPM system can be approximated as

$$PE \cong (M/2)Q(\sqrt{SNR}) \qquad (6.6.9)$$

where

$$SNR = \frac{K_s^2}{F(K_s + 2K_b) + 2K_n} \qquad (6.6.10)$$

This permits direct comparison between the binary digital systems in Section 6.2 and 6.3 and the block-coded PPM systems considered here.

6.7   CHANNEL CODING WITH PULSE POSITION MODULATION

The bit error probability performance of a digital link can often be improved by the insertion of channel coding. Channel coding is a form of additional coding placed on the transmitted data bits prior to digital transmission so that the subsequent receiver decoding is achieved with a better (lower) PE than if the bits were transmitted directly. In channel coding, the data bit sequence is converted (coded) into a sequence of channel symbols which are transmitted over the digital link. After receiver processing to recover the symbol sequence, the symbols are converted back (channel decoded) into the bit sequence with the improved PE.

A block diagram of the channel-coded optical system is shown in Figure 6.17. The source data bits are channel coded to the symbol sequence, which may also be binary symbols. The symbols are transmitted over the optical link using the optical-digital transmission methods previously discussed. The receiver photodetector output is processed to recover the transmitted symbols.

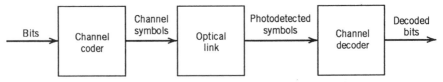

Figure 6.17. Optical link with channel encoding and decoding.

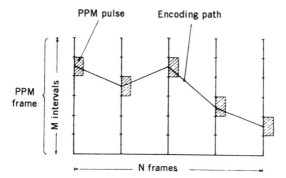

Figure 6.18. Coding tree paths in PPM decoding.

The channel decoder then converts the recovered symbols back to the data bits by inverse mapping of the channel coder. The structure of the bit-to-symbol coding permits the recovered bits to have less errors than occurred in the transmitted symbols. Note that the channel coding and decoding is accomplished entirely in the electronic domain, independent of the optical link itself.

In channel coding with PPM, the data bits are first separated into channel words, each word is considered as a sequence of symbols, and each symbol is $\log_2 M$ bits long. Thus, an N symbol word would correspond to block lengths of $N \log_2 M$ bits. Each symbol can now be transmitted as a PPM frame with M slots, and the entire block-encoded word is transmitted as a sequence of N PPM frames. Decoding is based on the photodetected slot voltages collected over the PPM frames. In *soft decisioning*, the entire word is collected before a decision is made. In *hard decisioning*, a PPM symbol decision is made each frame time, and the set of frame decisions is used to decode the word.

If we envision the PPM frame in Figure 6.14*a* as rotated to a vertical axis, and each of the N frames as placed in sequence, we generate the diagram in Figure 6.18. Placing an optical pulse in one slot in each frame defines a path through this N frame tree, as shown. It is clear that L distinct paths can be selected, where $L \leqslant M^N$. Channel coding therefore corresponds to mapping a block of $\log_2 L$ data bits into a specific path in the tree.

Decoding corresponds to deciding at the receiver which of the possible tree paths has been transmitted. The data bit sequence that was mapped into the decided tree path is then decoded. The decoder will correctly decode the $\log_2 L$ bit block if the correct path is decided. A soft decision receiver decides by computing a path *metric*, or score, for each possible path and selecting the path with the highest score after the N frames. The path score is obtained by summing the individual slot voltages that constitute the path. In hard decisioning, a PPM frame decision is made each frame time, with the set of N resulting frame decisions forming the decided path.

This PPM channel coding can be viewed directly in terms of coding theory [9–11]. The set of paths can be considered an M-ary code consisting of L

N-symbol words from an M-ary alphabet having minimum code *distance*, say d. Here the distance between two code words is equivalent to the number of frames where their paths do not use the same slots. Thus, d is the minimum distance among all code pairs in the set. Channel coding can then be viewed as a mapping of the integers $(1, 2, ..., L)$ into a word from this code set. Although the detailed study of such code constructions is beyond the scope here, we shall assume the use of such codes in our subsequent study.

Consider a shot-noise-limited direct detection PPM receiver. The transmitter sends the sequence of N PPM frames, each pulse encoded according to the channel-coding tree. The receiver photodetects and integrates over each slot time to generate the slot voltages. Let v_{ij} be the slot voltage from slot i in frame j. The soft decisioning decoder computes the score Λ_q for path q by summing

$$\Lambda_q = \sum_{\mathscr{P}(q)} v_{ij} \tag{6.7.1}$$

where $\mathscr{P}(q)$ is the specific set of slots constituting path q over the N frames. A running score is obtained for each of the L paths that could have been sent by the channel coder, and the path having the largest score is decided. The decision is correct if the transmitted path produces the largest score. A decoding eror is made in the N frame processing if any other path produces the highest score. An upper bound on the coded word error probability PWE can be obtained from the union bound, based on pairwise error probabilities between the correct path and any other path candidate. This yields

$$\text{PWE} = \sum_{v=d}^{N} L(v, N) P_v \tag{6.7.2}$$

where $L(v, N)$ is the number of paths having distance v from the correct path. The P_v are the probability of erring in attempting to decide between the correct path and one differing in v branches. For the shot-noise-limited Poisson channel, this is given by

$$P_v = \frac{\text{probability that a Poisson variable with mean } v(K_s + K_b)}{\text{exceeds a Poisson variable with mean } vK_b} \tag{6.7.3}$$

For the class of Reed-Solomon (RS) codes [9–11], it is known that $d = N - \log_M L + 1$, and the path enumerator is

$$L(v, N) = \sum_{j=N-v}^{N-d} (-1)^{j+v+N} \binom{j}{N-v} \binom{N}{j} (M^{N-d+1-j} - 1) \tag{6.7.4}$$

We are interested in the performance of this code set as a function of block code length N, the number of PPM frames in a code word. We note that the information rate of the channel-coded PPM system is

$$R_b = \frac{\log_2 L}{NMT_s} = \delta \left[\frac{\log_2 M}{MT_s} \right]$$ (6.7.5)

where δ is the channel code rate

$$\delta = \frac{\log_2 L}{N \log_2 M}$$ (6.7.6)

Because the bracket in Eq. (6.7.5) is the data rate for a standard single-frame uncoded PPM system, the code rate parameter δ indicates the loss in data rate due to the channel coding. If we fix the code rate for a fixed PPM frame size M, then the number of code paths L must increase as $L = M^{\delta N}$.

To convert the block coded PWE in Eq. (6.7.2) to the bit error probability PE, we use the fact that, when a path decoding error is made, it will most likely be made with a path a distance d away. The probability that a given symbol (PPM frame) error is made when the path is decoded incorrectly is approximately d/N. Likewise, given a PPM frame error, the probability is approximately $1/2$ that a given bit of the frame is in error, when M is large. Hence,

$$PE = \left(\frac{d}{2N} \right) PWE$$

$$= \left(\frac{N - \log_M L + 1}{2N} \right) PWE$$ (6.7.7)

$$\cong \frac{1}{2}(1 - \delta)PWE$$

The coded word error probabilty can be computed from Eq. (6.7.2), using Eqs. (6.7.3) and (6.7.4), and converted to PE using Eq. (6.6.7). The result is shown in Figure 6.19 for the code rates $\delta = 1/2$ and $1/3$, for the Poisson channel with noise count $K_b = 1$, $M = 100$, and various signal pulse counts K_s. The improvement in PE as the code length N is increased at fixed code rate is apparent. For the values of N shown, it is evident that a low-quality uncoded optical PPM system ($N = 1$) can be significantly improved by channel coding over multiple frames. This is particularly important in space systems where link background noise may prevent a standard PPM system from achieving satisfactory performance with a given laser power. Channel coding allows processing over multiple frames to improve the decoding performance.

The alternative to soft decision channel decoding is frame-to-frame hard decisioning. If hard decisioning is used, a PPM frame decision is made every frame time, with the set of frame decisions determining the path. A correct path will be decided if less than d frame errors are made, whereas d or more frame

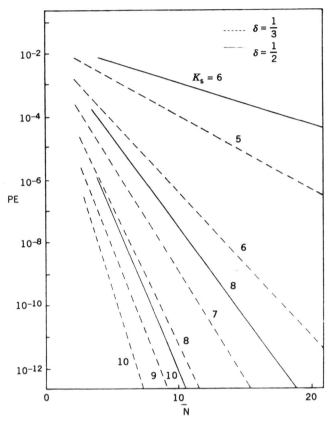

Figure 6.19. Coded bit error probabilities versus number of PPM frames. Poisson shot noise receiver. K_s = signal pulse count, δ = code rate, $M = 100$, and $K_b = 1$.

decision errors will produce block errors. Hence for hard decisioning,

$$\text{PWE} = \sum_{j=d}^{N} \binom{N}{j} (\text{PWE}_1)^j (\text{PWE}_1)^{N-j} \qquad (6.7.8)$$

where PWE_1 is the single frame PPM decision error in Eq. (6.6.3) or Eq. (6.6.7). Figure 6.20 compares the PWE performance of the hard decision decoder to the bound in Eq. (6.7.7) for fixed K_s and K_b, with several frame sizes M, and with $\delta = 1/2$. The results indicate the improvement achieved by channel coding even with hard decisioning and also shows the advantage of soft decision decoding, where the pulse energy is effectively integrated up for path decisioning as opposed to making individual frame decisions.

When APD or low-gain photodetectors are used, the decoding statistics change from Poisson counting to more generalized Gaussian-type noise, as we showed earlier, and the performance with channel coding must be recomputed.

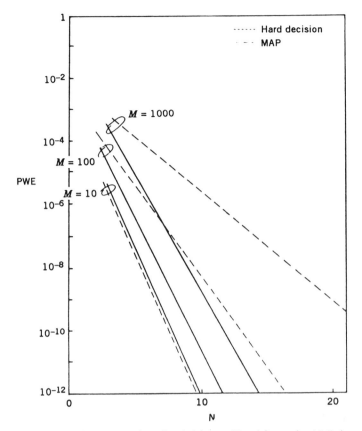

Figure 6.20. PPM PWE versus number of coded frames N and frame size M. Poisson shot noise receiver. $K_s = 10$, $K_b = 1$, $\delta = 1/2$.

One must replace the P_v in Eq. (6.7.3) by equivalent terms generated from Gaussian probabilities. Studies have shown that performance improvements roughly paralleling those in Figures 6.19 and 6.20 are possible. The interested reader may pursue these generalizations in references 12 and 13.

Another way to evaluate the effect of coding is to write the link data rate R_b (bits per second) instead as

$$R_b = \left[\frac{\text{bits}}{\text{photoelectrons}}\right] \cdot \left[\frac{\text{photoelectrons}}{\text{second}}\right]$$

$$= \rho_c n_s \tag{6.7.9}$$

Here n_s is the received signal count rate, and we call ρ_c, in units of bits per photoelectron, the channel *photon efficiency*. The achievable photon efficiency

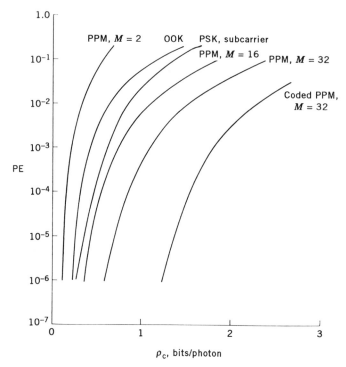

Figure 6.21. Channel photon efficiency versus PE for digital transmission.

depends on the encoding format and the desired PE of the decoder. Equation (6.7.9) factors the link bit rate into a product of an efficiency factor ρ_c, which depends purely on the encoding, and a term n_s, which depends on only the average power level. This interpretation can be used for all digital links-coded or uncoded. To determine ρ_c, it is necessary to determine the required number of photoelectrons per bit needed to achieve a desired PE. Figure 6.21 plots PE versus ρ_c for several of the encoding formats. If ρ_c is to be increased, when using a specific encoding, the PE is likewise increased. To reduce PE, either ρ_c must be decreased, or a more complex encoding and decoding must be inserted. If a desired PE is to be maintained, one can read off the achievable ρ_c for each format, and the resulting bit rate follows from Eq. (6.7.9). In general, ρ_c will be limited to $\rho_c \leqslant 1$, in which case n_s in counts per second is an upper bound to the achievable bit rate in bits per second.

6.8   TIMING ERROR EFFECTS IN PULSED OPTICAL SYSTEMS

Pulsed communications are designed to use highly peaked, narrow width, laser pulses to carry data, but require accurate pulse or slot integrations for ideal decoding. These integrations must be clocked by auxiliary timing circuits that provide the start and stop markers for the integrators, so that the integration variables are produced only over the exact slot times. In deriving the expressions for the PE for these systems, we have been assuming perfect timing. In reality, if the timing is not accurate, the integrations occur over offset intervals, leading to false integation values that can degrade the decoding performance. In this section, we attempt to assess the timing error effects on the pulsed systems previously described.

Consider first the OOK system, in which we redraw the receiver decoder block diagram as in Figure 6.22, showing the timing subcircuit. This timing subsystem must provide the clocking for the bit integrations by continually monitoring the photodetector outputs. It operates in parallel with the decoder and must continually correct for any bit time shifting or drifting that may occur during transmission. Methods for achieving this timing are considered in Section 6.9.

Assume a timing error of Δ exists, so that the decoder bit integration interval is displaced from the true bit interval, as shown in Figure 6.23. If a signaling pulse arrives (a "1" bit was sent in that bit time) the offset decoder will integrate over only a portion of that pulse energy, while integrating some of the adjacent bit time, which may or may not contain a pulse. (If Δ is positive, the following bit is involved, whereas, if Δ is negative, the previous bit is involved.) The possibilities that can occur for a given bit interval and its adjacent bit are summarized in Table 6.1. When a timing offset occurs, the tabulation shows how the integrated signal energy varies with the offset Δ for the OOK system. Because the integrator still integrates for T_b seconds, it collects the photodetector noise independent of Δ. The bit error probabilities, based on a threshold test with the signal energies in Table 6.1, will require averaging over all the various energy possibilities. Denoting $P(v \gtrless v_T | K_s)$ as the probability of the event $v \gtrless v_T$ when the signal pulse energy is K_s, the bit

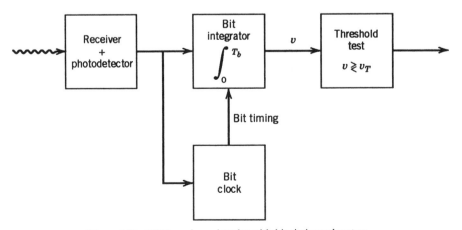

Figure 6.22. OOK receiver–decoder with bit-timing subsystem.

TABLE 6.1   Effect of Timing error offset on signal counts

Present bit	Adjacent bit	OOK signal bit count	Manchester signal count	
			correct slot	incorrect slot
1	0	$n_s(T_b - \Delta)$	$n_s\left(\dfrac{T_b}{2} - \Delta\right)$	0
1	1	$n_s T_b$	$n_s\left(\dfrac{T_b}{2} - \Delta\right)$	$n_s\Delta$
0	1	$n_s\Delta$	$n_s\Delta$	$n_s T_b/2$
0	0	0	$n_s\Delta$	$n_s\left(\dfrac{T_b}{2} - \Delta\right)$

n_s = signal count rate, cps
T_b = bit time, sec
Δ = timing offset, sec

error probability caused by a Δ offset is then

$$\begin{aligned}
\text{PE}\,|\,\Delta = \tfrac{1}{4}P(v < v_T\,|\,K_s) + \tfrac{1}{4}P(v > v_T\,|\,0) + \tfrac{1}{4}P(v < v_T\,|\,K_s') \\
+ \tfrac{1}{4}P(v > v_T\,|\,K_s'')
\end{aligned} \tag{6.8.1}$$

where $K_s = n_s T_b$, $K_s' = n_s(T_b - \Delta)$, and $K_s'' = n_s\Delta$. The first two terms are the same terms producing the OOK PE if their were no offset ($\Delta = 0$). The remaining terms account for the effect of the offset, and correspond to threshold tests with a PE degraded from the $\Delta = 0$ case. Hence, timing offsets always degrade the decoding OOK performance. Note also that $\text{PE}\,|\,\Delta$ is the same for both positive and negative values of Δ.

For the Poisson channel, the last two terms in Eq. (6.8.1) become

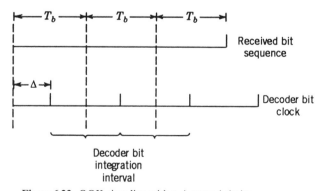

Figure 6.23. OOK time line with a Δ-second timing error.

$$P(v < v_T \mid K_s') = \sum_{k=0}^{v_T} \text{Pos}(k, K_s' + K_b)$$

$$P(v > v_T \mid K_s'') = \sum_{k=v_T}^{\infty} \text{Pos}(k, K_s'' + K_b) \tag{6.8.2}$$

If we let $\varepsilon = \Delta/T_b$ as the fractional offset, and denote $K_b' = K_b + K_s'' = K_b + K_s\varepsilon$, and rewrite $(K_s' + K_b) = K_s' + K_b' - K_s'' = K_s(1 - 2\varepsilon) + K_b'$, then Eq. (6.8.2) becomes

$$P(v < v_T \mid K_s') = \sum_{k=0}^{v_T} \text{Pos}[k, K_s(1 - 2\varepsilon) + K_b']$$

$$P(v > v_T \mid K_s'') = \sum_{k=v_T}^{\infty} \text{Pos}(k, K_b') \tag{6.8.3}$$

When combined together, these correspond to the OOK PE of a perfectly timed system with the noise count increased from K_b to K_b', and the signal count reduced from K_s to $K_s(1 - 2\varepsilon)$. Denote $PE(K_s, K_b)$ as the perfectly timed OOK PE with signal counts K_s and K_b. The offset PE for the Poisson channel can be written

$$PE \mid \Delta = \tfrac{1}{2} PE(K_s, K_b) + \tfrac{1}{2} PE[K_s(1 - 2\varepsilon), K_s\varepsilon + K_b] \tag{6.8.4}$$

Note that the timing errors are exhibited only in the second term of Eq. (6.8.4) and can be attributed to the last two terms in Eq. (6.8.1), where the adjacent bit is opposite from the true bit. These error probabilities depend on the choice of the threshold used for decisioning. When the threshold in Eq. (6.2.8) is used, PE in Eq. (6.8.4) is plotted in Figure 6.24 as a function of the fractional timing offset for several values of K_s and K_b. The curves indicate the rapid degradation that occurs as the timing offsets increase. At small offsets, increasing K_s decreases the PE, but at larger offsets the opposite is true, and a crossover is observed. An examination of Eq. (6.8.1) reveals that for $K_b \ll 1$, the first three terms tend to zero, and the resulting PE is directly attributable to the last term, corresponding to a zero sent and the adjacent bit is a one. In the limit, as $K_s \to \infty$, it follows that this last term becomes unity for any $\varepsilon \neq 0$. The overall $PE \mid \Delta$, therefore, becomes 0.25, and the result is plotted as the $K_s = \infty$ curve in Figure 6.24.

For the APD Gaussian receiver the last two terms in Eq. (6.8.1) must be separately evaluated. If we assume that the receiver is thermal noise-limited, and a threshold is set at $v_T = K_s/2$, then

$$\text{Prob}[v < v_T \mid K_s'] = \int_{-\infty}^{v_T} G[x, K_s(1 - \varepsilon), K_n] \, dx$$

$$= Q\left[\frac{K_s(1 - \varepsilon) - v_T}{\sqrt{K_n}} \right]$$

$$= Q[\sqrt{\text{SNR}(\tfrac{1}{2} - \varepsilon)^2}] \tag{6.8.5}$$

where $\text{SNR} = K_s^2/K_n$ is the photodetected SNR. Likewise,

$$\text{Prob}[v > v_T \,|\, K_s''] = \int_{v_T}^{\infty} G(x, K_s\varepsilon, K_n) \, dx$$

$$= Q[\sqrt{\text{SNR}(\tfrac{1}{2} - \varepsilon)^2}] \tag{6.8.6}$$

Thus the thermal noise-limited OOK receiver, with a fractional offset Δ, has the PE

$$\text{PE} = \tfrac{1}{2}Q(\sqrt{\text{SNR}}) + \tfrac{1}{2}Q(\sqrt{\text{SNR}(\tfrac{1}{2} - \varepsilon)^2}) \tag{6.8.7}$$

Note that when $\varepsilon = 0.5$ (the offset equals half the bit time) the last term dominates, and $\text{PE} \to 1/2$. The decoder is no longer performing, and the system is essentially ruined. Thus, timing errors must be maintained to a small fraction of the bit time for adequate decoding.

The effect of the timing error for the Manchester encoded system can be analyzed similarly to the OOK system. Table 6.1 lists the corresponding variation in signal energy from the integraton offsets for the various bit combinations. The bit $\text{PE} \,|\, \Delta$ must again be determined by computing PE for each combination and averaging over each. The bit decoding no longer performs a threshold test, but rather a integrator comparison test. Timing offsets cause loss in pulse energy in the correct interval, and a build-up of energy in the incorrect interval.

For the Poisson receiver, the Manchester decoder will have its PE modified from the perfectly timed $\text{PE}(K_s, K_b)$ in Eq. (6.3.6) to

$$\text{PE}\,|\,\varepsilon = \tfrac{1}{2}\text{PE}(K_s', K_b') + \tfrac{1}{4}\text{PE}(K_s'', K_b) + \tfrac{1}{4}\text{PE}(K_s', K_b') \tag{6.8.8}$$

where

$$K_s' = K_s(1 - 2\varepsilon)$$
$$K_s'' = K_s(1 - \varepsilon) \tag{6.8.9}$$
$$K_b' = K_s\varepsilon + K_b$$

Figure 6.25 plots Eq. (6.8.8), again exhibiting the relatively fast increase in PE as the offset is increased relative to the pulse time. Again we see that timing must be maintained to within a small fraction of the integration interval. Because this interval is itself only half the bit interval, the timing requirements for the Manchester signals, relative to the bit time, are tighter than for the OOK system. Hence, Manchester systems are more susceptible to timing errors than the OOK system.

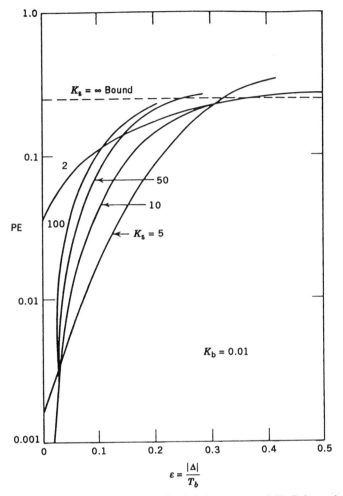

Figure 6.24. OOK decoding with normalized timing error $\varepsilon = \Delta/T_b$, Poisson channel.

For the APD Gaussian receiver with thermal noise dominating, the Manchester PE with timing error is now

$$PE\,|\,\varepsilon = \frac{1}{2}Q\left[\sqrt{\frac{SNR}{2}\,(1-2\varepsilon)^2}\right] + \frac{1}{2}Q\left[\sqrt{\frac{SNR}{2}\,(1-\varepsilon)^2}\right] \qquad (6.8.10)$$

where $SNR = K_s^2/FK_b + K_n$. The result shows a similar behavior as for the OOK system.

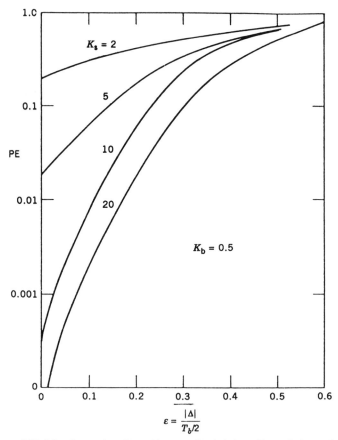

Figure 6.25. Manchester decoding with normalized timing offset ε, Poisson channel.

In PPM decoding, timing offsets, as shown in Figure 6.26, reduce signal energy in the signal slot, increase signal energy in the adjacent slot, and leave all other slots unaffected (we neglect end effects where end slots may be affected by adjacent frames). We see that slot timing offsets produce a correct slot with signal energy $K_s(1 - \varepsilon)$, an incorrect slot with signal energy $K_s\varepsilon$, and $M - 2$ remaining slots with noise energy only. We must then sum over the probabilities of all situations in which a decoding word error can occur.

For the Poisson channel, the integrations produce count variables for each slot. Thus,

$$
\text{PWE} \mid \varepsilon = 1 - \sum_{k=0}^{\infty} \text{Pos}[k, K_s(1 - \varepsilon) + K_b] \sum_{j=0}^{k} \text{Pos}(j, K_s\varepsilon + K_b)
$$

$$
\cdot \left[\sum_{j=0}^{k} \text{Pos}(j, K_b) \right]^{M - 2}
\tag{6.8.11}
$$

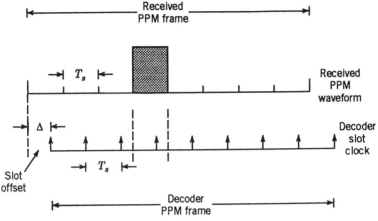

Figure 6.26. PPM timing diagram with slot clock offsets.

This PWE is plotted in Figure 6.27. Again we see a rapid degradation in performance as the offset becomes a significant portion of the slot time. Thus the slot timing capability places a lower limit on the slot times (pulse widths) that can be used in PPM systems.

For the APD receiver, the PPM PWE can be estimated from the union bound approximation in Eq. (6.6.6). For the thermal noise-limited receiver,

$$\text{PWE}|\varepsilon \cong Q[\sqrt{\text{SNR}(1 - 2\varepsilon)^2}] + (M - 2)Q[\sqrt{\text{SNR}(1 - \varepsilon)^2}] \qquad (6.8.12)$$

where now $\text{SNR} = K_s^2/FK_b + 2K_n$, and the Ks refer to counts collected over a slot time.

6.9 CLOCK SYNCHRONIZATION WITH PULSED OPTICS

We have shown that pulsed digital communication requires accurate timing to properly set the photodetector integrations used in the decoding. Since performance degrades rather quickly as timing errors increase, receiver decoding clocks must be accurate to within a fraction of the pulse time. In this section, we consider the circuitry usually used for pulsed timing in OOK, Manchester, and PPM systems.

Digital clocking is achieved in a parallel subsystem to the decoder, as was shown in Figure 6.22. The slot clocking objective can be described by redrawing Figure 6.22 as in Figure 6.28. The transmitted laser data pulse arrives at the receiver as shown, its location is time dependent on the laser pulsing and propagation delays. The receiver decoder requires a pulse clock (or slot clock in PPM) that is time coherent with the pulse edges to set the start and stop times of the decoder integration. This clock can be represented by a

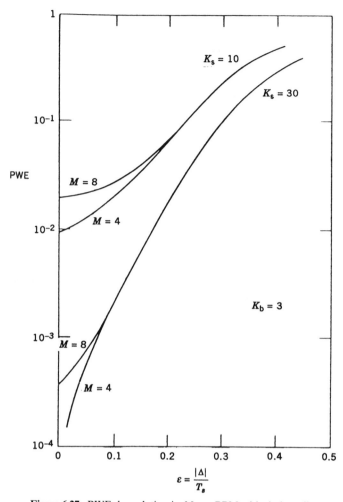

Figure 6.27. PWE degradation in M-ary PPM with timing offsets ε.

periodic square wave, as shown, whose leading edges time the decoder integrations. This clock must remain time coherent with the arriving pulse edges in spite of variations in the laser pulsing time and propagation delays, which may drift in time from pulse to pulse. This clock tracking can be accomplished by attempting to measure the time offset Δ between an arriving pulse and the clock edges every pulse or frame time and using the measured error to update (shift) the local clock.

A system for automatically accomplishing this clock tracking is shown in Figure 6.29. We consider PPM clocking here for an M-slot PPM frame, but the results can be applied to Manchester clocking ($M = 2$) or OOK clocking ($M = 1$) as well. The periodic slot clock [assumed a binary $(+1, -1)$ square

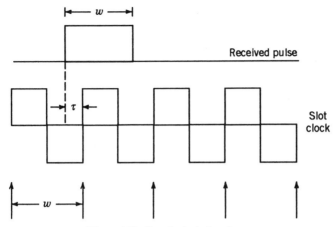

Figure 6.28. Slot clock timing diagrams.

wave at the slot rate] is multiplied with the photodetector output and integrated (filtered) over each clock period to generate a correction voltage. This voltage is then loop filtered and used to correct the clock timing. Because only the multiplication occurring over the true pulse slot is useful, a running integration over each slot time is used to "gate" the multiplier outputs. This produces a sequence of weighted slot integrations over each frame, which are then accumulated according to the loop bandwidth.

The objective of the gating is to turn off the multiplier except when the true arriving pulse is occurring. The upper arm therefore serves as a modulation *wipe-off* for the tracking operation of the closed loop. The gating signal can be obtained by simply filtering at the slot bandwidth and using either threshold comparison or bias limiters for extracting the gate.

The system can be modeled by the block diagram in Figure 6.30. The input is the photodetector output $y(t)$ containing the optically detected pulse field and the detector noise. The correction signal is obtained by multiplying and integrating (correlating) over each receiver slot time. The wipe-off gate uses a slot integration followed by a threshold test to select a binary $(0, 1)$ gate voltage. The loop correlates the slot clock with the photodetector output each clock period and uses the gate signal to weight the instantaneous correlation. The tracking signal so generated over a particular PPM frame is then

$$e = \sum_{i=1}^{M} b_i e_i \tag{6.9.1}$$

where

$$e_i = \int_{(i-1)w}^{iw} y(t)c(t)\ \mathrm{d}t \tag{6.9.2}$$

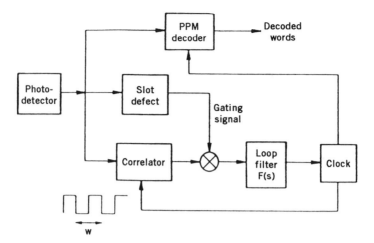

Figure 6.29. PPM clocking subsystems.

and M is the number of slots per frame, $c(t)$ is the clock square wave, and w is the slot time. The variable b_i generated from the threshold test is given by

$$b_i = \left.\begin{matrix}1\\0\end{matrix}\right\} \text{ if } \int_{(i-1)w}^{iw} y(t)\, dt \gtrless \Upsilon \tag{6.9.3}$$

where Υ is the selected threshold. It should be emphasized that b_i is a binary decision made independently on each slot and differs from the PPM decisioning, which is an M-ary decision from among all slots.

The timing performance of the tracker can be directly related to the instantaneous timing error between the clock and the received pulse. Because $y(t)$ is the photodetector output, $e(t)$ evolves as an integrated shot noise process. The mean tracking error can be determined by inserting the mean of the shot noise and tracing it through the loop. The mean value of e in Eq. (6.9.2) is

$$\begin{aligned}
\overline{e(t)} &= [1 - P(0\,|\,1)]R_{cp}(\tau) - P(1\,|\,0)R_{cp}(\tau)\\
&= [1 - P(0\,|\,1) - P(1\,|\,0)]R_{cp}(\tau)
\end{aligned} \tag{6.9.4}$$

where

$$\begin{aligned}
P(0\,|\,1) &= \text{Prob[slot decision is 0 when pulse is present]}\\
P(1\,|\,0) &= \text{Prob[slot decision is 1 when pulse is not present]}
\end{aligned} \tag{6.9.5}$$

and $R_{cp}(\tau)$ is the clock correlation signal with the photodetected pulse $p(t)$,

$$R_{cp}(\tau) = \int_0^w c(t)p(t + \tau)\,dt \tag{6.9.6}$$

Because $c(t)$ is a binary clock with period w, this is equivalent to

$$R_{cp}(\tau) = \int_0^{w/2} p(t + \tau)\,dt - \int_{w/2}^w p(t + \tau)\,dt \tag{6.9.7}$$

corresponding to early–late integrations over each half-slot time. For a rectangular optical pulse,

$$\begin{aligned} p(t) &= n_s & 0 \leqslant t \leqslant w \\ &= 0 & \text{elsewhere} \end{aligned} \tag{6.9.8}$$

where n_s is the laser photodetected count rate, proportional to the laser received peak power during the pulse. The resulting correlation in Eq. (6.9.6) is shown in Figure 6.31 as a function of offset τ. If the correct slot generates a correlation $R_{cp}(\tau)$, then the adjacent slot (previous slot for positive τ and subsequent slot for negative τ) will generate $-R_{cp}(\tau)$. This would cancel the correct correlation if both slots integrations are allowed to accumulate. This is why the gating is needed in Figure 6.30 and shows the importance of accurately distinguishing the correct signaling slot.

If τ is confined to the linear range of $R_{cp}(\tau)[|\tau| \leqslant w/2]$, and if photodetected SNR are suitably high, then each P function in Eq. (6.9.5) is negligibly small, and the loop is modeled by the equivalent linear loop in Figure 6.32. This loop has a clearly defined loop gain function, from which a loop noise bandwidth

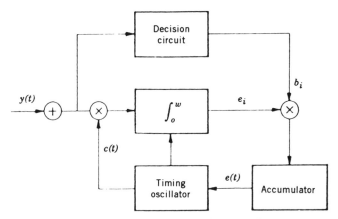

Figure 6.30. PPM clocking block diagram.

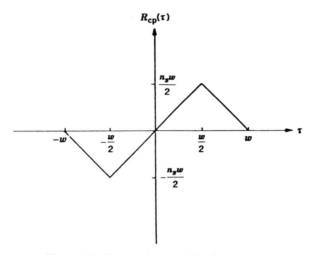

Figure 6.31. Clock–pulse correlation function.

B_L can be determined. As in any feedback system, this loop bandwidth must be wide enough to track the expected time variations in τ occurring during transmission.

The loop tracking error variance can be obtained by applying an effective noise spectral level of

$$N_{Od} = \frac{\text{var}(e)}{1/T_f} \tag{6.9.9}$$

Here var(e) is the variance of e in Eq. (6.9.1) and T_f is the frame time ($T_f = Mw$). The loop error variance is then obtained by inserting this noise into the equivalent loop in Figure 6.32 and applying standard feedback theory [16, 17]. Thus,

$$\sigma_\tau^2 = \left[\frac{N_{Od}/2}{n_s^2}\right] 2B_L \tag{6.9.10}$$

where B_L is the loop noise bandwidth. Because b_i is a binary variable and we assume zero mean tracking error, the variance of e follows as

$$\text{var}(e) = \sum_{i=1}^{M} (\text{var } e_i)(\text{Prob } b_i = 1) \tag{6.9.11}$$

where

$$\begin{aligned}
\text{Prob}[b_i = 1] &= [1 - P(0|1)] \quad \text{for pulse slot} \\
&= P(1|0) \quad\quad\;\; \text{for nonpulse slot}
\end{aligned} \tag{6.9.12}$$

The variance of e_i is the result of integrating the photodetector output over a clock period. Hence

$$\text{var}(e_i) = [n_s + n_b]w \qquad \text{for pulse slot}$$
$$= n_b w \qquad \text{for nonpulse slot} \qquad (6.9.13)$$

where n_s is the pulse count rate and n_b is the noise count rate. Thus,

$$\text{var}(e_i) = (n_s + n_b)w[1 - P(0|1)] + (M - 1)n_b wP(1|0) \qquad (6.9.14)$$

As long as $P(0|1)$ is small relative to $1(\leqslant 10^{-1})$, the resulting tracking loop error variance in Eq. (6.9.14) simplifies to

$$\sigma_\tau^2 = \frac{B_L T_f(\text{var } e)}{n_s^2}$$
$$= \frac{B_L T_f}{n_s^2}[(n_s + n_b)w + (M - 1)n_b wP(1|0)] \qquad (6.9.15)$$

As a fraction of the slot time w, this can be written as

$$\left(\frac{\sigma_\tau}{w}\right)^2 = \frac{B_L M}{n_s}\left[\frac{n_s + n_b[1 + (M - 1)P(1|0)]}{n_s}\right] \qquad (6.9.16)$$

The above represents the fractional mean-square tracking error of the slot clock. Note that the bracket plays the role of a squaring loss, while the term $B_L M/n_s$ represents the theoretical quantum limit of the clocking capability in tracking an unmodulated periodic pulse train. The term n_s/M corresponds to the average rate of detected laser counts and is, therefore, proportional to the average received laser power. Note that the modulation is effectively wiped off

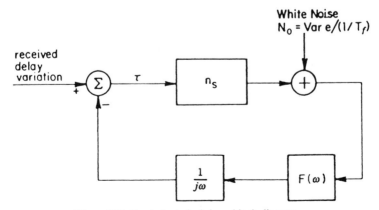

Figure 6.32. Equivalent clock-loop block diagram.

only if the P terms in Eq. (6.9.14) can be neglected. In general these conditions require only enough slot signal energy for slot thresholding with P values on the order of 10^{-2} or 10^{-3}. Typically, the required pulse energy is set by the decoding PE, for which these threshold conditions are almost always satisfied. The latter can be estimated from standard OOK PE, although it must be remembered that the gating probabilities are themselves functions of the time error τ.

When the term $(M - 1)P(1 \mid 0)$ is small, Eq. (6.9.16) reduces to

$$\left(\frac{\sigma_\tau}{w}\right)^2 = \frac{B_L M}{n_s} \left[\frac{n_s + n_b}{n_s}\right] \tag{6.9.17}$$

Thus with accurate gating, the clock tracking accuracy (neglecting squaring loss) will be independent of the number of slots per frame M and the slot size w, as long as the laser is average power limited. The squaring loss, however, depends on the peak power n_s, which increases with M under an average power constraint. The advantage of high-M PPM systems is that the slot clock will operate with squaring losses closer to 1, which improves tracking accuracy. The ultimate (quantum-limited) clocking accuracy is therefore given by the reciprocal of the average number of detected photoelectrons in a time period of $1/B_L$ seconds. Figure 6.33 plots Eq. (6.9.17) as a function of n_s/B_L for several M and n_b/B_L values. The results show that timing errors well below 10 percent can be achieved with count accumulations of approximately several hundred.

If the laser is peak power limited (n_s is constrained instead of n_s/M), the clocking performance in fact degrades with M (root-mean-square timing error increases as $M^{1/2}$ as higher M values are used). This means the transition between peak and average power constraints on laser sources is critical to the slot clocking.

Some alternative clock loop designs have been suggested [14, 15]. Since adjacent slots contain the pulse overlap and will produce negative correlation values, a possibility is to use $(+1, -1)$ gating signals, using the negative gate on all zero signaling slots to pick up the additional correlation. Repeating the analysis of Eqs. (6.9.9) to (6.9.17) will derive instead

$$\left(\frac{\sigma_\tau}{w}\right)^2 = \frac{B_L M}{2n_s} \left[\frac{n_s + M n_b}{n_s}\right] \tag{6.9.18}$$

We see that we have indeed doubled the effective tracking power, but we have increased the interfering noise in a frame. Clearly, the advantage of this negative gating will depend on the relative values of n_s and n_b. If the noise is

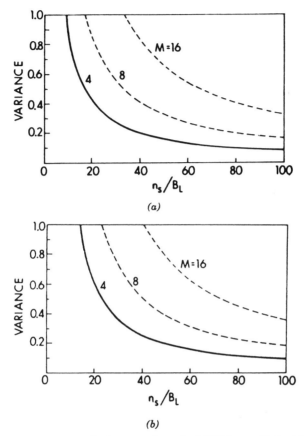

Figure 6.33. Tracking-loop normalized error variance for MPPM vs. n_s/B_L for (a) $n_b/B_L = 1$ and (b) $n_b/B_L = 10$.

weak, the negative gating will double the effective signal count and improve tracking. With strong noise, the noise integration will outweigh the signal doubling.

Clocking systems that eliminate the gating operation have been proposed. If the correlation is made over an entire frame, a zero signal correlation will be generated unless the correlating signal is reshaped. For example, a slot clock at one half the clock rate (clock period is two slot widths) eliminates the negative correlation value but introduces an ambiguity in the correlation voltage.

Another procedure is to use transmitted slot dead time so that the slot width w is wider than the laser pulse width. If the timing error is less than the slot dead time, a correlation will only be generated from the slot integration corresponding to the true pulsed slot (no negative correlation will be obtained from the adjacent slot). An integration must still be made over all slots but no gating is required. The system however integrates up noise in all M slots, as in

Eq. (6.9.18). This system also has the disadvantages that the inserted dead time slows the data rate, and the linear tracking operation is limited to the dead time width. Digital (sampled versions) of the above tracking loops have also been proposed [14].

PROBLEMS

6.1   An OOK direct detection system with a Poisson channel sends pulses producing an average count K_s over a bit time. The background produces an average count K_b. Let k be the observed bit count at the decoder, with the count probability $P(k|i)$ if bit i was sent. Optimal decoding theory states that bit decoding should be based on the comparison test

$$\log P(k|1) \gtrless \log P(k|0)$$

evaluated at the observed k. Show that the optimal Poisson channel test reduces to the comparison test

$$k \gtrless \frac{K_s}{\log\left(1 + \dfrac{K_s}{K_b}\right)}$$

6.2   **(a)** Use the rule for differentiating summations:

$$\frac{d}{dx}\left[\sum_{k=0}^{x} f(k)\right] = f(x)$$

to show that the threshold value m_T minimizing PE in Eq. (6.2.6) occurs at the value in Eq. (6.2.7).

(b) Solve for m_T in part (a) to produce the threshold in Eq. (6.2.8).

(c) What conclusion can be obtained by comparing to Problem 6.1?

6.3   Use the definition of the incomplete gamma function:

$$\Gamma(n, x) = \int_0^x e^{-t} t^{n-1}\, dt$$

and the identity

$$\sum_{n=c}^{\infty} \frac{x^n e^{-x}}{n!} = \frac{\Gamma(c, x)}{\Gamma(c, \infty)}$$

to show that the OOK Poisson PE with threshold m_T can be written as

$$PE = \frac{1}{2}\left[1 + \frac{\Gamma(m_T, K_s) - \Gamma(m_T, K_s + K_b)}{\Gamma(m_T, \infty)}\right]$$

6.4 It is desired to send 10^5 bps with OOK encoding over a 5-km fiber with a 1-dB/km loss and achieve a PE $= 10^{-4}$ with Poisson detection.

(a) What pulse energy is required at $\lambda = 1\,\mu m$ from a light source, assuming no input coupling loss and a PIN detector quantum efficiency of 0.5? Express the answer in joules (1 joule $=$ 1 watt/ second).

(b) What pulse power is needed to produce this energy value?

6.5 Use the rule for differentiating integrals:

$$\frac{d}{dx}\left[\int_0^x f(u)\,du\right] = f(x)$$

to show that PE in Eq. (6.2.11) is minimized at the value of v_T that satisfies Eq. (6.2.12).

6.6 (a) Use the result in Eq. (6.2.12) to show that under the Gaussian model, PE in Eq. (6.2.18) is minimized at the threshold value in Eq. (6.2.19).

(b) Show that when Eq. (6.2.19) is used in Eq. (6.2.18), PE simplifies to that given in Eq. (6.2.20).

6.7 The Marcum Q-function is

$$Q(a, b) = \int_b^\infty \exp\left[-\frac{a^2 + x^2}{2}\right] I_0(ax)\,dx$$

where $I_0(x)$ is the imaginary Bessel function,

$$I_0(x) = \sum_{k=0}^\infty \frac{(x^2/4)^k}{(k!)^2}$$

The Q function has the identity

$$Q(a, b) = \sum_{k=0}^\infty \frac{(b^2/2)e^{-b^2/2}}{k!} \sum_{j=0}^{k-1} \frac{(a^2/2)e^{-a^2/2}}{j!}$$

Show that the Manchester PE in Eq. (6.3.1) can be written as

$$PE = 1 - Q(\sqrt{2K_b}, \sqrt{2(K_s + K_b)}) - \tfrac{1}{2}I_0(2\sqrt{(K_s + K_b)K_b})e^{-K_s - 2K_b}$$

6.8 Use the fact that for Gaussian integrals

$$\int_{-\infty}^{\infty} Q\left(\frac{x - m_2}{\sigma_2}\right) G(x, m_1, \sigma_1) \, dx = Q\left(\frac{m_1 - m_2}{\sqrt{\sigma_1^2 + \sigma_2^2}}\right)$$

to show that the Manchester PE in Eq. (6.3.2) can be written as in Eq. (6.3.3).

6.9 Given a Manchester coded direct detection system with additive receiver thermal noise of one-sided level N_{0c} and assuming Gaussian integrator output variables v_1 and v_2, write an expression for the bit error probability based on the decoding test

$$(v_1 - v_2) \gtrless 0$$

in terms of the mean and variances of the v variates.

6.10 Assume a desired SNR value in Eq. (6.3.7). Find the value of K_s that produces this SNR value in terms of the remaining parameters in the equation. (Be careful with the multiple roots.)

6.11 In a BPSK heterodyne systems with RF carrier referencing, the ratio (CNR_L) of the RF carrier to noise power in the tracking loop bandwidth B_L must be least 20 dB. Derive an expression for the CNR_L at the photodetected output in Figure 6.12b in terms of the decoding E_b/N_{0d}. Assume the RF filter bandwidth is larger than the B_L bandwidth and the strong local laser condition is valid.

6.12 An optical M-ary PPM system uses a source of average power P_s and a specified PRF (pulses per second). Derive the equation for the peak pulse power in a system sending R_b bps.

6.13 Consider M-ary PPM in a Poisson channel with pulse count K_s and background slot count K_b. Let $\mathbf{k} = (k_1, k_2, \ldots, k_M)$ be the sequence of observed independent count variables over a PPM frame. Let $P(\mathbf{k} \mid i)$ be the probability of the k sequence given a pulse in slot i. Optimal (minimal PWE) decoding requires that the decoding decision be made that

$$[\text{pulse is in slot } j] \quad \text{if} \quad \left[\max_i P(\mathbf{k} \mid i) \text{ occurs for } i = j\right]$$

Show that, for the Poisson channel, this is equivalent to selecting the slot i corresponding to the maximum slot count.

6.14 A 4-bit PPM system is desired with a word error probability of 10^{-4} with a noise count of $K_b = 3$ in each slot. How much pulse energy (in joules) must occur in the received pulse? Assume $\alpha = 10^{18}$.

REFERENCES

1. M. Abramowitz and I. Stegun, *Handbook of Mathematical Functions*, National Bureau of Standards, Washington DC, 1964.

2. J. Pierce, Optical channels—practical limits in photon counting, *IEEE Trans Commun.* 26, December 1968.

3. S. Karp, E. O'Neill, and R. Gagliardi Communication theory for the free space optical channel, *Proc. of IEEE*, 58, October (1970).

4. A. Viterbi, *Principles of Coherent Communications*, McGraw Hill, New York, 1966.

5. J. Proakis, *Digital Communications*, 2nd ed., McGraw Hill, New York, 1989.

6. R. Ziemer and R. Peterson, *Introduction to Digital Communications*, McGraw Hill, New York, 1992.

7. R. Gagliardi, *Introduction to Communication Engineering*, 2nd ed., Wiley, New York, 1988.

8. J. Wozencraft and I. Jacobs, *Principles of Communication Engineering*, Wiley, New York, 1965.

9. W. Peterson and E. Weldon, *Error Correcting Codes*, MIT Press, Cambridge, MA, 1972.

10. S. Lin and D. Costello, *Error Control Coding*, Prentice Hall, Englewood Cliffs, NJ, 1983.

11. G. Clark and J. Cain, *Error Correction Coding for Digital Communications*, Plenum Press, New York, 1988,

12. G. Prati and R. Gagliardi, Block coding and decoding for the optical PPM channel, *IEEE Trans. Info, Theory*, 28, January, 1982.

13. J. Massey, Capacity, cut-off, and coding for the direct detection optical channel, *IEEE Trans. Commun*, 20, November 1981.

14. G. Ling and R. Gagliardi, Slot synchronization in optical PPM, *IEEE Trans. Commun*, 34, December 1986.

15. R. Gagliardi, Synchronization using pulse edge tracking in optical PPM, *IEEE Trans. Commun*, 22, October 1974.

16. F. Gardner, *Phaselock Techniques*, Wiley, New York, 1968.

17. W. Lindsey, *Synchronization Systems in Communication and Control*, Prentice Hall, Englewood Cliffs, NJ, 1972.

7

FIBEROPTIC COMMUNICATIONS

In Section 1.5, the fiber and some of its key communication parameters were introduced. This provided some understanding of field propagation and power flow in a fiber channel. In this chapter, further details of the fiber communication channel are explored, permitting both system analysis and design for this type of link.

7.1 FIBER POWER FLOW AND DISPERSION

In Section 1.5, it was stated that the optical fields that were launched into fibers cores propagated at ray line angles θ (Figure 7.1) within the critical angle θ_p determined by the fiber indices of refraction. This critical angle was given as

$$\theta_p = \cos^{-1}\left(\frac{n_2}{n_1}\right) \qquad (7.1.1)$$

where n_1 and n_2 are the indices of the core and cladding. A closely related parameter is the numerical aperature, NA, that could be written in any of the equivalent forms,

$$
\begin{aligned}
NA &= n_1 \sin \theta_p \\
&= [n_1^2 - n_2^2]^{1/2} \\
&\cong n_1[\sqrt{2\Delta}]
\end{aligned}
\qquad (7.1.2)
$$

where $\Delta = (n_1 - n_2)/n_1$ is the fractional index difference. The number of field modes within the fiber was given in Eq. (1.5.8) as

$$\text{number of fiber modes} \cong 2\left(\frac{\pi d}{\lambda}\right)^2 (NA)^2 \qquad (7.13)$$

233

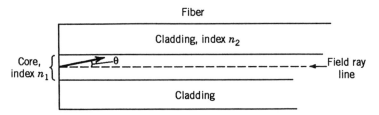

Figure 7.1. Fiber construction with field flow.

where d is the core diameter. A single mode fiber occurs only if $\theta_p \cong 0$, or if Eq. (7.1.3) is equal to or smaller than one, which places a limit on the core size. Otherwise, it is a multimode fiber.

The power P_f coupled into a fiber depends on the coupling mismatch losses between the emitting area and angles of the source and the fiber core, as in Eq. (1.5.12). The fiber attenuation depends on the type of fiber and on the field wavelength and can be obtained from curves similar to Figure 1.19. When the fiber attenuation coefficient α_f is given in decibels/length, then a fiber length Z will produce an output power of

$$P_r = P_f 10^{-\alpha_f Z/10} \tag{7.1.4}$$

The above equations can be used to obtain an estimate of the power levels delivered by a fiber to its output. As we have seen in Chapters 4 and 5, these power values directly determine the performance of the receiving systems in the link.

In describing propagating guided fields as a communication channel, however, not only is power flow important but field dispersion as well. Field dispersion is basically a relative time displacement of the propagating optical field and leads to an effective filtering or waveform distortion in the channel. This filtering is equivalent to having a bandwidth limitation placed on the fiber channel, as we shall see.

There are two main causes of dispersion in a fiber: material dispersion and mode dispersion. *Material dispersion* occurs within a mode and is due to the fact that the fiber core material causes the different frequencies that make up a mode waveform to travel at different velocities within the mode. Material dispersion is given in terms of time differences in propagating a unit length between two wavelengths $\Delta\lambda$ apart in the vicinity of wavelength λ. This dispersion is proportional to $\Delta\lambda$ and is usually normalized to the percent wavelength difference $\Delta\lambda/\lambda$. Material dispersion is usually stated in units of nanoseconds per kilometer for a fixed percentage of wavelength bandwidth, as a function of λ. Typical dispersion is approximately 1 nsec/km, multiplied by the percent bandwidth. An optical bandwidth of 1000 Å at $\lambda = 0.86 \, \mu$m would, therefore, produce approximately 0.1 nsec/km, whereas a 10-Å bandwidth would produce approximately 10^{-3} nsec/km. Because these are extremely small

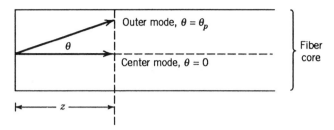

Figure 7.2. Two propagatory fiber modes at the center and maximum propagation angle.

dispersion values, a single mode should have negligible effect on communication performance.

Mode dispersion is caused by dispersive interaction between two separate modes. Two modes propagating at different ray lines within the core, but at the same velocity, will arrive at a point down the fiber at different times. Consider the ray diagram in Figure 7.2, showing an outer mode (maximum propagating angle) traveling a different path than the center ($\theta = 0$) mode. While the center mode travels a distance z, the outer mode travels a distance $z/\cos\theta_p$, both at the fiber wave velocity c/n_1, with c the speed of light. The difference between the arrival times of the modes at the plane at $z = Z$ is then

$$t_d = \left(\frac{Z}{\cos\theta_p} - Z\right)\frac{n_1}{c} \tag{7.1.5}$$

in seconds. Using Eq. (7.1.2), this corresponds to a time differential per unit length of

$$\frac{t_d}{Z} = \left(\frac{n_1}{n_2} - 1\right)\frac{n_1}{c} \cong \frac{n_1\Delta}{c} \tag{7.1.6}$$

Hence, worst-case dispersion, as predicted from basic ray line theory, depends only on the fiber index difference. A fiber with a Δ of one percent will have a mode dispersion of approximately 50 nsec/km. That is, light impulses launched into the fiber at the same time in these inner and outer two modes will arrive 50 nsec apart for each kilometer of fiber length.

The derivation of Eq. (7.1.6), using Figure 7.2, immediately suggests that the t_d/Z value for these two modes can be decreased by reducing the time differential. This requires speeding up the travel time of the outer ray relative to the inner ray. To do this, it is necessary to reduce the core index n_1 at the core edges. Thus, the core would have to exhibit a variable index along its radial direction, from n_1 at the center to a decreased value of index at the edges. The index grading is exhibited as an index profile. Figure 7.3 shows some examples of index profiles observed radially. Figure 7.3a is the standard

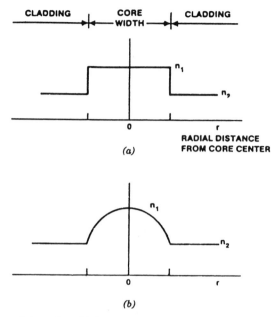

Figure 7.3. Fiber radial profiles with index grading. (*a*) Step-index fiber. (*b*) Graded-index fiber.

nongraded fiber with the core index maintained at n_1 throughout the core radius. (This is often called a *step-index* fiber due to the shape of the profile.) Figure 7.3*b* is an example of a graded profile, showing the gradual reduction of index along the radius. In essence, the index profile acts like a continuous lensing action that continually refocuses the light beam along the fiber path. A common form of this profile, describing the grading along the fiber radius, is

$$n_1(r) = n_1 \left[1 - \Delta \left(\frac{r}{d/2} \right)^2 \right] \tag{7.1.7}$$

where r is the radial distance from the core center, d is the core diameter, and Δ is the index differential. This profile has the maximum delay spread of approximately

$$\frac{t_d}{Z} \cong \left(\frac{1}{\cos \theta_p} \right) \frac{n_2(1 + \Delta^2/2)}{c} - \frac{n_1}{c}$$

$$\cong \frac{n_1 \Delta^2}{2c} \tag{7.1.8}$$

This is approximately $2/\Delta$ times smaller than that for the step-index profile in Eq. (7.1.6). For Δ equal to a fraction of a percent, this corresponds to reduction in dispersion by several orders of magnitude. Thus, although a step-index fiber might have a dispersion of approximately 50 nsec/km, graded fibers may be well below 1 nsec/km. We emphasize that if the single mode condition can be maintained, mode dispersion will not occur, and field dispersion is limited to only that of the core media. The tradeoff is then a higher quality, thinner, low-NA fiber, making splicing and fiber field control more difficult, versus a multimode fiber that lessens these latter problems, but has increased field dispersion.

The use of the previous dispersion analysis employing only central and outer ray lines may raise some questions as to the extent to which this dispersion is really harmful to the propagating fields when many modes are involved. With many modes, the field propagates at all ray angles out to the maximum ray angle, and their contribution to dispersion has not been included. In addition, the power of the central angles may be significantly greater than that of the maximum angle, so the effect of the outer angle may be overemphasized. To perform a more exact analysis of field dispersion, it is necessary to consider the power distribution across all propagating angles, and to take into account power conversion that may occur between modes. This is done in the next section.

7.2 DISPERSION AND PULSE SHAPING IN FIBERS UNDERGOING DIFFUSION

The dispersion discussed in Section 7.1 was purely in terms of isolated ray lines and field modes. A more exact analysis must account for possible diffusion and mode regeneration as the field propagates. To develop this approach, the optical field in the fiber must be treated as undergoing diffusion in which propagation at one ray line can couple, or diffuse, into another ray line. This can be handled by considering the ray angle θ as a continuous variable and allowing diffusion over this angle. This inherently implies a multimode condition—a large number of modes exist so that all angles between $\theta = 0$ (central ray angle) and θ_p (maximum ray angle) are occupied.

The study of fiber power flow based on diffusion theory has been well developed [1]. The analysis begins with the basic diffusion equation in cylindrical coordinates that describes the ray line power in the fiber field $P(t, z, \theta)$ at time t, distance z down the fiber, in ray line direction θ. The total power in the fiber at time t and position z is obtained by a circular integration over all propagation angles θ. Hence we denote

$$Q(t, z) = \left[\begin{matrix} \text{integrated power over all angles at time } t \text{ and position } z \\ \text{due to an initially launched source power distribution} \end{matrix} \right]$$

$$= 2\pi \int_{\text{all}\,\theta} P(t, z, \theta)\, d\theta \qquad (7.2.1)$$

Thus $Q(t, z)$ defines the total integrated power that can be collected over the fiber cross-sectional area at distance z and time t. If the initial source power injected into the fiber was an impulse in time, then $Q(t, z)$ is the impulse response of the fiber at point z.

A complete solution for $Q(t, z)$ was derived by Gloge [1], for a fiber with a given attenuation and diffusion coefficient, the latter coefficients dependent on the fiber material. The general solution is somewhat complicated, but has some interesting limiting cases suitable for our purposes. For a condition of a short fiber, defined by the condition

$$z < z_f \qquad (7.2.2)$$

the solution behaves as

$$Q(t, z) \cong (P_f e^{-z/z_f})(e^{-t/\tau_f z}) \qquad (7.2.3)$$

where P_f is the initial source power, and z_f and τ_f are constants dependent on the fiber material. This short fiber solution is plotted in Figure 7.4. The parameter z_f, besides defining the short fiber condition in Eq. (7.2.2), plays the role of an average attenuation factor and is related to both attenuation and diffusion coefficients in the fiber, whereas τ_f is similar to a fiber time constant associated with the exponential spreading. Note that as z increases (we move further down the fiber), the peak power decreases, and the impulse response spreads in time according to the time constant.

For a long fiber, now defined by the condition $z > z_f$, the solution behaves instead as shown in Figure 7.4. We see that for a long fiber the impulse response is more pulselike, exhibiting both an inherent delay and pulse spreading. As the point z is further increased, the pulse delay increases, and the pulse shape widens into a bell-shaped response. An exact expression for the pulse widths in Figure 7.4 has been computed [1] using the general solution for

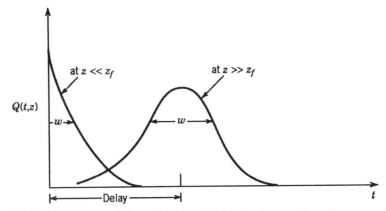

Figure 7.4. Impulse response of fiber undergoing diffusion, for short and long fibers. z_f = equilibrium distance, w = pulse width.

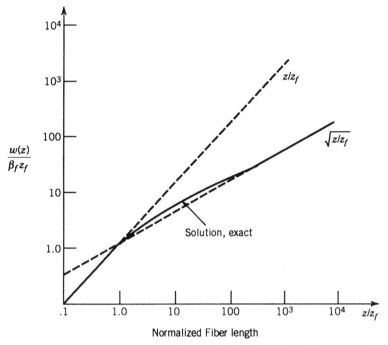

Figure 7.5. Impulse dispersion widths in fiber undergoing diffusion with asymptotic forms. $w(z)$ = pulse width at distance z, z_f = fiber equilibrium length.

$Q(t, z)$, and is shown in Figure 7.5, showing how the diffused pulse spreads as it propagates down the fiber. This pulse-width curve has the limiting forms [1]

$$w(z) = \begin{cases} \beta_f z & \text{for } z < z_f \\ \beta_f z_f \sqrt{z/z_f} & \text{for } z > z_f \end{cases} \tag{7.2.4}$$

where β_f is given by

$$\beta_f = \frac{(NA)^2}{2cn_1} \tag{7.2.5}$$

Here NA is fiber numerical aperture, n_1 is the core index, and c is the speed of light. Note that for short distances, the pulse spreading is proportional to distance; but, for long distances, the spreading eventually becomes proportional to the square root of z. That is, the spreading increases at a slower rate at the long distances. The transition occurs approximately where the two asymptotes cross in Figure 7.5, which is at the point $z = z_f$. For this reason the parameter z_f is often called the *equilibrium* distance of the fiber. Because pulse spreading can be readily measured, z_f can be determined empirically. Typically, z_f will have a value on the order of about one kilometer for most fibers.

7.3 PULSE STRETCHING AND BANDWIDTH LIMITATIONS IN MULTIMODE FIBERS

The results of the previous section can now be directly applied to model the pulse response that occurs in a multimode fiber. Let the fiber have length Z and assume a time impulse of power (short burst of light) is launched into the fiber uniformly over all angles within. The power distribution at the fiber output can be obtained, as a function of time, by using the earlier equations with $z = Z$. In particular, the pulse width will be obtained from Eq. (7.2.5) as

$$w(Z) = \begin{cases} \beta_f Z & Z < z_f \\ \beta_f z_f \sqrt{Z/z_f} & \text{for } Z > z_f \end{cases} \tag{7.3.1}$$

with β_f given in Eq. (7.2.6). Thus the amount of pulse spreading that occurs will depend on the fiber length, its equilibrium distance, and the coefficient β_f.

Since β_f determines the spreading, let us examine it in more detail. If we substitute for the numerical aperture, we have

$$\beta_f = \frac{n_1^2 - n_2^2}{2cn_1}$$

$$\approx \frac{n_1 - n_2}{c} \tag{7.3.2}$$

$$= \frac{n_1 \Delta}{c}$$

If we relate this to our earlier result in Eq. (7.1.6), we see that the spreading coefficient β_f obtained from diffusion theory, is identical to the differential time delay per length, t_d/Z, obtained from ray line analysis. Indeed, the maximum angular ray line contribution to the pulse dispersion is significant, and dispersion reduction techniques via fiber grading are theoretically justified. However ray line analysis always predicts a linear increase in spreading with fiber length, and will therefore produce oversized spreading estimates in long fibers, as the diffusion theory result has shown.

The spreading of a pulse during fiber propagation can be interpreted as a form of channel filtering. Because the Fourier transform of a pulse of width w has a frequency bandwidth of approximately $1/w$ in hertz, the pulse spreading can be attributed to an effective low pass filtering on the pulse intensity, caused by the fiber having this bandwidth. Hence, a fiber of length Z can be modeled

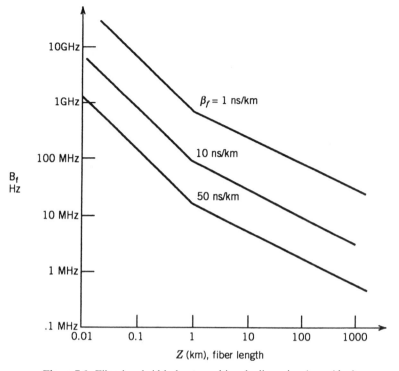

Figure 7.6. Fiber bandwidth due to multimode dispersion ($z_f = 1$ km).

as having the intensity bandwidth

$$B_f = \frac{1}{w(Z)} \tag{7.3.3}$$

in hertz, where $w(Z)$ is the pulse width in Eq. (7.3.1). Thus,

$$B_f = \begin{cases} \dfrac{1}{\beta_f Z} & \text{for } Z < z_f \\[2ex] \dfrac{1}{\beta_f z_f \sqrt{Z/z_f}} & \text{for } Z > z_f \end{cases} \tag{7.3.4}$$

This fiber bandwidth is plotted in Figure 7.6 as a function of normalized fiber length Z for several values of β_f. As the fiber length increases, the fiber bandwidth is reduced, thereby reducing the ability to transmit undistorted, high bandwidth information (narrow pulses) over this channel.

7.4 COMMUNICATION LINK MODELS FOR THE FIBER CHANNEL

The fiber characteristics of the previous sections can now be used to formulate a basic fiberoptic channel model. The procedure is to integrate the key parameters of the communication link with the individual characteristics of the fiber itself. When viewed in this context, the only role of the fiber is to carry the modulated light field from transmitter to receiver. The properties of the channel will therefore depend on the manner in which the light is effected by the fiber.

A direct detection fiberoptic communication system is shown in Figure 7.7. The light source emits an intensity modulated optical field, with modulated power $P_f(t)$, into the fiber. The fiber having length Z, is characterized by its attentuation loss and dispersion, which effectively filter the light modulation. The field power at the fiber output can then be modeled by the fiber low pass filtering

$$P_r(t) = P_f(t) \otimes h_f(t) \tag{7.4.1}$$

where $\otimes$ denotes time convolution, and $h_f(t)$ is the effective impulse response of the fiber. For a single mode fiber, $h_f(t)$ is a fairly wideband response, limited only by the core material dispersion. For a multimode fiber undergoing diffusion, $h_f(t)$ is given by

$$h_f(t) = Q(t, Z) \tag{7.4.2}$$

where $Q(t, Z)$ is the response in Eq. (7.2.1). Following our discussion in Chapter 4, a direct detection receiver, collecting all fiber modes, will produce the photodetected shot noise current, whose mean time variation (signal component) is

$$i_s(t) = \bar{g}\alpha[P_r(t) \otimes h_d(t)] \tag{7.4.3}$$

where $\bar{g}$ is the photodetector mean gain, $\alpha = \eta/hf$, and $h_d(t)$ is the filtering response produced by the detector itself on the intensity modulation. In the

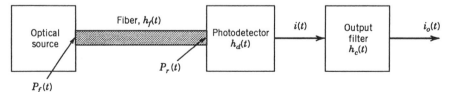

Figure 7.7. Fiber communication model.

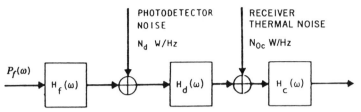

Figure 7.8. Fiberoptic filter model.

frequency domain, Eq. (7.4.1) and (7.4.3) yield

$$I_s(\omega) = \bar{g}\alpha[P_f(\omega)H_f(\omega)H_d(\omega)] \qquad (7.4.4)$$

where the caps denote Fourier transforms. The photodetected output current is then filtered by any postdetection filtering, $H_c(\omega)$, that is inserted. The resulting output signal current from an initial modulation $P_f(t)$ has the transform

$$I_o(\omega) = \bar{g}\alpha[P_f(\omega)H_f(\omega)H_d(\omega)H_c(\omega)] \qquad (7.4.5)$$

This establishes the baseband equivalent link model shown in Figure 7.8, which describes the way in which intensity (power) modulation is transmitted over the fiber to the receiver output. For typical modulating bandwidths (say megahertz or less), $H_d(\omega)$ is relatively wideband, and the principle filtering in Eq. (7.4.5) will come from the fiber itself and the postdetection circuitry. The use of baseband equalization filters following photodetection to compensate for the fiber filtering can be directly applied here. That is, one would attempt to design postdetection filters to compensate for the filtering produced by the fiber and detector.

To the fiber filtering model, we can add the photodetector noise, as shown in Figure 7.8, which has the combined spectral level from the shot noise and dark current of

$$N_d = [\overline{g^2}(e^2\alpha)P_r + eI_{dc}] \qquad (7.4.6)$$

where e is the electron charge. To this we add the output thermal noise having the spectral level N_{0c}. The filtered version of this combined noise, therefore, appears at the receiver output.

Communication performance for the fiber link can now be determined from these channel models. For example, the output SNR can be computed by the standard power flow analysis. If we assume that the power modulation $P_f(t)$ imposed at the transmitter in a direct detection system is narrowband relative to the overall link filtering, the SNR computation follows the analysis in Section 4.3. The output detected signal power is then $[\beta \bar{g} e \alpha \bar{P}_r]^2$, where β is the intensity modulation index and $\bar{P}_r$ is the average fiber power at the fiber output. If the output postdetection filter has a noise bandwidth B_c, the output noise power is then $2[N_d + N_{0c}]B_c$. The detected SNR is then

$$\text{SNR} = \frac{(\beta \bar{g} e \alpha \bar{P}_r)^2}{[F(\bar{g}e)^2 \alpha \bar{P}_r + e I_{dc} + N_{0c}]2B_c} \tag{7.4.7}$$

Note that, with the fiber and detector filtering neglected, the output SNR is similar to the SNR of a direct detection free-space optical space link with background noise eliminated. A fiber link can therefore be shot noise limited, thermal noise limited, or quantum limited, depending on the values of these parameters.

7.5   RADIO FREQUENCY CARRIER TRANSMISSION OVER FIBER

In RF wireline communications, a modulated RF carrier is routed over coaxial cable or microwave waveguides. The RF carrier is generally in the mega- or gigahertz range and, therefore, well within the bandwidths of optical fibers. An important topic of interest is that of replacing the RF cabling by optical fiber for routing the RF carrier. This would have the packaging advantage of the narrower fiber, as well as the lower attenuation losses. The approach is to intensity modulate the RF carrier directly on the laser and transmit the optical field over the fiber to the receiver. After photodetection, the RF carrier is recovered. The objective is for the optical link to have as little effect as possible on the RF carrier routing. The conditions for this to occur can be obtained directly from the previous analysis.

Consider the RF-fiber system in Figure 7.9. A modulated RF carrier is filtered by the RF bandpass prefilter centered at the carrier frequency. The

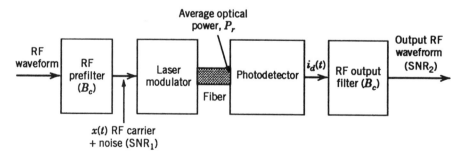

Figure 7.9. RF carrier transmission over fiber.

output of the filter is the sum of the modulated RF carrier $c(t)$, with power P_c, and the filter output noise $n(t)$, with power $P_n = N_0 B_c$, where N_0 is the prefilter RF noise spectral level and B_c is the prefilter noise bandwidth. The prefilter output has the RF SNR given by

$$\text{SNR}_1 = \frac{P_c}{P_n} = \frac{P_c}{N_0 B_c} \qquad (7.5.1)$$

The combined output waveform $x(t)[= c(t) + n(t)]$ is then used to intensity modulate the laser source driving the fiber link. The laser, with average power P_f, would then be intensity modulated by $x(t)$ to produce the power variation at the fiber input of

$$P_f(t) = P_f[1 + \beta x(t)] \qquad (7.5.2)$$

where β is again the modulation index determined by the linear modulation range of the laser. Since $\beta x(t) \leqslant 1$, we require the condition $\beta^2 P_x \leqslant 1$, or

$$\beta \leqslant \frac{1}{\sqrt{P_x}} \qquad (7.5.3)$$

That is, the RF prefilter output power P_x must be adjusted to satisfy this condition. This, therefore, limits the RF gain that can be used in the prefilter. The input fiber field with the intensity in Eq. (7.5.2) is fed into the fiber for propagation. Neglecting the filtering effects of the fiber, and taking the fiber transmission losses (coupling and attenuation) as L_f, the photodetected output current is

$$\begin{aligned}
i(t) &= (\bar{g} e \alpha) P_r(t) + i_d(t) \\
&= (\bar{g} e \alpha) P_r \{ 1 + \beta[c(t) + n(t)] \} + i_d(t)
\end{aligned} \qquad (7.5.4)$$

where $P_r = P_f L_f$ and $i_d(t)$ is the combined detector noise current from Figure 7.8. Note that the signal portion of $i(t)$ [the first term in Eq. 7.5.4) contains a component due to both the carrier and the input RF noise. The latter will be photodetected and appear as additional noise in the detector output. The output is filtered by the postdetection bandpass filter, which is assumed to have the same carrier bandwidth B_c as the input prefilter. The detected output carrier power is then

$$P_{c2} = (\beta \bar{g} e \alpha L_f P_f)^2 P_c \qquad (7.5.5)$$

Neglecting the dark current, the total output RF filtered noise has power

$$P_{n2} = (\beta \bar{g} e \alpha L_f P_f)^2 P_n + [F(\bar{g}e)^2 \alpha P_r + N_{0c}]2B_c \qquad (7.5.6)$$

The output SNR is

$$\mathrm{SNR}_2 = P_{c2}/P_{n2}$$

$$= \frac{P_c}{P_n + \left[\dfrac{F}{\alpha \beta^2 P_r} + \dfrac{N_{0c}}{(\bar{g}e\alpha P_r \beta)^2}\right]2B_c} \qquad (7.5.7)$$

Rewriting in terms of the input SNR_1 in Eq. (7.5.1) and inserting the condition in Eq. (7.5.3) with equality, yields

$$\frac{\mathrm{SNR}_2}{\mathrm{SNR}_1} = \frac{1}{1 + \left[\dfrac{F(\mathrm{SNR}_1 + 1)}{\alpha P_r} + \dfrac{N_{0c}(\mathrm{SNR}_1 + 1)}{(\bar{g}e\alpha P_r)^2}\right]2B_c} \qquad (7.5.8)$$

Equation Eq. (7.5.8) shows the extent by which the fiber link has degraded the RF input SNR during the carrier routing. For negligible degradation, it is necessary that the denominator bracketed term be much less than one. The first term in the bracket is due solely to the shot noise and requires

$$\frac{\alpha P_r}{2B_c} \gg F(\mathrm{SNR}_1 + 1) \qquad (7.5.9)$$

The left-hand side can be interpreted as the average signal count in the time interval $1/2B_c$ seconds produced at the fiber output. Thus, Eq. (7.5.10) requires that this count should be larger than a value dependent on the SNR_1 itself. The second term in the denominator bracket requires

$$\left(\frac{\alpha P_r}{2B_c}\right)^2 \gg \left(\frac{n_c}{2B_c}\right)(\mathrm{SNR}_1 + 1) \qquad (7.5.10)$$

where n_c is the equivalent count rate caused by thermal noise

$$n_c = N_{0c}/(\bar{g}e)^2 \qquad (7.5.11)$$

Thus, the transparency of the fiber interconnecting link in Figure 7.9 depends directly on the optical power levels that can be maintained. This means laser power, fiber coupling losses, and attenuation losses become critical design parameters in RF-fiber links.

7.6 OPTICAL AMPLIFIERS IN FIBER LINKS

Significant advances in fiber technology has occurred because of the development of on-line optical amplifiers. As stated in Section 1.8, optical amplifiers
provide field intensity amplification that directly increases the power level of
the fiber signal. This permits the amplifier to help overcome fiber-propagation
loss, thereby increasing power levels at the detector input. At the same time,
optical amplifiers produce spontaneous emission noise that adds to the
detector input signal, and it is not immediately obvious how the overall link is
affected.

Optical amplifiers can be used in fibers in two basic arrangements, as shown
in Figure 7.10. In Figure 7.10a, the amplifier is placed at the fiber output, acting
as a receiver preamplifier prior to photodetection. In Figure 7.10b, the amplifier
is used at the fiber input to amplify the source power inserted into the fiber. If
the amplifier was an ideal noiseless device, with gain values independent of
input power levels, the systems would be identical in terms of source power
delivered to the detector. The nonideal gain, and the addition of the amplifier
spontaneous emission noise, means that each system must be separately
examined.

Consider the link in Figure 7.10a. Let P_f be the fiber input power, L_f the
fiber propagation loss, P_r the fiber output power, and assume the optical
amplifier has gain G and spontaneous noise coefficient n_{sp}. This means the
amplifier inserts at its output the spontaneous emission noise with spectra in
Eq. (1.8.3),

$$N_{sp} = (G - 1)(hf)n_{sp} \tag{7.6.1}$$

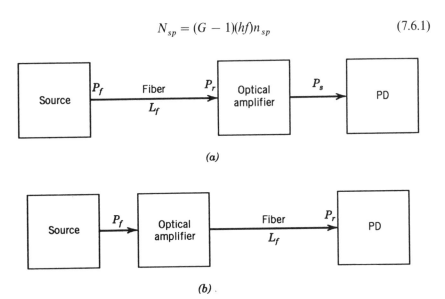

(a)

(b)

Figure 7.10 Optical amplifier arrangements (a) as a predetection amplifier and (b) as a source
amplifier.

The photodetector input signal power is then $P_s = GP_f L_f$. The direct detection SNR follows our discussion in Section 4.1 with $N_0 = N_{sp}$. The last two terms in (4.1.26) become $(\alpha \bar{g})^2 (N_{sp}^2 B_o + 2GP_r N_{sp})$. Since the last term is usually larger,

$$SNR = \frac{(e\bar{g}\alpha GP_r)^2}{[(\bar{g}e)^2 F\alpha GP_r + 2(\alpha e\bar{g})^2(GP_r N_{sp}) + N_{0c}]2B_m}$$

$$= \frac{\alpha GP_r}{\left[F + 2(G-1)n_{sp} + \dfrac{N_{0c}}{(e\bar{g})^2 \alpha GP_r}\right]2B_m} \tag{7.6.2}$$

where B_m is the detector output bandwidth. The amplifier gain has increased the detector input signal power, and helps to overcome the receiver thermal noise. Note that the optical gain G acts identical to the photodetector gain $\bar{g}$ in this regard. With the thermal noise negligible, the detector achieves

$$SNR = \frac{\alpha GP_r}{[F + 2(G-1)n_{sp}]2B_m}$$

$$\cong \frac{\alpha P_f L_f / 2B_m}{2n_{sp}} \tag{7.6.3}$$

The amplifier, therefore, has produced quantum-limited detection, but has diluted the theoretical SNR_{QL} by the factor $2n_{sp}$. Hence this latter term plays a role similar to a preamplifier noise figure, just as in electronic circuitry. For typical values of n_{sp}, the reduction factor can approach 3 to 10 dB.

Now consider the system in Figure 7.10b, with amplification applied at the fiber input. The photodetector input signal power is again $P_s = L_f GP_f = GP_r$, but the photodetector input noise will now have the attenuated spectral level $N_{sp}L_f$. The photodetected SNR is now

$$SNR = \frac{(\bar{g}e\alpha GP_r)^2}{[(\bar{g}e)^2 \alpha FGP_r + 2(\bar{g}e\alpha)^2 GP_r N_{sp}L_f + N_{0c}]2B_m} \tag{7.6.4}$$

The gain G increases the signal level and produces a quantum-limited SNR that is now

$$SNR \cong \frac{\alpha P_r / 2B_m}{2n_{sp}L_f}$$

$$= \frac{\alpha P_f / 2B_m}{2n_{sp}} \tag{7.6.5}$$

This is higher than the result in Eq. (7.6.3) due to the fiber attenuation of the spontaneous noise. Because $P_r = P_f L_f$, Eq. (7.6.5) depends on only the fiber input power, and the amplifier has effectively eliminated the fiber loss. The power input to the optical amplifier is significantly larger; therefore, the amplifier gain value G may not be as high as when used as a detector preamplifier because of the possible saturation effect of typical amplifiers. This reduced gain may limit the capability of the amplifier to eliminate the thermal noise term at the detector output.

7.7 DIGITAL COMMUNICATIONS OVER FIBERS

One of the more important uses of fiber communications is the transmission of digital data in the form of high-speed light pulses. The ability to pulse light at extremely high rates (≈ 1 Gbps) leads to an obvious potential for high-bit-rate fiberoptic channels. In this section, we examine the parameters needed to determine the extent to which this potential is achieved.

We have seen that the fiber channel can cause transmitted light pulses to be attenuated and spread during propagation. This spreading can be interpreted as an effective channel filtering, as discussed in Section 7.3. This means that if an idealized square pulse of light of width τ is launched into the fiber, it will appear at the output with the intensity spread to a width w, as shown in Figure 7.11 with the amount of spreading dependent on the fiber length. If this pulse corresponded to a digitally modulated OOK pulse, and if a sequence of such pulses were sent, the pulse spreading will cause (a) a pulse energy loss, because a portion of the pulse is spread outside its original width, and (b) interpulse distortion, because the tails of a given spread pulse will overlap onto adjacent pulses. To reduce these effects, it is necessary to use transmitted pulse intervals greater than the expected spreading. Thus the minimal transmitted pulse width equals the spread pulse width, and $\tau = w$. The maximum laser pulse rate is therefore limited to

$$R_p = \frac{1}{w} \quad \text{pulses/sec} \tag{7.7.1}$$

The right-hand side of Eq. (7.7.1) was previously defined as the fiber bandwidth in Eq. (7.3.3), and Eq. (7.7.1) limits the maximum source pulse rate to the fiber

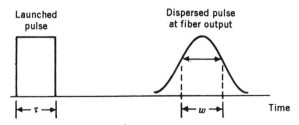

Figure 7.11. Pulse spreading with data pulse transmission.

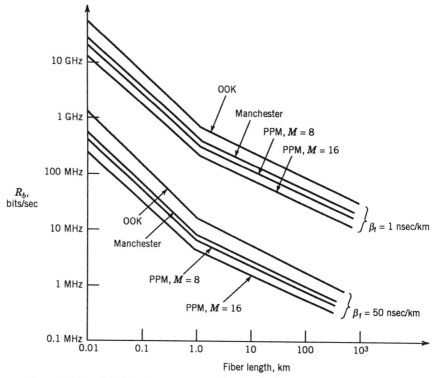

Figure 7.12. Band-limiting bit rates versus fiber length. β_f = fiber dispersion coefficient.

bandwidth. Because the fiber bandwidth depends on fiber length, the allowable transmitter pulse rate is determined by this length and can be determined directly from Figure 7.6 by simply rescaling the ordinate axis as pulses-per-second instead of hertz.

The actual transmitted bit rate depends on the encoding format. For OOK signaling, each pulse carries one data bit, and the bit rate R_b equals the transmitted pulse rate R_p. Hence the fiber bandwidth determines the bit rate, and Figure 7.6 can be related directly in terms of OOK bit rate in bits per second, producing the bit rate curve in Figure 7.12. In Manchester signaling, a bit requires two pulse widths, and the maximum bit rate is 1/2 of the fiber bandwidth. In PPM, the fiber bandwidth determines the minimum pulse width, and the system sends $\log_2 M$ bits in M pulse times. Hence the bit rate is $\log_2 M/M$ times the fiber bandwidth. These results are summarized in Table 7.1, and the bit rates are included in Figure 7.12.

As the coded pulse propagates down the fiber, it is attenuated as well as dispersed. The degraded pulse power lowers the decoding pulse energy and therefore lowers the decoding bit error probability (PE). The required pulse energy to achieve a specified PE with a particular encoding format can be obtained from the equations and curves in Chapter 6. The use of these curves

TABLE 7.1 Fiber Digital Encoding Parameters

Fiber Encoding Format	Bit Rate for Fiber Bandwidth (B_f)	Peak Pulse Power at Average Power (P_a)
OOK (pulsed)	B_f bps	$2\,P_a$
Manchester (pulsed)	$\dfrac{1}{2}B_f$	$2\,P_a$
M-ray PPM (pulsed)	$\left(\dfrac{\log M}{M}\right)B_f$	MP_a

first requires determining the photodetected noise levels, either as a noise variance, or converted to equivalent noise counts (K_n) in a specified pulse interval. With this, PE can be directly related to pulse signal counts (K_s), which in turn can be related to fiber pulse power.

Consider the OOK noncoherent fiber system, perhaps the simplest and most popular type of digital fiber link. The pulse time is equivalent to the bit time T_b. The detector thermal noise count collected in a bit time in the OOK decoder integrator is, from Eq. (6.3.7)

$$K_n = n_c T_b \qquad (7.7.2)$$

where n_c is again the effective noise count rate in Eq. (7.5.11). Using Figure 6.7, the required value Υ of SNR to achieve the desired PE can be determined, which can then be related to the signal pulse count $K_s = \alpha P_p T_b$ and the noise count K_n by

$$\begin{aligned}
\Upsilon &= \frac{(\alpha P_p T_b)^2}{F(\alpha P_p T_b) + K_n} \\
&= \frac{K_s^2}{F K_s + n_c T_b}
\end{aligned} \qquad (7.7.3)$$

If a link amplifier was used, with coefficients (G, n_{sp}), Eq. (7.7.3) would be modified by replacing K_s by $K_s = \alpha G P_p T_b$, and replacing F by $(F + 2n_{sp})$, as in Eq. (7.6.2). Equation (7.7.3) can be rewritten as

$$\frac{(\alpha P_p)^2}{F(\alpha P_p) + n_c} = \frac{\Upsilon}{T_b} = \Upsilon R_b \qquad (7.7.4)$$

where $R_b = 1/T_b$ is the bit rate. Equation (7.7.4) can be solved as a universal curve (recall Problem 6.10) of the form $y = x^2/(x + 1)$, where $x = \alpha P_p/(n_c/F)$ and $y = R_b/(n_c/\Upsilon F^2)$. This leads to the two important ranges of solution $x = y$, when $y \gg 1$, and $x = \sqrt{y}$ when $y \ll 1$, or

$$\alpha P_p = R_b[FY] \qquad \text{if } F^2 R_b Y/n_c \gg 1 \tag{7.7.5a}$$

$$= (R_b Y n_c)^{1/2} \qquad \text{if } F^2 R_b Y/n_c \ll 1 \tag{7.7.5b}$$

Because Eq. (7.7.5a) is usually true, required receiver power tends to be directly proportional to required bit rate.

Equation (7.7.5) relates required fiber output pulse power to fiber bit rates at a specified PE for the OOK system. The fiber output pulse power P_p is related to its input power through its attenuation loss, as was given in Eq. (7.1.4). Because fiber attenuation loss increases with fiber length, the available data rate decreases with the fiber length, when the fiber input power is fixed. Thus fiber length limits data rate because of its power attenuation as well as its bandwidth. Combining Eqs. (7.7.5a) and (7.1.4), the bit rate can be related to the fiber length Z by

$$R_b = Q(10^{-\alpha_f Z/10}) \tag{7.7.6}$$

where Q is the parameter

$$Q \triangleq \frac{\alpha P_f}{FY} \tag{7.7.7}$$

with Y the value of SNR producing the desired OOK PE. Note that the fiber input power and desired performance (PE) are all incorporated into the one parameter Q in Eq. (7.7.7). Typically, Q will have values in the range 10^{10} to 10^{16}. Equation (7.7.6) specifies the data rate allowed by the link power limitations for a fiber length Z and attenuation factor α_f. Note that this data rate can be increased for any value for fiber length by increasing the parameter Q, that is, by increasing source power or accepting a lower PE (lower Y).

A similar result can be developed for Manchester or M-ary PPM pulsed signaling. The corresponding SNR equation is now

$$\begin{aligned} Y &= \frac{(\alpha P_p T_p)^2}{F(\alpha P_p T_p) + 2n_c T_p} \\ &= \frac{K_s^2}{FK_s + 2K_n} \end{aligned} \tag{7.7.8}$$

where T_p is the pulse time, and Y is now the value of SNR for a specified PE for these systems. The pulse rate is related to the bit rate as shown in Table 7.1, which produces $Y/T_p = YMR_b/\log_2 M$, with $M = 2$ for Manchester signaling. This again produces the approximate result in Eq. (7.7.6) with Q modified to

$$Q = \begin{cases} \dfrac{\alpha P_f \log_2 M}{F \Upsilon M} & \text{for PPM} \\[2ex] \dfrac{\alpha P_f}{2 \Upsilon F} & \text{for Manchester} \end{cases} \qquad (7.7.9)$$

The data rates specified by Eq. (7.7.6) correspond to the maximum power-limited rates (i.e., the rates limited by the available power down the fiber). This must be balanced against the data rate permitted by the bandlimiting of the fiber in Figure 7.12. Equation (7.76) is plotted in Figure 7.13 as a function of the fiber length for a fixed set of Q values and loss coefficients. The OOK dispersion curves from Figure 7.12 are superimposed, as shown. The results combine to give the limiting data rates at each fiber length Z. Note that a fiber data rate can be either power limited or bandwidth limited, depending on the link parameters and the length of the fiber.

Once the components of a fiber link have been selected (power levels, fiber coefficients, receiver, and encoding format), the analysis will produce one of the rate–length set of curves in Figure 7.13. Assume this resulting curve is redrawn as in Figure 7.14. This will then indicate the maximum data rate at each fiber length for the selected parameters. If we consider (R_b, Z) as an arbitrary operating point in the two-dimensional space of the figure, then the point can

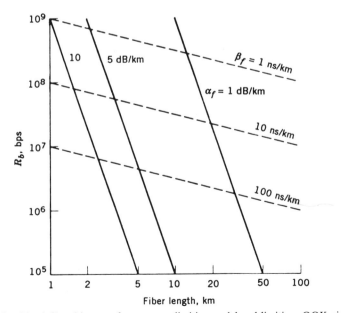

Figure 7.13. Combined fiber bit rates from power limiting and band-limiting. OOK signaling. PE = 10^{-9}. $Q = 10^{10}$.

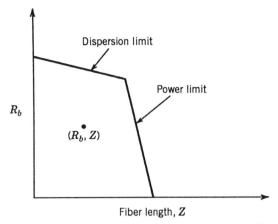

Figure 7.14. Example of power band-limited rate curve for fibers.

be supported by the link design only if it falls within the curve region. In fact, the rate can be increased to the point obtained by projecting vertically to the curve, or the fiber length could be increased to the value obtained by projecting horizontally to the curve. The inner region of the plot therefore defines the set of data rate-length values achievable by the link. Any operating point outside the curve will necessarily require redesign, or the insertion of repeaters or parallel data lines.

7.8 COHERENT COMMUNICATIONS OVER FIBERS

A coherent communications system can be used over a fiber channel with a heterodyning receiver operated at the fiber output. The overall fiber would be similar to that shown in Figure 7.15. Ideally, a heterodyning fiber link would have a lower threshold SNR (lower sensitivity) advantage over a direct detection (noncoherent) system. Of particular importance is the use of the heterodyning laser power at the receiver to overcome the detector thermal noise that limits noncoherent operation. This would permit operation at lower source power levels, or at higher data rates, then in direct detection links. The analysis in Chapter 5 can be directly applied here to determine demodulated SNR or decoding PE when heterodyning is applied.

The difficulty in coherent fiber operation is in maintaining the proper conditions for achieving these advantages. Often auxiliary hardware and higher quality components are needed to insure the necessary heterodyning matching. A laser source must be carefully selected to ensure a suitable narrow linewidth around the carrier wavelength. This is necessary to control the phase noise that will be superimposed on the heterodyned carrier frequency. This means the relatively inexpensive LEDs and laser diodes normally used in pulsed non-

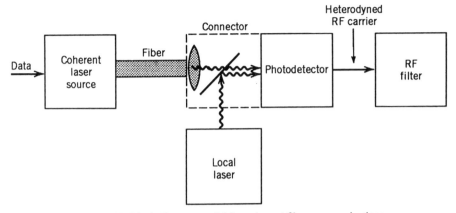

Figure 7.15. Block diagram model for coherent fiber communications.

coherent systems (where the laser linewidth is not important), are generally not suited for the heterodyned system. Likewise the wavelength stability of the source is also important to be sure that the heterodyned carrier remains within the RF bandwidth in Figure 7.15. This can be partially compensated by separate frequency locking loops that monitor the instantaneous heterodyne carrier frequency and keep the local laser wavelength at the proper separation. However, this now requires tunable lasers for the local source, with tuning bands wide enough to cover the source instability.

It is almost mandatory that a heterodyning fiber system use single-mode fiber. Because focusing and alignment hardware must be used at the fiber output, the fiber Airy pattern must be carefully controlled to prevent local field mismatches. When the fiber field is spread over many modes, either a more complicated local field pattern is required or a substantial heterodyning loss will occur due to power loss to other modes. This can only be corrected by multimode heterodyning, as was discussed in Section 5.5.

One of the more serious problems in coherent fiber systems is that of polarization matching, even with single-mode fibers (recall Eq. 5.4.14). Although the source polarization can be adequately predicted (generally a linear polarized field), a problem occurs during the fiber propagation. The birefringance (asymmetrical polarization rotations) within the fiber, produced by nonuniform polarization shifts due to bending and temperature effects, causes uncertainty in the field polarization rotation in the fiber. Because this rotation is unknown and random in nature, the local field polarization at the receiver cannot maintain alignment at all times with the fiber output polarization. This time variable mismatching causes power fading during communications.

To combat this polarization problem, there has been significant research in developing fibers that maintain polarization states during propagation. Such an approach requires careful control of the birefringance effect in fibers. This is achieved by techniques such as modifying the design of the fiber core (elliptical instead of circular) or modifying the cladding construction. Such design modifications add to the cost of the fiber and are still in the laboratory development stage.

It appears that as long as high-gain low-cost optical amplifiers are readily available to augment the noncoherent systems, the added complexity of using heterodyning in fiber communications may be difficult to justify in fiber links.

PROBLEMS

7.1   A 0.1-mW laser Lambertian source inputs light into a fiber with a numerical aperture of $NA = 0.1$. The fiber has a 10-μm glass core diameter and a total propagation loss of 10 dB. A photodetector ($\alpha = 10^{18}$) is directly connected to the fiber output. Write an expression for the probability that a count of 100 or less photoelectrons will be produced in a 10^{-9}-sec interval.

7.2   Given a short fiber with coefficients $z_f = 1$ km and $\tau_f = 10^{-5}$ sec/km. Sketch the impulse response of the fiber at the distances (a) $z = 0.01$ km, (b) $z = 0.1$ km, (c) $z = 0.9$ km.

7.3   Estimate the dispersion coefficient for a long fiber with $NA = 0.1$ and a core index of 1.5.

7.4   What is the approximate transmission bandwidth of a fiber with a dispersion coefficient of 5 nsec/km and an equilibrium distance of $z_f = 1$ km, having lengths (a) $z = 0.1z_f$, (b) $z = 10z_f$.

7.5   A fiber is modeled as a low-pass filter having the first-order response $H_f(\omega) = L_f/[1 + j(\omega/2\pi B_f)]$.
(a) If a pulse of width τ sec is fed into the fiber, compute the photodetected mean output current time response. Neglect detector filtering.
(b) If the fiber filtering is to be equalized (canceled out) describe the required output circuit filtering needed.

7.6   A noisy RF carrier with $SNR_1 = 10$ dB and carrier bandwith 10 MHz is transmitted over a fiber-coupling link. The detector has $F = 3$ and a thermal noise count rate of $n_c = 10^7$. If the output SNR_2 is to be within 3 dB of the input SNR, estimate the required received signal count rate.

7.7   A fiber has a dispersion coefficient of 50 nsec/km.
(a) If an OOK encoding system is used, what fiber length will permit a bit rate of 10 Mbps.
(b) If the dispersion is reduced to 1 nsec/km, what bit rate can be achieved?

7.8   The fiber system parameters are $n_c = 10^5$ cps, $F = 2$, $\alpha = 10^{18}$, and $\bar{g} = 100$. We desire a OOK bit rate of 500 Mbps at $PE = 10^{-9}$. Estimate the required received pulse power P_p.

7.9 A laser has an average power of 1 mW and is to be modulated with OOK data at a bit rate of 100 Mbps and fed into a fiber. The fiber has an attenuation coefficient of 5 dB/km, and the system is to maintain a $PE = 10^{-4}$. How long can the fiber be?

7.10 A fiber communication link uses noncoherent OOK signaling with a laser producing a received pulse power of 10^{-11} W. The link has a bandwidth limitation that permits pulses no shorter than 10^{-7} sec. A $PE = 10^{-4}$ is to be maintained. Assume a quantum-limited receiver with $\alpha = 10^{18}$.

(a) What is the maximum bit rate possible in the link?

(b) Repeat with the received laser power increases to 10^{-8} W.

7.11 A fiber has an attenuation of 4 dB/km. A digital OOK format of 1 Gbps is used with a laser source rated at 1 mW. A power coupling loss of 5 dB occurs at both the fiber input and output. The detector is quantum limited, has a gain of 1, an $\alpha = 10^{18}$, and must produce an electron count of 20 in each bit time. How long can the fiber be?

7.12 A 4-km fiber with a dispersion coefficient of 5 ns/km and attenuation of 2.5 dB/km transmits 16-slot PPM data. The detector requires a pulse count of 25 to decode. Determine the transmit laser diode power at which the dispersion-limited bit rate equals the power-limited bit rate. Assume $\alpha = 10^{18}$.

7.13 A fiber system has $\beta_f = 1$ ns/km, $\alpha_f = 5$ dB/km, and a equilibrium length of 1 km. The source can support a quantum-limited bit rate of 10^{12} bps over a 10-km length.

(a) Sketch the dispersion-limited and attenuation-limited bit rates versus fiber length.

(b) How many repeater sections are needed to support 1 kbps over a 100-km fiber pathlength?

7.14 Given a cascade of N optical amplifiers, each with gain G and spontaneous emission coefficient n_{sp}. Show that in terms of output signal and noise power, the cascade is equivalent to a single optical amplifier with gain G^N and coefficient $(G^{N-1} - 1/G - 1)n_{sp}$.

REFERENCES

1. D. Gloge, Impulse response of clad optical multimode fibers, *BSTJ*, 52, 801–816, March 1973.

2. D. Marcuse, *Theory of Optical Waveguides*, Academic Press, New York, 1974.

3. S. Karp, R. Gagliardi, S. Moran, and L. Stotts, *Optical Channels*, Plenum Press, New York, 1988, Appendix D.

4. C. Sandbank, *Optical Fiber Communication Systems*, Wiley, New York, 1980.

8

FIBER NETWORKS

The ability to transmit high data rates over thin fiber cable lines has made optical fibers important in communication networking. In typical optical communication networks, multiple sets of optical transmitters and receivers are interconnected by optical fiber to permit independent information flow to be simultaneously dispersed among the set.

Fiber networks may involve interconnections among devices that can be (1) confined to a specific location such as a single building (local area network, LAN), (2) spread between several buildings, say, within a city (metropolitan area networks, MAN), or (3) spread over an entire continent (wide area networks, WAN). Thus a network may involve individual links that may span meters, hundreds of meters, or thousands of meters. In this chapter, we examine some of the communication characteristics of general forms of fiber networks, and examine the relationships between the optical parameters and the key network performance parameters. The most important communication parameters of any network are (1) the size (number of simultaneous links) that can be sustained, (2) the data rate of each link, and (3) the accuracy of the transmitted data. Secondary aspects, such as complexity, latency (delays), and cost issues may also become important, depending on the application of the network, but are not pursued here.

8.1 FIBER NETWORK INTERCONNECTION ELEMENTS

The development of fiber networks has been spurred by the rapid achievements in specific fiber interconnection elements. Figure 8.1 shows two of the basic optical elements for fiber networking. Figure 8.1a shows a fiber splitter, an interconnection device that permits light from one fiber to be split into separate output fibers. A splitter is usually constructed by tapering down the fiber ends and fusing them together into a Y-shaped connection. The tapering causes the input fiber to focus its output field to illuminate the input ends of both output fibers. Assuming a constant light brightness in the tapered output field, each

Figure 8.1. Fiber interconnections. (a) Splitter and (b) combiner.

output fiber collects one half of the available fiber output power. (recall Problem 2.15.) Thus if the input fiber has an output power of P_r, each output fiber collects a power of $P_r/2$. The output power may also be reduced by any losses of light (for example, light lost into the cladding or leakage light) during the splitting. This latter power loss is called the excess loss (L_e). Hence each output fiber in Figure 8.1a collects a power of

$$P_o = (P_r/2)L_e \qquad (8.1.1)$$

A fiber combiner, shown in Figure 8.1b, is the opposite of a splitter, and permits input fibers to combine their light fields into a single output fiber. As in the splitter, combiners are formed by tapering and fusing fiber ends. This permits two separate fiber fields to be superimposed noncoherently, so the power in the output fiber is the sum of the input fiber powers, reduced by any excess loss factor. Thus if P_1 and P_2 are the input powers, the output power of the combiner is

$$P_o = (P_1 + P_2)L_e \qquad (8.1.2)$$

Often the output fiber is constructed with a larger core area than the input fibers to reduce the excess loss term (i.e., single-mode input fibers combined into a single-multimode output fiber). In this case, $L_e \approx 1$, and the output power approaches the direct sum of the input field powers.

An optical *directional coupler* is a two port device that interconnects light among four fibers (Figure 8.2). In typical operation, arriving light into one port (port 1) is coupled to a desired output port (port 2), whereas a fraction of the light is tapped off at another port (port 3), with no light permitted to exit at port 4. Directional couplers are constructed as light cavities with internal reflectors (mirrors or prisms) to direct the fields. By controlling the internal light paths, the throughput light and tapped light power levels can be adjusted, whereas the undesired port can be adequately isolated.

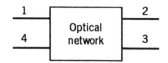

Figure 8.2. Directional coupler, 4 ports.

The important parameters of a directional coupler are the power ratios describing the routing at a particular wavelength. If P_1 is the input fiber power at port 1 and P_2, P_3, and P_4 are the coupler output powers into the connecting fiber at each corresponding output terminal, then the following ratios are often defined.

$$\text{throughput ratio } (L_T) = P_2/P_1$$
$$\text{tap ratio } (L_s) = P_3/P_2$$
$$\text{leakage ratio } (L_L) = P_4/P_1 \tag{8.1.3}$$
$$\text{excess loss ratio } (L_e) = (P_2 + P_3 + P_4)/P_1$$

These loss ratios are directly related, and it is straightforward to show that

$$L_e = L_T + L_T L_s + L_L$$
$$= L_T(1 + L_s) + L_L \tag{8.1.4}$$

Note that the tap ratio indicates how the input power is split between the throughput port and the tap port. If the tap ratio L_s is 1, and if the isolation is perfect ($L_L = 0$), then $L_T = (1/2)L_e$. Thus the directional coupler would act as a perfect splitter, as in Eq. (8.1.1). For other tap ratios, the throughput ratio becomes

$$L_T = L_e/(1 + L_s) \tag{8.1.5}$$

Thus for a $-10\,\text{dB}$ tap ratio ($P_3 = 0.1P_2$), with no excess loss ($L_e = 1$), the throughput power is $P_T = 0.9P_1$, corresponding to a $0.4\,\text{dB}$ throughput loss.

Directional couplers can be operated in a reverse direction as well. In Figure 8.2, power inserted into ports 2 and 3 will combine and appear at port 1 as an output. The output power would then be

$$P_1 = [L_T P_2 + L_T L_s P_3]L_e \tag{8.1.6}$$

and the coupler functions as a combiner.

The concept of a splitter and combiner in Figure 8.1 can be extended to multiple output ports, as shown in Figure 8.3. A single input fiber can be split into N output fibers by generalizing the 1×2 splitter in Figure 8.1a. Such N order splitters can be constructed by fusing multiple fibers (Fig. 8.3a) or by cascading 1×2 splitters (Fig. 8.3b). Again the input fiber power will be divided down and distributed over the output fiber set. If P_r is the input power, then the power coupled into any output fiber is

$$P_o = \left(\frac{P_r}{N}\right)L_e \tag{8.1.7}$$

where L_e is again the excess loss factor.

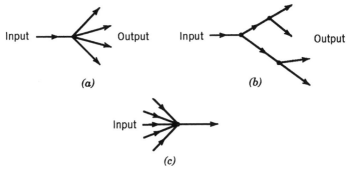

Figure 8.3. Nth-order fiber interconnects. (*a*) Splitter. (*b*) Splitter with cascaded (1 × 2) splitters. (*c*) Combiner.

Likewise an Nth order combiner (Fig. 8.3*c*) can be constructed as an extension of the 2 × 1 combiner. Again the output power is the sum of the N input fiber powers. Of most concern is the possible limit to the total light power that be inserted into a single output fiber before the nonlinearities of the fiber become important [1, 2]. Generally fiber powers must combine to over a watt before these effects must be considered.

8.2   NETWORK ARCHITECTURES

The fiber interconnect devices can be used to distribute multiple source data over an optical network. Figure 8.4 shows a generalized block diagram representing a basic fiber network architecture that may be designed into a local, metropolitan, or wide area network. A set of N optical transmitters are to be interconnected to N separate receivers through a fiber interconnection system. Any receiver should have the capability of receiving any transmitter, and conversely any transmitter should be able to transmit to any particular receiver. These distribution requirements can be achieved in several ways. The most basic is to route the transmissions from all transmitters to all receivers, and use multiple accessing techniques to separate the transmissions. *Multiple accessing* is a communication format that uses a form of transmitter signal

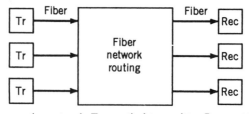

Figure 8.4. Fiber routing network. Tr = optical transmitter, Rec = optical receiver.

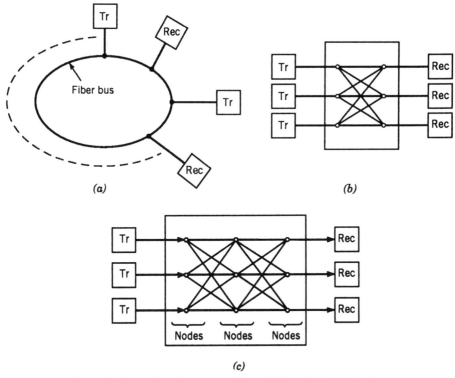

Figure 8.5. Network architectures. (*a*) Bus. (*b*) Star. (*c*) Layered nodes.

addressing that permits its transmission to be distinguished from all others. Multiple accessing in optical systems is discussed in Section 8.5.

Another network interconnection procedure is to connect each transmitter to a specific receiver and periodically, or upon command, switch the interconnections so that over a given time period any transmitter has a direct link to a specific receiver. This is often called *circuit switching* or *spatial switching*.

Data routing networks take on several basic architectures, as shown in Figure 8.5. In the bus system, a common fiber channel, or optical bus, interconnects all transmitters and receivers. Each transmitter couples its modulated light field onto the bus, and each receiver taps off the bus, using the multiple-accessing formats to identify the transmitted fields. In the star arrangement (Fig. 8.5*b*) the transmitter fields are carried by fiber to a central distribution point, where the fields are split and distributed to output fibers that connect to each receiver. Thus, each receiver collects the sum transmissions from all the transmitters. Again, multiple accessing permits a receiver to separate out the individual transmitters. The bus and star architectures are commonly used in local area networking.

The layered node network in Figure 8.5*c* generally models wide area networking. The network is composed of fiber-connected nodes, or stations,

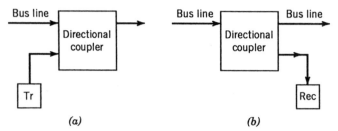

Figure 8.6. Bus network terminals. (*a*) Transmitter. (*b*) Receiver.

that have the capability of receiving and/or retransmitting its input fields. The transmitters identify the desired receiver node and, in-between nodes, serve as relays of the source field until the desired destination is reached. Information is generally sent as digital data packets, or word blocks, in which the destination information is directly appended in the packet (usually as header bits that are added to the date to form the packet). At a node, the data is decoded, the header is read, and the packet data is optically retransmitted to the next node, until the destination is reached.

Bus networks are constructed with fiber and directional couplers, as shown in Figure 8.6. The couplers permit each source to insert its modulated field into the fiber and allow any receiver to tap off the fiber field for demodulation. In each case, the field flowing in the fiber can pass through the coupler. By the use of the source addressing, the receiver can recover any source, while allowing the fiber field to continue to propagate through to the subsequent receivers. By properly designing the coupler, both a transmitter and a receiver can be combined at a particular node.

The power flow equations between any transmitter and receiver in a bus can be directly computed from the coupler and fiber parameters. Consider the bus link in Figure 8.7 showing a laser source with average power P_a inserting its light field into a fiber via a directional coupler having thoughput loss L_T, tap loss L_s, and excess loss L_e, as defined in Eq. (8.1.3). The inserted field passes through $N - 1$ identical couplers before being tapped off at the Nth coupler. Neglecting input fiber losses, the average received power tapped off at the

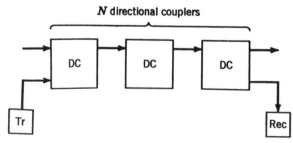

Figure 8.7. Bus link from transmitter (Tr) to receiver (Rec) through N couplers.

receiver is then

$$P_r = P_a[L_T(L_T)^{N-2}L_s^2]L_e^N 10^{-\alpha_f Z/10} \tag{8.2.1}$$

where α_f is the fiber attenuation coefficient in decibels per length, and Z is the bus length covered. The overall loss factor of the source, when expressed in decibels, is then

$$\text{dB loss} = (N-1)(L_T)_{dB} + 2(L_s)_{dB} + N(L_e)_{dB} - \alpha_f Z \tag{8.2.2}$$

Note that the bus link loss increases directly proportional to the number of couplers it must pass through. Even if the throughput loss $(L_T)_{dB}$ and excess loss $(L_e)_{dB}$ were limited to a fraction of a decibel, a sizable loss can still occur with a network having as few as 10 to 20 nodes. Note also that the received power distribution will vary over the receiver set, depending on its location relative to the source. That is, some receivers will receive higher power levels than others. Hence the multiple-accessing format at any receiver may correspond to the presence of strong unwanted signals combined with weaker desired signals. The bus system has the advantage of simplicity and requires only a single fiber to interconnect all nodes.

The star architecture in Figure 8.5b is redrawn in Figure 8.8, showing N sources interconnected to N receivers. Here an individual source couples to a receiver via the Nth order splitter that distributes the combined light fields from all sources to any receiver. The power flow from a source of power P_a to a receiver involves the fiber loss and the Nth order splitting loss. Thus,

$$P_r = P_a[10^{-\alpha_f Z}](L_e/N) \tag{8.2.3}$$

where Z is the overall fiber length. Expressed in decibels, the loss is

$$\text{dB loss} = -\alpha_f Z - (N)_{dB} + (L_e)_{dB} \tag{8.2.4}$$

Note that the loss now varies only logarithmically with the network size N rather than linearly as in the bus system. Also, the power from equal sources

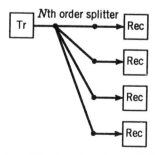

Figure 8.8. Star network using Nth-order splitters.

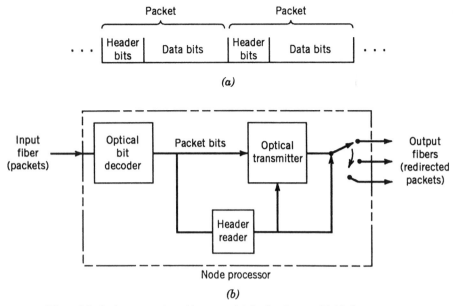

Figure 8.9. Packet network architectures. (*a*) Packet format. (*b*) Node processors.

is equally distributed at all receivers. Hence, star networks are less lossy and more power equivalent than bus networks. The star systems, however, require more fiber lengths to complete the network, essentially maintaining a separate fiber link for each source–receiver pair.

The distributed node system in Figure 8.5*c* represents dense networks involving widely separated stations. Figure 8.9 shows how a node terminal may be formatted. Packets composed of both destination (header) bits and data bits (Fig. 8.9*a*) are transmitted to the nodes, say as OOK optical pulses. The node decodes the bits of the packets, storing the data bits and separating out the header bits, which are electronically read to identify the destination. The node electronics determines the best output route (output fiber) to retransmit the packet. The entire packet is then retransmitted (Fig. 8.9*b*) optically through the selected output fiber. This operation is repeated at all nodes until each packet reaches its destination. Nodal relays have the advantage of power regeneration and dynamic route selection, but problems with packet collisions at a node, node delays during retransmission, and decoding bit errors must be considered in overall network evaluation. These latter topics are beyond the scope of our study.

8.3 OPTICAL MULTIPLE ACCESSING

Multiple accessing refers to the communication operation of having multiple signals propagate over the same link and yet be individually separable at a receiver. The separability is provided by transmitting the signals with inherent addressing formats superimposed on the signal in some manner. A specific source signal can be recovered by tuning, or aligning, the receiver to the addressing format. All other signals, having an undesired address, will be rejected (partially or totally) by the address recovery procedure. The resulting recovered signals containing the proper address will then be separated out, in the presence of detector noise and any interference from the signals that were not totally rejected. This latter interference is referred to as "multiple access" noise, or simply "crosstalk." Crosstalk can take on various forms, can appear noiselike in nature, or can appear with definate signal structure.

Optical signal addressing can take on various forms. The most popular are summarized below.

8.3.1 Wavelength Division Multiplexing

Each source is addressed by assigning it a unique wavelength band. All information is transmitted by a laser source using a modulation bandwidth confined to the wavelength assigned to it. The modulation can be noncoherent (intensity modulation) or coherent and involve digital or analog data. Receiver separability is achieved by frequency tuning (predetection optical filtering) to the desired band. Signals in other bands will be rejected by the optical filtering at the receiver. Because each source uses a separate band for transmitting, the transmission can be completely independent and nonsynchronous. Note that each source uses only a portion of the total available wavelength band of the network. The optical wavelength division multiplexing (WDM) system exactly parallels the frequency division systems used in radio communications and, therefore, is the most fundamental type of multiple accessing.

8.3.2 Time Division Multiplexing

Each source is addressed by assigning it a specific time interval. A given source transmits only during its assigned time period, and no other source is allowed to transmit during that interval. An allotted time interval for a particular transmitter is generally repeated periodically or interleaved with others, so that its communication is bursty in nature. A receiver recovers the transmission from a specific source by time gating to only the intervals of that source. Note that TDM requires complete network synchronization, so that all transmitters and receivers are accurately aligned to the proper time slots. Hence TDM is inherently a synchronized network. Because only one transmitter is transmitting at one time, it theoretically can use the entire wavelength bandwidth available to the entire network.

8.3.3 Code Division Multiplexing

Each source is assigned a unique optical pulse code sequence that can be recognized at the receiver. A digital laser source sends its data by superimposing its data bits onto its own code. All sources transmit their modulated pulse codes asynchronously over the same channel. Separability is achieved at the receiver by looking for (correlating with) the proper pulse code sequence and decoding the superimposed data. Ideally all other pulse code sequences arriving at the receiver will be correlated out. Code Division Multiplexing (CDM) systems are almost always noncoherent and are primarily designed for digital transmission formats.

8.4   WAVELENGTH DIVISION MULTIPLEXED NETWORKS

A WDM network [3] composed of N sources sending information simultaneously to N receivers over an interconnecting fiber system is shown in Figure 8.10. Each source is assigned a separate wavelength band within a specified tuning range. Tuned lasers are used as the optical source for each transmitter, and then data modulated within its assigned band for transmission. The distribution network routes the WDM signals from all sources to each receiver terminal. The receiver isolates a transmitter by tuning an optical filter to the desired band, while filtering out all others. This can be accomplished by the use of tunable optical filters followed by photodetectors and decoders. The prime disadvantage in optical WDM is the fairly expensive and sophisticated hardware required (wavelength-controlled tunable lasers and high-quality narrowband tunable filters) for each channel.

The wavelength spectrum of the transmitted and received WDM network is shown in Figure 8.11. Each modulated transmitter occupies a fixed wavelength band, depending on the data rate and laser line width, with the set of all bands

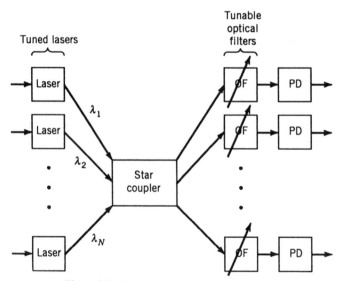

Figure 8.10. WDM network architecture.

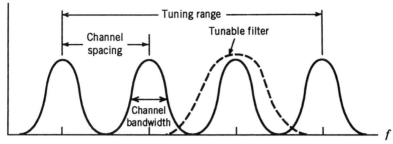

Figure 8.11. WDM spectral diagram.

spanning the available tuning range. The tuned filter at the receivers must be able to tune over the entire range to select a specific band. Tunable optical filters were discussed in Section 2.5, the most popular types being the Fabry-Perot [4, 5] and Mach-Zehnder interferometer filters. Other types of optical tuning filters use acoustic-optic [6] and electrooptic mode coupling, or can be obtained from narrowband laser amplifiers. Table 8.1 summarizes some basic characteristic of these filters.

Power and crosstalk analysis for a WDM system can be obtained by referring to the wavelength diagram in Figure 8.11. The figure shows adjacent WDM band spectra (all assumed to be equal), and the filtering effect of a single receiver-tuned filter aligned to one of the bands. The signal power collected is that of the desired band falling within the filter bandwidth. The crosstalk is due primarily to the adjacent spectra falling within the same filter band. This crosstalk, therefore, acts as an additive input noise field impinging on the receiver photodetector. If we let $S(f)$ be the modulated optical carrier spectrum, and let $H_f(f)$ be the tuning filtering function expressed in the frequency domain, then the signal power P_s is

$$P_s = \int_{-\infty}^{\infty} S(f) |H_f(f)|^2 \, df \qquad (8.4.1)$$

TABLE 8.1 Typical Optical Filter Parameters

Filter Type	Tuning Range	Bandwidth (nm)
Fiber Fabry–Perot	50 nm	<0.01
Two-stage		<0.01
Waveguide	4.5 nm	~0.04
Mach–Zehnder		(5 GHz)
Acoustooptics	400 nm	1
Electrooptics	16 nm	0.6
DFB Filter Amplifier	0.5 nm	0.1
2-section DFB Amplifier	0.6 nm	0.085

whereas the crosstalk power is

$$P_{CT} = \int_{-\infty}^{\infty} S(f - f_d) |H_f(f)|^2 \, df \qquad (8.4.2)$$

where f_d is the channel separation between band centers. Let us assume a second-order modulation spectrum and filter functions (centered at $f = 0$, for convenience),

$$S(f) = \frac{2P_r/\pi B_s}{1 + (2f/B_s)^2} \qquad (8.4.3)$$

$$H_f(f) = \frac{1}{1 + (2f/B_f)^2} \qquad (8.4.4)$$

where P_r is the received signal power, and B_s and B_f are the spectra and filter 3-dB bandwidths. The resulting integrations in Eqs. (8.4.1) and (8.4.2) now yield (see Problem 8.4) $P_s = P_r L_f$, and $P_{CT} = P_r L_{CT}$, where the loss functions from the filtering are

$$L_f = \left[\frac{B_f}{B_f + B_s} \right]^2 \qquad (8.4.5)$$

$$L_{CT} = \left\{ \frac{1}{1 + [2f_d/(B_f + B_s)]^2} \right\}^2 \qquad (8.4.6)$$

Thus the signal and crosstalk powers depend on the spectral widths and the channel spacings. Because the crosstalk appears at the receiver as a relatively low spectral level noise field (its spectral levels are well below the signal spectrum), it can be modeled as a weak noise field at the photodetector input. Its primary effect is that its power level increases the shot noise level of the photodetector. Thus, the photodetected SNR derived in Chapters 4 and 5 are now reduced by the crosstalk addition to the shot noise term, effectively replacing the background power term P_b.

In a digital WDM system using OOK direct detection, the crosstalk level also adds to the mean detector output, which can further effect the OOK decoding, even if the receiver thermal noise dominates the shot noise. The offset mean values caused by the crosstalk during decoding can shift the bit integration output relative to the threshold. Repeating our OOK decoding analysis in Secton 6.2, the off-bit and on-bit mean values in Eq. (6.2.17), when crosstalk is present, now become

$$\begin{aligned} m_1 &= e\bar{g}K_s + e\bar{g}(\alpha P_{CT} T_b) && \text{with probability one-half,} \\ &= e\bar{g}K_s && \text{with probability one-half,} \end{aligned}$$

$$m_0 = e\bar{g}(\alpha P_{CT} T_b) \quad \text{with probability one-half,}$$
$$= 0 \qquad\qquad \text{with probability one-half,} \tag{8.4.7}$$

where $K_s = \alpha P_s T_b$. For a thermal-noise limited receiver, $\sigma_1^2 = \sigma_0^2 = \sigma_n^2$, where σ_n^2 is the bit integrated thermal noise variance in Eq. (6.2.10). With a decoding threshold set at $e\bar{g}K_s/2$, the channel bit error probability PE, averaged over equal likely possibilities of the presence and absence of the crosstalk from the adjacent WDM channel, is then

$$\text{PE} = \frac{1}{2}\left\{\frac{1}{2} Q\left(\frac{K_s/2 + \alpha P_{CT} T_b}{\sigma_n/e\bar{g}}\right) + \frac{1}{2} Q\left(\frac{K_s/2}{\sigma_n/e\bar{g}}\right)\right.$$
$$\left. + \frac{1}{2} Q\left(\frac{K_s/2 - \alpha P_{CT} T_b}{\sigma_n/e\bar{g}}\right) + \frac{1}{2} Q\left(\frac{K_s/2}{\sigma_n/e\bar{g}}\right)\right\} \tag{8.4.8}$$

Grouping terms and rewriting,

$$\text{PE} = \tfrac{1}{2}Q(\tfrac{1}{2}\sqrt{\text{SNR}}) + \tfrac{1}{4}Q[\tfrac{1}{2}\sqrt{\text{SNR}}(1 + \varepsilon)] + \tfrac{1}{4}Q[\tfrac{1}{2}\sqrt{\text{SNR}}(1 - \varepsilon)] \tag{8.4.9}$$

where $\text{SNR} = K_s^2/K_n$, $K_n = \sigma_n^2/(e\bar{g})^2$, and $\varepsilon = 2P_{CT}/P_s$. The first term is the OOK PE if there were no crosstalk term, whereas the remaining terms account for crosstalk during decoding. The last term is the most important, always producing a degradation (increase) in PE. To insure that this degradation is negligible, we generally want $\varepsilon \gtrsim 0.02$, which, from Eq. (8.4.6) with $B_f = 2B_s$, requires $(2f_d/3B_s)^2 \approx 15$, or

$$f_d/B_s \cong \sqrt{15}(3/2) \cong 5.8 \tag{8.4.10}$$

Thus center-to-center band spacings of about six times the spectral bands are necessary to negate the crosstalk effect on decoding in OOK WDM.

With crosstalk eliminated by band spacing, the PE of a channel link reduces to the standard power-bandwidth tradeoff of as single fiber link, as discussed in Chapter 7. The assigned channel bandwidths determine the narrowest OOK pulse width and sets the maximum link bit rate. In general, this is independent of the network size N.

The bit rate is also limited by the available received channel pulse power needed to achieve a desired pulse count K_s for a specific PE. Recall our discussion in Chapter 7. An optical OOK pulse, sent from a particular transmitter and confined to the channel band, suffers the network transmission losses discussed in Section 8.2. For a star network, the received pulse power due to a source pulse power P_p is obtained from Eq. (8.2.3),

$$P_r = P_p\left(\frac{L}{N}\right) \tag{8.4.11}$$

where L is the combined splitter excess loss, fiber transmission loss, and tuning filter loss. The L term may or may not be significant, depending on the type of network and associated fiber losses. Setting $P_r T_b = K_s/\alpha \triangleq E_b$, we see that $P_p T_b L/N = E_b$ or

$$R_b = \frac{P_p L}{N E_b} \tag{8.4.12}$$

Here an inherent inverse relationship exists between network size N and allowable channel bit rate R_b, because of the star splitting of the source power. We see now that achievable data rates in networking can be limited by either bandwidth, source power, or network size. This also shows why optical amplification, which can be inserted to overcome the expected splitting losses, are important elements in dense (high-N networks).

Lastly, the network size N may be limited by the number of WDM channels permitted by the optical tuning ranges in Figure 8.11. Because each receiver filter must have the capability of tuning to any transmitter band, the tuning ranges of the filters can limit the number of allowable channel bands. With the individual channel band set by the modulating bit rates and laser source linewidths, the number of channels N that can be encompassed will be directly related to the overall range. Let f_d again denote the channel spacing and let Δf be the laser linewidth in hertz. The channel bandwidth B_s can be approximated as the sum of the linewidth and the channel modulating bit rate. This means that the channel spacing should be approximately

$$f_d \cong 6(\Delta f + 2R_b) \tag{8.4.13}$$

according to our discussion in Eq. (8.4.10). If ΔF is the available tuning band in hertz, then

$$\Delta F = N f_d \approx 6N(\Delta f + 2R_b) \tag{8.4.14}$$

The maximum channel bit rate R_b and the number of channels N permitted by the tuning range is then related by

$$N \leqslant \frac{\Delta F}{6(\Delta f + 2R_b)} \tag{8.4.15}$$

Note that when $R_b > \Delta f$, N and R_b are inversely related. When $R_b < \Delta f$, N is independent of R_b and is limited to the asymptotic value of $N = \Delta F/6\Delta_f$.

In summary, the OOK bit rate per channel in a WDM network of size N can be limited by the power levels, bandwidth, or filter tuning ranges. At a specified N, the smallest of the bit rates permitted by each of these effects will determine the overall network link capability.

8.5 TIME DIVISION MULTIPLEXED NETWORKS

In TDM networks, multiple accessing is achieved by separating all source transmissions in time and using time gating at the receiver for tuning to the data of a desired source. The required timing for the transmitters and receivers must be accurately controlled by a master clock that is available to all stations. Thus TDM networks are inherently a time synchronized system.

A TDM network can be generated in one of two basic formats. Figure 8.12 shows a time-slotted format, in which a periodic time frame is partitioned into N time slots, and each of the N transmitters are assigned to one time slot in each frame. With all transmitters aligned to the same frame clock, a source transmits only during its own slot. Thus only one source is producing waveforms at any one time. A receiver synchronized to the same frame clock (allowing for transmission delay) gates only during the slot time of the desired source.

Source modulations can be coherent or noncoherent, intensity modulated or digital, but the source must burst on periodically during its slot times. Each transmitter, however, has available the entire optical bandwidth during its transmission time. A transmitter may have to buffer (store-and-forward) its data to properly align its data rate with its source transmission time. Crosstalk from other channels will only occur if timing errors cause time slot overlap between adjacent time channels. By inserting guard spaces between slot times to compensate for possible timing errors, the crosstalk can be effectively removed.

If the source sends OOK bits at rate R_s bits per second during its slot time, and if each time slot is T_s seconds long, then each source sends $R_s T_s$ bits during each frame time. The transmitted bit rate of a single channel is then

$$R_b = R_s T_s / N T_s = R_s / N \qquad (8.5.1)$$

in bits per second. Thus a source must transmit OOK bits during its slot time at a rate R_s that is N times faster than its desired bit rate. Conversely, if the maximum OOK transmission rate of a source is fixed at R_s bits per second, the channel data rate is lower by the factor N. Again we see an inverse relation

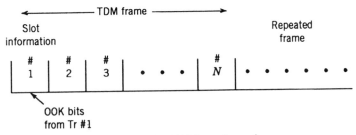

Figure 8.12. TDM frame format.

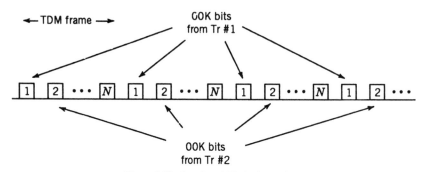

Figure 8.13. Interlaced TDM frame format.

between the channel data rate and the network size N, just as with WDM. Note that the receiver decoder must operate at the channel rate R_s rather than the date rate R_b, and must reestablish bit synchronization during the proper slot time of each frame. However, network timing need only be accurate to a fraction of a slot time, which may be on the order of microseconds for typical frame structures.

A second way to produce TDM is via direct pulse interlacing, as shown in Figure 8.13. In this format, each source transmits at its desired channel data rate R_b, but its OOK bit pulses are assigned to specific periodic pulse slots in the frame. The bit pulses of all sources are then interleaved on a bit-by-bit basis to generate the TDM frame. Thus in the bit time of one source, $N - 1$ other sources interleave their bits. If T_c is the laser pulse width, $T_b = NT_c$, or

$$R_b = \frac{1}{NT_c} \tag{8.5.2}$$

Thus each source must produce pulses of width T_c at the bit rate R_b satisfying Eq. (8.5.2), and all such pulses must be timed to interleave into the TDM frame in Figure 8.13. If $1/T_c$ corresponds to the bandwidth of the channel, then Eq. (8.5.2) is identical to Eq. (8.5.1). Thus, for fixed bandwidth networks, the slotted and interleaved TDM system operate at the same channel bit rate. However, the interleaved system requires no data buffering, and the laser sources need only pulse at the bit rate instead of the higher rate R_s in Eq. (8.5.1).

A receiver for the interlaced system must be accurately synchronized to within the narrow pulse time of the transmitters (allowing for propagation delay) so as to only gate during the desired bit pulse times. Thus the receiver timing must be on the order of optical pulse widths (picoseconds) instead of the wider slot intervals (microseconds) for the slotted system.

An important element in the interleaved system is the optical clocks. These must produce narrow-width optical pulses exactly timed to the bit location of the sources. Such clocks are needed at both the transmit and receiver ends. Optical clocks can be obtained by electronically triggering laser pulses at a lower pulse rate and multiplying the rate to the optical rate needed, as shown

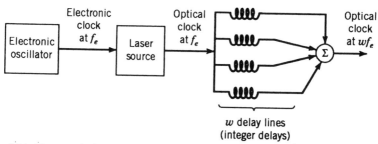

Figure 8.14. Electrooptical clock. Output optical clock pulses at rate qf_e driven by electronic clock at rate f_e.

in Figure 8.14. The electronic clock at rate f_e produces the triggers for the laser pulses at this same rate. Rate multiplication is achieved by splitting the laser pulses into parallel fiber lines having integer time delays and summing the outputs. The summed output appears as a sequence of delayed echoes of the input laser pulse, forming a regenerating optical pulse sequence at the higher rate. If w is the number of delay lines, the output pulse sequence rate is $f_o = wf_e$, and the output rate is a multiplied version of the electronic trigger rate.

The optical delay lines in Figure 8.14 correspond to sections of fiber cut to the exact length to produce the desired delay. For example, a z meter length of glass core fiber with index $n_1 = 1.5$ will have a delay of $z/(c/n_1) = (z/2) \times 10^{-8}$ seconds. A delay of 1 nsec (10^{-9} sec) would therefore require a fiber length of approximately 0.2 m or 8 in. Because we are dealing with delay accuracies on the order of optical pulse widths, delay differentials through the various parallel paths must also be controlled to within fractions of this accccuracy.

The advantage of the electrooptical pulse clock in Figure 8.14 is that the clock laser need only pulse at the electronic rate f_e, while the output optical pulse rate can achieve the desired data rate. The clock pulses, however, suffer an inherent power loss during the splitting into the delay lines, and the output optical pulse power is $1/w$ of the laser pulse power.

The optical pulse clock must be regenerated at the receivers to time gate to the desired bit pulses of the data. A receiver optical clock, synchronized to the same electronic clock at rate f_e (allowing for transmission delay), can produce the desired receiver optical pulse clock by properly selecting the delay lines. The receiver clock can then be aligned with the bits of the desired transmitter (this may require an additional delay shift). The optical clock and the TDM bit stream can then be summed and photodetected. A threshold detection circuit responding only to the summed light field can then decode only the OOK bits to which the clock is aligned. The resulting bit error probability PE can be determined from the OOK analysis in Section 6.2, and will depend on the transmitter data pulse power and on the network transmission losses discussed in Section 8.2.

8.6 CODE DIVISION MULTIPLEXED NETWORKS

Code division multiplexing (CDM) [7, 9] is an alternative to WDM and TDM for multiple accessing, and its value is that it involves simple, nonsynchronous pulsed laser modulation for its operation. Each transmitting source is assigned a unique optical pulse sequence for its address. Digital data is encoded onto these assigned pulse sequences, and decoded at a receiver by a pulse sequence correlation. The system has the advantage of using simple pulsed lasers (without wavelength control) and standard wideband photodetectors (without narrow optical filters). All sources operate independently, and no clock is needed to align transmitters.

In OOK CDM, each OOK bit is sent as the sequence of that transmitter or its absence, a shown in Figure 8.15a. The code sequence is produced at the encoder from the OOK bit pulse and can be generated either actively or passively. In the passive encoder (Fig. 8.15b) the sequence is generated by parallel fiberoptic delay lines, whose delays are adjusted to the pulse sequence separations, so that the summed output forms the sequence as the echoed version of the initial OOK pulse. In a active encoder, Figure 8.15c, the laser is directly modulated by the code sequence. All transmitters asynchronously superimpose their OOK pulse sequences for distribution by the network.

An individual receiver observes the sum of all such pulse transmissions, and recovers the data from a particular transmitter by using a matched pulse code correlator. This correlator can be accomplished optically by the use of parallel fiber delay lines, similar to Figure 8.15b, except the delays are matched to the desired sequence. If the sequence is sent during a bit time (an OOK "1" bit is sent) the code correlator will produce a high correlator signal at the end of the bit, when the received pulse sequence exactly fills the correlator. Received sequences from other transmitters that do not match the correlator delays will correlate to a lower value. Photodetection following the correlator detects the optical correlation peak (or its absence) to decode the OOK bit. Other arriving pulse sequences having pulses that happen to overlap a pulse of the desired sequence will produce correlation crosstalk, which can degrade the decoding.

There have been numerous reported studies of optical pulse code sequences that produce low levels of pairwise cross correlations. The selection of candidate sets can be viewed as a coding problem involving (0, 1) sequence sets and has been rigorously formulated as such [9, 10]. Sets of sequence having no more than one pulse overlap in the pairwise cross correlations have been derived, and their construction is now fairly well understood. An important property of a set of N such sequences, each having w pulses (called the code weight), is that they require a code length l (number of 0, 1 symbols) given approximately by

$$l \cong Nw(w - 1)/2 \qquad (8.6.1)$$

Thus, the code lengths must increase with the number of such sequences in the

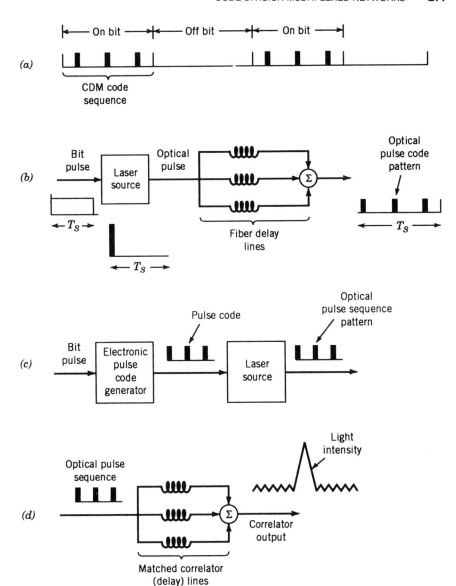

Figure 8.15. CDM diagram. (*a*) Coded pulse format. (*b*) Passive code generator. (*c*) Active code generator. (*d*) Optimal matched correlator.

set, for a specific code weight if the minimal correlation is to be achieved. Because the code assigned to each transmitter is sent to represent an OOK bit, the bit rate of a set of transmitters using this code set can be determined. If T_c is the laser pulse widths used in the sequence, a bit time must be $T_b = lT_c$. The minimal pulse widths T_c will be determined by the link bandwidths of each channel. Hence the channel bit rate for an OOK CDM system is related to the bandwidth B_c by

$$R_b = \frac{2B_c}{Nw(w-1)} \qquad (8.6.2)$$

This shows that the number of transmitters N and the channel bit rate are again inversely related for a fixed bandwidth; the actual value depends on the code weight. The code weight, however, is important in determining the decoding bit error probability of each receiver, because it determines the pulsed bit energy of a code.

The laser source of any transmitter must produce sufficient pulse power to produce a pulse sequence whose power levels will be strong enough to overcome the channel and network losses. The delay line splitting during encoding produces a splitting loss of $1/w$. (This loss is not incurred if an active source is used). At the receiver, the optical correlator produces a second splitting loss of $1/w$. The total loss from laser pulse output to the corresponding photodetector is then

$$\text{pulse power loss} = L/Nw^2 \qquad (8.6.3)$$

where $1/N$ is the network splitting loss and L is again the combined excess, coupling, and fiber transmission losses. If P_p is the peak laser power, then the peak received single pulse power is

$$P_r = P_p \left(\frac{L}{Nw^2} \right) \qquad (8.6.4)$$

The optical correlator sums w of the correct sequence pulses to form a correlation peak of wP_r when the OOK one bit is sent. The receiver photodetects the correlator output and decodes the bit by comparing to the OOK threshold. Other sources transmitting different sequences will produce correlation values that add as crosstalk during bit decoding.

Consider again the OOK decoder in Figure 6.2, and let m_1 and m_0 be the bit integrator mean values and σ_n^2 be the variance due to thermal noise. If there are k overlapping crosstalk pulses (that is, k pulses from the other codes happen to overlap any of the pulses of the desired code, and therefore will be accumulated in the optical correlator), then the mean values become

$$m_1 = e\bar{g}(\alpha w P_r T_c) + e\bar{g}(\alpha k P_r T_c)$$
$$m = e\bar{g}(\alpha k P_r T_c) \qquad (8.6.5)$$

Because the parameter k is dependent on the random time relation between the codes and on the data modulation (a code may be on or off during a bit time) it evolves as a random variable in Eq. (8.6.5), with a probability density $P(k)$.

The bit error probability PE for an OOK channel in the CDM network is obtained by averaging over all k values. Hence

$$PE = \sum_k P(k) \left[\frac{1}{2} Q\left(\frac{m_1 - z}{\sigma_n}\right) + \frac{1}{2} Q\left(\frac{z - m_0}{\sigma_n}\right) \right] \qquad (8.6.6)$$

where z is the decoding threshold. This can be rewritten as

$$PE = \sum_k P(k) \left[\frac{1}{2} Q\left(\frac{1}{2}\sqrt{w\mathrm{SNR}_1}\left(1 + \frac{2k}{w}\right)\right) + \frac{1}{2} Q\left(\frac{1}{2}\sqrt{w\mathrm{SNR}_1}\left(1 - \frac{2k}{w}\right)\right) \right]$$
$$(8.6.7)$$

where we have introduced the single pulse SNR_1

$$\mathrm{SNR}_1 = K_s^2 / K_n \qquad (8.6.8)$$

with $K_s = \alpha P_r T_c$, and we assumed a threshold value of $K_s/2$. Equation (8.6.7) shows that code weight w effects PE in two ways. It multiplies up the single-pulse SNR but can also decrease SNR due to the delay line splitting in the passive encoder and matched correlator.

Crosstalk degrades the CDM performance by altering the signal mean relative to the threshold values. Deriving accurate statistical models for the crosstalk probability P(k) based on the code design becomes an important aspect of this analysis. For the minimal correlation codes, only one pulse overlap can occur between any two codes, and this occurs with probability $w^2/2l$ (see Problem 8.9). Because one interfering code can produce at most one overlap, $N - 1$ interfering codes can produce at most $N - 1$. Thus, for these codes, k in Eq. (8.6.5) is in the range $(0, N - 1)$, and has the probability

$$P(k) = \binom{N-1}{k} \left(\frac{w^2}{2l}\right)^k \left(1 - \frac{w^2}{2l}\right)^{N-1-k} \qquad (8.6.9)$$

The CDM PE in Eq. (8.6.7) always approaches a PE floor from the crosstalk, even with high values of pulse SNR_1. This can be seen by letting $\mathrm{SNR}_1 \to \infty$ in Eq. (8.6.7) and using the facts

$$\begin{aligned} Q(x) &\to 0 & \text{as } x \to \infty \\ Q(x) &\to 1 & \text{as } x \to -\infty \end{aligned} \qquad (8.6.10)$$

This shows that

$$PE \text{ floor} = \sum_{k=w/2}^{N-1} P(k) \qquad (8.6.11)$$

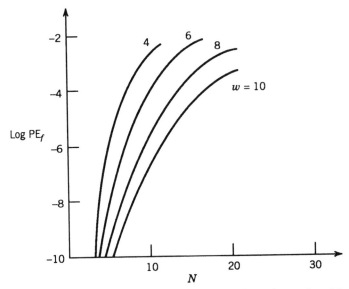

Figure 8.16. CDM PE floor versus network size N for various code weights w.

This PE floor is plotted in Figure 8.16 as a function of N for several values of code weight w. It is evident that high enough weights (enough code pulses) must be used, for a given network size N, to lower the PE floor to the desired value to prevent the link from being dominated by the code crosstalk. With w selected, we can then determine if the link SNR_1 is high enough to support a specified bit rate at the desired PE.

8.7 OPTICAL CIRCUIT SWITCHED NETWORKS

The alternative to addressed waveforms for achieving multiple accessing in networks is the use of circuit switching. Rather than depend on recognizeable addresses to separate channels, the network instead directly switches the optical paths to complete a link between a receiver and transmitter.

A circuit switched network is obtained by placing on–off switches (sometimes called optical shutters) in the fiber paths interconnecting the receiver to the star distribution, as shown in Figure 8.17. The switch can be physically located at the star output or at the receiver input. By externally switching off the undesired channels, a direct link is maintained to the desired channel, ideally with no crosstalk interference. The switching can be converted upon command so that any receiver can be connected to any transmitter, while blocking off the other channels.

Each shutter in Figure 8.17 is separately accessible by a control signal.

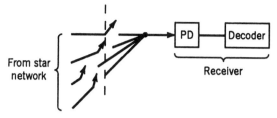

Figure 8.17. Implementation of shutter switches in space switching network.

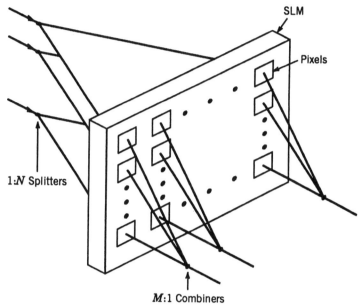

Figure 8.18. Spatial light modulator (SLM) shutter switch.

Several popular ways can be used to accomplish the shutter operation. One is by the use of spatial light modulators (SLM) in Figure 8.18. These devices correspond to a matrix of pixels, in which each pixel can be separately controlled and designed to perform as a shutter. The shuttering is achieved by polarization control, in which the input light polarization is shifted by the control to either match or be orthogonal to an output polarizer. A variety of materials can be used to accomplish the voltage-controlled polarization rotation, such as ferroelectric liquid crystals or other magnetooptical materials. The shutters can also be implemented with optical amplifiers that are biased into either the off or on state.

The key parameter in a shutter system is the leakage light, that is, the amount of light that passes through a shutter when it is in the off state. Any such leakage appears as crosstalk in the desired channel and can accummulate from all blocked channels. If P_L is the leakage light, then an N channel network can produce a crosstalk of

$$P_{CT} = (N - 1)P_L \qquad (8.7.1)$$

The effect can be analyzed similar to an additive input light noise field in the WDM system. The PE is again given by Eq. (8.4.9) now with $\varepsilon = (N - 1)\rho_L$, when ρ_L is the fractional light leakage. Clearly, the leakage will limit the number of channels in the network, depending on the accuracy degradation that can be tolerated. For $\varepsilon \leqslant 0.02$, N will be restricted to the value $N \gtrsim 1/50\rho_L$, and leakages less than $-30\,\text{dB}$ are needed to support a 20-channel shutter network.

PROBLEMS

8.1   A 1×2 splitter with excess loss of $7\,\text{dB}$ feeds into fibers having attenuation loss coefficient of $0.5\,\text{dB}$ per kilometer. How long must the output fiber be before the propagation loss exceeds the splitter loss?

8.2   Use Eq. (8.1.5) and plot a curve showing throughput loss L_T in decibels versus tap loss L_s in decibels. Assume zero excess loss.

8.3   A network uses identical directional couplers having a 10-dB tap ratio and a 1-dB excess loss. Neglect fiber loss. Determine the total transmission loss increase in going from 10 to 15 terminals for both a bus and a star network.

8.4   Use the integration identity

$$\int_{-\infty}^{\infty} \left(\frac{1}{1 + a^2 f^2} \right) \left(\frac{1}{1 + b^2 f^2} \right) df = \frac{\pi}{a} \left(\frac{1}{1 + (b/a)^2} \right)$$

to verify that Eqs. (8.4.5) and (8.4.6) follow from using Eqs. (8.4.3) and (8.4.4) in Eqs. (8.4.1) and (8.4.2).

8.5   Sketch a curve showing how the crosstalk relative power ratio P_{CT}/P_r decreases as the channel spacing parameter f_d/B_s is increased. Assume $B_f = 2B_s$ in Eq. (8.4.4).

8.6   Determine the accuracy to which a fiber delay line must be cut to insure that the time delay it produces is accurate to within one tenth of a 100-psec pulsewidth.

8.7   (a) Show how a frequency offset in the electronic clock ($f_e \rightarrow f_e + \Delta f$) in Figure 8.15 translates to an offset delay error in the optical clock.

(b) If timing errors are to be constrained to a fraction of a bit pulse, show how the result in (a) limits the channel bit rates in a TDM system.

8.8 Show that the OOK bit decoding PE for the interlaced TDM system in Figure 8.14 will not depend on the receiver clock pulse power, provided that the clock power is higher than the received data pulse power.

8.9 Consider the two (0, 1) pulse sequence codes below.

$$1\ 0\ 0\ 1\ 0\ 0\ 0\ 0\ 0\ 0\ 0\ 0\ 1\ 1$$

$$1\ 0\ 0\ 0\ 1\ 0\ 0\ 1\ 0\ 0\ 0\ 1\ 0\ 0$$

(a) Compute the cross correlation function of the pair by shifting one relative to the other, multiply, and sum.

(b) Prove that the maximum cross correlation value of one will occur if and only if the distance between ones in one code is not repeated in the other code.

(c) Show that if (b) is true, the number of ones in the correlation function is always equal to w^2, the code weight.

(d) Assume the ones in (c) are uniformly distributed over the correlation range $(0, l)$, and show that the probability of a two ones overlapping is w^2/l.

(e) Using (d), show that, with N such codes in the set, the probability that $N - 1$ codes will produce k overlapping ones is given by Eq. (8.6.9).

8.10 Consider a CDM code (similar to those in Problem 8.9) having w pulses, code length l, and a pulse in the first slot. A matched correlator is to be designed that produces a peak value at the end of the code word.

(a) Determine the delays that must be used in the parallel delay line correlator in terms of the pulse separations in the code.

(b) Show that this correlator, as the sequence slides through, adds only the code symbols in the past corresponding to the pulse positions of the matched code.

8.11 A general CDM code has length l.

(a) For an OOK CDM system, relate the time delay of the longest delay line that may have to be used in the matched correlator in terms of the bit rate of the channel.

(b) Determine the maximum length of the glass core fiber needed for a 30-Mbps rates.

(c) For 200 Mbps.

(d) For 1Gbps.

8.12 Write the expression for the SNR of a shot-noise-limited photodetector for the shutter system in Figure 8.17, with signal power P_r, leakage loss coefficient of ρ_L, and bandwidth B_o.

REFERENCES

1. J. Palais, *Fiberoptic Communications*, 2nd ed., Prentice Hall, Englewood Cliffs, NJ, 1988, Chapter 5.
2. G. Kaiser, *Optical Fiber Communications*, 2nd ed., McGraw Hill, New York, 1991, Chapter 2.
3. C. Brackett, Dense wavelength division multiplexing networks—principles and applications, *J. Selected Areas Commun.*, 8, August (1990).
4. P. Humblet and W. Hamby, Crosstalk analysis of single and double cavity Fabret-Perot filters, *J. Selected Areas Commun.*, 8, August 1990.
5. S. Mallinson, Wavelength selective filters for WDM using Fabret-Perot interferometers, *Appl. Opt.* 26, February 1987.
6. K. Cheung, Acousto-optical tunable filters in WDM networks, *J. Selected Areas Commun.* 8, August 1990.
7. J. Salehi, Code division multiple-access techniques in optical fiber networks—Part I: Fundamental principle, *IEEE Trans. Commun.*, 37(8), 824–833 (1989).
8. J. Salehi and C. Brackett, Code division multiple-access techniques in optical fiber networks—Part II: Systems performance analysis *IEEE Trans. Commun.*, 37(8), 834–842 (1989).
9. F. Chung and J. Salehi, "Optical orthogonal codes: Design, analysis and applications, *IEEE Trans. Info. Theory*, 35(3), 595–604 (1989).
10. H. Chung and P. V. Kumar, Optical orthogonal codes—new bounds and optimal construction, *IEEE Trans. Info. Theory*, July 1990.

9

THE ATMOSPHERIC OPTICAL CHANNEL

In our earlier discussion of the optical space communication system, it was assumed, for the most part, that operation occurred over the free-space channel. The source field due to the modulated laser was considered a plane wavefield having beamwidths and power levels set by the transmitter parameters. The received field was then considered a spatially coherent field over the beamfront area, with a power level reduced by the free-space loss over the propagation distance.

When the space link involves transmission through the atmosphere, additional effects may have to be considered. In particular, the atmosphere produces additional space losses as well as possible beam distortion. System designers for space–atmospheric links must therefore be aware of these effects to properly assign power budgets, perform link analyses, and produce efficient system designs. In this chapter, we summarize some of the recent studies on optical channels and show how these results can be used to modify earlier analysis.

9.1 THE ATMOSPHERIC CHANNEL

The atmosphere is composed of collections of gases, atoms, water vapor, pollutants, and other chemical particulates that are trapped by the Earth's gravity field; it extends to approximately 400 miles in altitude. The heaviest concentration of these particles is near Earth in the troposphere level (Fig. 9.1), with particle density decreasing with altitude up through the ionosphere. Actual particle distributions depend on the atmospheric conditions. The upper levels of the ionosphere contain ionized electrons that form radiation belts that surround the Earth. These atmospheric particles interact with all radiation fields that propagate through the radiation belts, with the primary effects being power losses and wavefront distortion.

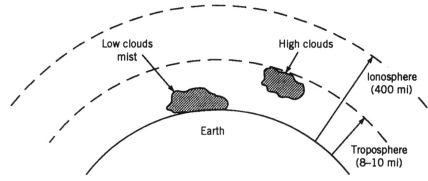

Figure 9.1. Earth's atmosphere.

Power losses and distortion are caused by the absorption and scattering of the radiation fields by the particulates. These effects become most severe as the radiation wavelengths begin to approach the cross-sectional dimension of the particulates. Because particle sizes in the atmosphere range from centimeters down to micrometers, the atmosphere is especially deleterious to optical transmissions.

The communication effect also depends on the type of link. Figure 9.2 shows two of the basic space link types. Figure 9.2a shows a vertical link, with either up-link or down-link transmission, characterizing Earth-to-space communications. The atmosphere therefore produces an integrated effect over the altitudes involved. In addition, since the atmosphere is relatively close to the Earth, long-range vertical space links may have different characteristics for the up-link than the down-link. That is, in one case the source is closer to the atmosphere, whereas in the other the receiver is closer. Figure 9.2b shows a horizontal link, which can sustain the severest effects if close to the Earth, where the highest particle densities exist. Horizontal links above the Earth's atmosphere act as free-space channels.

Atmospheric conditions can be roughly classified into three basic types: clear air, clouds, and rain. The clear-air channel is the most benign, character-

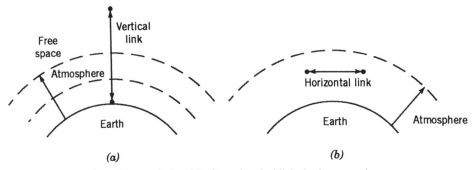

Figure 9.2. Vertical and horizontal optical links in the atmosphere.

ized by long-range visibility, clear weather, and relatively low attenuation. Clear air, however, can still contain eddies and temperature gradients (clear-air turbulence) that can produce changes in the index of refraction of impinging fields. These index changes act as optical lenses that refocus and reorient beam propagation. Cloudy atmospheric conditions can range from mist and fog to heavy cloud cover extending from near Earth to upper altitudes and is characterized with increasing water vapor accumulation and higher levels of attenuation. Rain represents the presence of water droplets of significant sizes, and can produce the most severe effects, depending on rainfall rate and rain cloud extent.

As field radiation impinges on an atmospheric particulate, a portion of its energy is absorbed, and the angle of the remainder is redirected. The particle absorption produces a field power loss, and the angle redistribution produces scattered energy. Surrounding particles further absorb and scatter the redistributed field, producing an aggregate field attenuation and scattering mechanism on the overall field as it propagates. It is precisely this scattering within a sunlit cloud, for example, that makes it appear uniformly illuminated when observed from below.

Figure 9.3 shows a plot of field transmittance observed at various optical wavelengths in propagating through the clear air atmosphere. The significant attenuation observed at selective wavelengths indicates field absorption from specific particles and shows the importance of properly selecting wavelengths in the high transmittive bands for space optics systems.

Various studies have attempted to model this mechanism [1]. The important parameters are the particle size (cross-sectional dimension relative to wavelength), and the particle density (volumetric concentration of particles) that represent the atmospheric conditions. Low densities of large particles tend

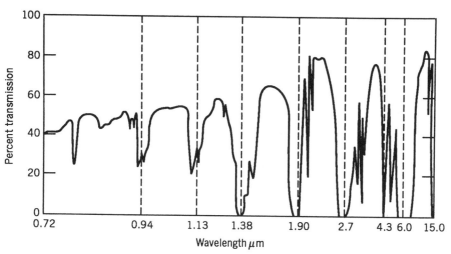

Figure 9.3. Transmissitivity of the atmosphere.

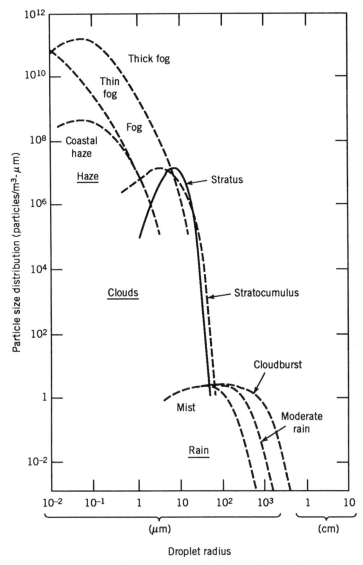

Figure 9.4. Cloud and precipitation drop size distribution. Reprinted from Karp et al. [1].

to obey single-scatter theory [2, 3], and produce primarily attenuation with mostly forward scattering at angles close to the normal. High densities of small particles tend to obey multiple-scatter theory [4, 5], which produces elemental beam scattering and random phase shifting across the wavefront as it propagates. This is referred to as beam breakup.

Figure 9.4 shows typical values for average particle size and corresponding particle densities for various cloud and rain conditions. We see that conditions vary from high-density small particles during mist and fog to low-density large-particle concentrations during heavy rain. Depending on the communication wavelength relative to the particle dimension and on the type of link, certain conditions may produce more severe effects than others. It should be emphasized that data as in Figure 9.4 represent average parameters, whereas the real atmosphere undergoes continual temporal changes in conditions from winds, thermal heating, Earth rotations, and so on. Hence, the atmosphere is actually a dynamic channel that continually changes its characteristics throughout a day, varying also from day to day and from season to season.

9.2 EFFECT OF THE ATMOSPHERE ON OPTICAL BEAMS

The discussion in Section 9.1 indicates that radiation propagation at optical wavelengths through the atmosphere are susceptible to additional power losses, beam spreading, and possible beam breakup, depending on the channel conditions. These conditions are also dependent on the beam itself, as shown in Figure 9.5. Figure 9.5a shows the free-space condition in which an optical source produces a transmitter beam of angle θ_b and a power P_a transmitted over a distance Z. As discussed in Section 1.4, the free-space beam generates a beamfront area at Z of $A_f = \pi(\theta_b Z)^2/4$, and a corresponding field intensity inside the beam of P_a/A_f watts/area. A receiver area in the beam collects this field intensity.

Figure 9.5b shows beam propagation in a clear-air channel, exhibiting possible eddies and temperature gradients as field turbulence. As long as the beamfront area is smaller than the turbulence dimension (shown as a flat plane) the field beam is attenuated by the clear-air transmittance but is undistorted, except for possible beam redirection. This causes beam drift and beam defocusing, at the receiver plane, similar to mispointing. The beamfront intensity for uniform beams is now $P_a L_a/A_f$, where L_a is the wavelength-dependent clear-air transmittance loss and A_f is the wavefront area including spreading. For shaped beams (say Gaussian beams), this beam drift may cause receiver operation at beam edges, producing further power losses even though the transmitter is accurately pointed. As the turbulence layer changes (slowly moving up and down and perhaps tilting) the beam drift causes the received beam pointing to wander over the receiver plane producing time varying power fades. In a long range uplink, the receiver is far from the turbulence, and a point source will appear to have a slightly spread (defocused) beam at the receiver, with power density fluctuations. On a long range downlink, the transmitted beam is wider when it reaches the atmosphere, and different points on the beam front may observe different turbulence conditions. This can cause both power fluctuations and beam breakup.

Figure 9.5c shows the situation for a diffracted beam passing through the clouds or rain. The channel appears as a dense collection of water vapor particles that cause beam absorption and multiple scattering. The beam is

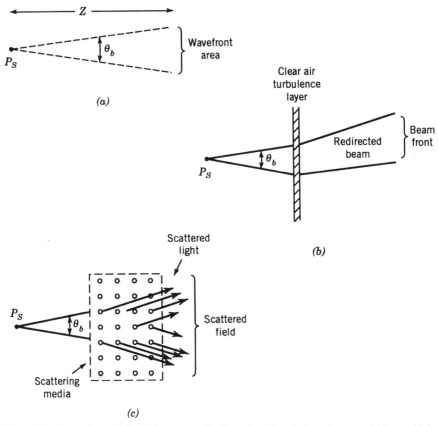

Figure 9.5. Channel models in (*a*) free space, (*b*) clear air with turbulence layer, and (*c*) a multiple scattering medium.

highly attenuated, and the particle distribution can cause localized beam phase shifts across the beamfront as it propagates through. A collimated beam will broaden due to the multiple scattering, decreasing the power density at the receiver. In addition the source will no longer appear as a point, and scattered radiation will arrive at the receiver from many angles (modes) giving the appearance of a spread source. (This is often called the "shower glass" effect.) As the receiver field of view is widened, more of the scattered light can be collected.

Power loss in atmospheric propagation over a distance Z is accounted for by writing the additional atmospheric power loss L_a (over that which would occur in free space) as

$$L_a = e^{-\alpha_e Z} \tag{9.2.1}$$

where α_e is the extinction loss coefficient of the channel in reciprocal distance units. The extinction coefficient is composed of the two parts

$$\alpha_e = \alpha_a + \alpha_{sc} \tag{9.2.2}$$

where α_a and α_{sc} are the absorption and scattering coefficients, respectively. Note that this factors the extinction loss L_a into separate loss terms due to both absorption and scattering. In some channels, α_e may be due entirely to absorption (which is always present), and in this case, $\alpha_e = \alpha_a$. Other channels may contain additional loss from scattering, which requires the addition of α_{sc}.

It is common to define

$$z_e = 1/\alpha_e \tag{9.2.3}$$

as the extinction length of the channel. Propagation over one extinction length causes an additional loss of $e^{-1} = 1/2.7 = -4.4\,\text{dB}$. this allows us to describe propagation distances in terms of multiples of z_e. That is, distance Z would contain Z/z_e extinction lengths, and have a total atmospheric loss of $(Z/z_e)(4.4)$ decibels.

Figure 9.6 shows some representative extinction lengths, as a function of wavelength, for several cloud conditions. Table 9.1 lists some specific cloud extinction lengths, and corresponding average cloud thickness, measured in extinction lengths. Clearly, the total extinction loss in a specific communication

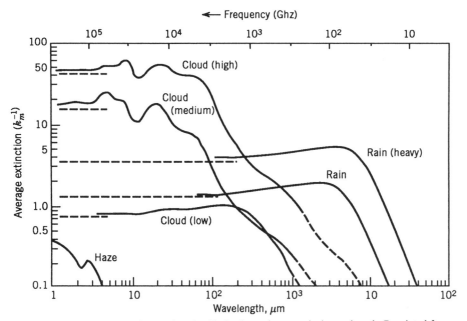

Figure 9.6. Extinction coefficients for cloud/rain channels vs. optical wavelength. Reprinted from Karp et al. [1].

TABLE 9.1 Some Typical Cloud Parameters

Type of Cloud	Average Altitude ($\times 1000$ ft)	Layer	Extinction Length (m)	Average Thickness (ext. lengths)
Cumulonimbus	2.5–18	middle	25	120
Nimbootratus	6.5–10	middle	55	51
Altocumulus	15–20	middle	65	18
Altostratus	14–22	middle	45	30
Stratocumulus	2.5–3.0	low	20	27
Stratus	2–5	low	15	62
Cumulus	2–5	low	50	10
Cirrosfratus	30–35	high	350	5
Cirrocumulus	22–26	high	350	5
Cirrus	20–30	high	350	3

link will depend on wavelength, propagation distance into the cloud, and altitude (type of cloud encountered).

Beam breakup from extensive turbulence or scattering produces a randomization of the field phase across the beamfront. This effect can be described by the field mutual coherence function defined in Section 1.7. Let $f(t, r)$ be the complex random source field at time t and point $\mathbf{r}$ in the wavefront after propagating a distance Z. Assume the field can be described by a separable, stationary, and homogenious coherence function

$$R_f(t_1, t_2, \mathbf{r}_1, \mathbf{r}_2) = R_t(\tau) R_s(\rho) \tag{9.2.4}$$

Here $R_t(\tau)$ is the temporal correlation function in terms of $\tau = t_2 - t_1$ and $R_s(\rho)$ is the spatial coherence function in terms of the scalar distance $\rho = |\mathbf{r}_1 - \mathbf{r}_2|$ between points $\mathbf{r}_1$ and $\mathbf{r}_2$. The parameter $R_t(0)$ is the field intensity at Z given by

$$R_t(0) = \bar{I} e^{-\alpha_a Z} \tag{9.2.5}$$

where $\bar{I}$ is the intensity for free space and the exponential is the absorption loss in Eq. (9.2.1).

Various studies have been used to derive models for the spatial coherence function for scattering channels [1]. For the clear-air turbulent channel, the theory of Rytov [2, 3] generates the approximate form

$$R_s(\rho) = e^{-3.44(\rho/r_o)^{5/3}} \tag{9.2.6}$$

where r_o is called the coherence length in the transverse plane at distance Z. The coherence length is roughly a measure of the spatial distance about a point in the wavefront plane over which surrounding points are spatially coherent (phase aligned). The larger the coherence length, the more the wavefront appears as a plane wavefield, as in free space. The shorter the coherence length, the more the channel has broken up the plane wave phase.

Equation (9.2.6) indicates that the field coherence falls off exponentially with a spatial width of about r_o. For a receiving aperture with dimension much smaller than r_o, the wavefront appears coherent over the aperture, and the only degradation is field attenuation from extinction caused by absorption and scattering. If the aperture is many times the coherence length, the wavefront no longer appears phase coherent over the entire aperture, and points separated by distances exceeding several times r_o are now phase incoherent. The aperture is effectively collecting a spatially random field, composed of elemental coherent cells. This can be directly related to the concept of random-field modes discussed in Section 1.6.

Values for r_o depend on the channel conditions, the wavelength, and type of link. Figure 9.7 shows a published [6–8] nomogram for estimating r_o for the clear-air turbulent channel for the vertical and horizontal links. For a vertical link of given height and wavelength, r_o is shown for both daytime and nighttime operation. Figure 9.7b shows the corresponding values of r_o for horizontal links of a given range and wavelength, in terms of the refractive index parameter C_n^2 of the atmosphere. The latter depends on the altitude of the horizontal link, as shown in Figure 9.8, and is again related to the particle density of the channel.

The model for $R_s(\rho)$ in Eq. (9.2.6) is strictly not valid for multiple-scattering channels (clouds and rain). A more accurate small angle approximation [8] for these channels is obtained from

$$R_t(0)R_s(\rho) = \bar{I}e^{-\alpha_a Z}e^{-(\rho/r_o)^2} \qquad \rho \ll r_o$$

$$= \bar{I}e^{-(\alpha_a+\alpha_{sc})Z} \qquad \rho \gg r_o \tag{9.2.7}$$

where r_o is now given by

$$r_o = \frac{\lambda}{(\overline{\theta^2})^{1/2}\sqrt{\alpha_{sc}Z}} \tag{9.2.8}$$

Here $(\overline{\theta^2})^{1/2}$ is the root-mean-square forward scattering angle of the medium. The forward scatter angle is the angle from the propagation direction of the inserted field caused by particle redirection in the atmosphere. This angle is random, with a specified angular distribution, and Eq. (9.2.8) uses its root-mean-square value. The root-mean-square angle is a function of wavelength and particle density and typically has values in the range of 0.4 to 1.0 rad. Note that Eq. (9.2.8) indicates an r_o that decreases with propagation distance and is almost always less than a wavelength. Equation (9.2.7) indicates that for propagation distances much less than a scattering length ($Z \ll 1/\alpha_{sc}$) the field is still approximately coherent, with only absorption loss. After many scatter lengths, the field coherence is reduced, and the coherence function actually falls

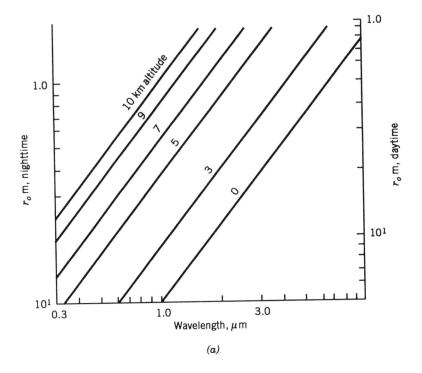

(a)

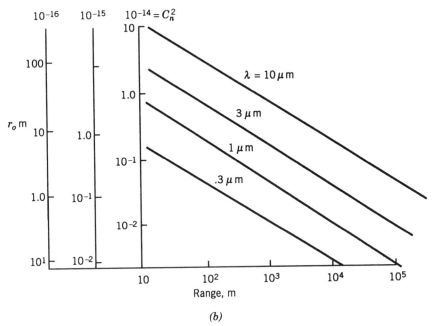

(b)

Figure 9.7. Coherence distance r_o. (a) Vertical link. (b) Horizontal link. $C_n^2 =$ refractor index parameter.

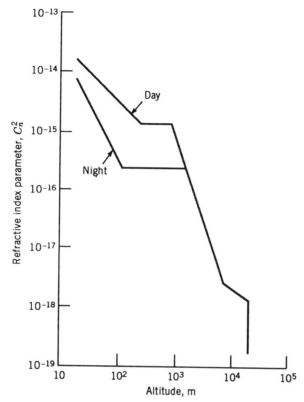

Figure 9.8. Refractive index parameter C_n^2 versus altitude, daytime and nighttime. Reprinted from Karp et al. [1].

off slightly faster in the multiple-scatter media than the clear-air channel. At small coherence distances, $(\rho \ll r_o)$, the field still appears coherent with only the absorption loss, but at large distances $(\rho \gg r_o)$ the field coherence has been reduced to a minimal limiting value. This minimal value can be interpreted as a residual plane wave that has propagated through the media but has been reduced by the total extinction (absorption plus scattering) loss of the pathlength.

9.3   EFFECT OF ATMOSPHERE ON DIRECT DETECTION RECEIVERS

When a direct detection optical link is operated over the atmospheric channel, the communication analysis in Chapter 4 must be modified to take into account the channel effects. Since the received signal field can now be random from the scattering, the collected signal power at the receiver aperture will evolve as a random variable. If the received random signal field is denoted $f(t, \mathbf{r})$, with field intensity $I(t, \mathbf{r}) = |f(t, \mathbf{r})|^2$, then the average received signal power over an aperture area A is

$$\bar{P}_r = \int_A \bar{I}(t, \mathbf{r}) \, d\mathbf{r} \tag{9.3.1}$$

where the overbar denotes averaging. This average power can be written in terms of the received-field coherence function, because

$$\bar{I}(t, \mathbf{r}) = \overline{|f(t, \mathbf{r})|^2} = R_f(t, t, \mathbf{r}, \mathbf{r}) \tag{9.3.2}$$

For the class of stationary and homogenious coherence functions given in Eq. (9.2.4) for the scatter channel,

$$\bar{I}(t, \mathbf{r}) = R_t(0) R_s(0) = \bar{I} e^{-\alpha_e Z}$$

and

$$\bar{P}_r = (\bar{I} A) e^{-\alpha_e Z} \tag{9.3.3}$$

Thus the average power is the free space average power $(\bar{I} A)$, reduced by the additional loss due to the extinction loss of the channel.

An indication of the variation of this power around its mean value is given by the power variance. The mean-square power value is

$$\overline{P_r^2} = \overline{\left(\int_A I(t, \mathbf{r}) \, d\mathbf{r} \right)^2}$$

$$= \int_A \int_A \overline{I(t, \mathbf{r}_1) I(t, \mathbf{r}_2)} \, d\mathbf{r}_1 \, d\mathbf{r}_2 \tag{9.3.4}$$

$$= \int_A \int_A R_I(t, t, \mathbf{r}_1, \mathbf{r}_2) \, d\mathbf{r}_1 \, d\mathbf{r}_2$$

where

$$R_I(t, t, \mathbf{r}_1, \mathbf{r}_2) = \overline{I(t, \mathbf{r}_1) I(t, \mathbf{r}_2)} \tag{9.3.5}$$

is the spatial coherence function of the field intensity. The variance of the power would follow as

$$\mathrm{var}(P_r) = \overline{P_r^2} - (\bar{P}_r)^2 \tag{9.3.6}$$

The above requires knowledge of the intensity coherence. When the signal fields are Gaussian and homogenious, the intensity coherence function is obtained directly from the field coherence function (recall Problem 4.4) as

$$R_I(t, t, \mathbf{r}_1, \mathbf{r}_2) = R_t^2(0) R_s^2(0) + 2R_t^2(0) R_s^2(\rho) \qquad (9.3.7)$$

For the atmospheric channel modeled by Eq. (9.2.7)

$$R_I(t, t, \mathbf{r}_1, \mathbf{r}_2) = (\bar{I}e^{-\alpha_a Z})^2 + 2(\bar{I}e^{-\alpha_a Z})^2 e^{-2(\rho/r_o)^2} \qquad (9.3.8)$$

and

$$\mathrm{var}(P_r) = 2 \int_A \int_A (\bar{I}e^{-\alpha_a Z})^2 e^{-2(\rho/r_o)^2} \, d\mathbf{r}_1 \, d\mathbf{r}_2 \qquad (9.3.9)$$

A measure of the relative magnitude of the signal power variance is obtained by normalizing with respect to the squared mean signal power. Thus we define

$$(\Delta P_r) = \frac{\mathrm{var}(P_r)}{(\bar{P}_r)^2} \qquad (9.3.10)$$

The integral in Eq. (9.3.9) can be evaluated by inserting a variable change $\boldsymbol{\rho} = \mathbf{r}_1 - \mathbf{r}_2$ and rewriting as

$$\int_A \int_A e^{-2(\rho/r_o)^2} \, d\mathbf{r}_1 \, d\mathbf{r}_2 = \int_A e^{-2(\rho/r_o)^2} H(\boldsymbol{\rho}) \, d\boldsymbol{\rho} \qquad (9.3.11)$$

where $H(\boldsymbol{\rho})$ is the aperture optical transfer function (OTF)

$$H(\boldsymbol{\rho}) = \int_A w(\mathbf{r}_1) w(\mathbf{r}_1 + \boldsymbol{\rho}) \, d\mathbf{r}_1 \qquad (9.3.12)$$

associated with the window aperture function $w(\mathbf{r}) = 1$ for $\mathbf{r} \in A$. For the circular aperture of diameter d, this integrates to

$$H(\boldsymbol{\rho}) = 2d^2 \left[\cos^{-1}\left(\frac{\rho}{d}\right) - \left(\frac{\rho}{d}\right)\sqrt{1 - \left(\frac{\rho}{d}\right)^2} \right] \qquad \frac{\rho}{d} \leq 1$$

$$= 0 \qquad\qquad\qquad\qquad\qquad\qquad\qquad \frac{\rho}{d} > 1 \qquad (9.3.13)$$

where $\rho = |\boldsymbol{\rho}|$, and Eq. (9.3.10) becomes

$$\Delta P_r = \frac{2}{d^2} \int_0^1 e^{-(ud/2r_o)^2} [\cos^{-1}(u) - u\sqrt{1 - u^2}] u \, du \qquad (9.3.14)$$

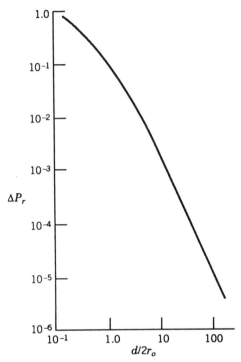

Figure 9.9. $\Delta P_r = \text{var } P/(\bar{P})^2$ versus $d/2r_o$, where d is the aperture diameter and r_o is the coherence radius.

The above is directly integrable and is plotted in Figure 9.9 as a function of the normalized aperture diameter $d/2r_o$. Note that the normalized variance rapidly decreases ($\ll 1$) as the receiver aperture is increased beyond the coherence length of the channel. Increasing d beyond $2r_o$ increases the number of coherence cells within the aperture, and the power randomness of each cell is effectively averaged out as more cells are collected. This effect is referred to as *aperture averaging* [9] and is a prime reason for increasing receiver apertures in direct detection systems with random signal fields. Larger apertures exhibit less power scintillation after propagating through scatter channels.

After photodetection, the direct detection SNR can be computed, taking into account the fluctuations of the signal power, as well as the shot noise and background noise terms. Because the power variation around the mean signal power appears as an additive "noise" the power variance must be considered a contributor to the detector noise. This produces the modified direct detection shot-noise-limited SNR in a bandwidth B_c of

$$\text{SNR} = \frac{(\alpha \bar{P}_r)^2}{\alpha(P_r + P_b)F2B_c + \alpha^2 \text{var}(P_r)} \tag{9.3.15}$$

Note that the atmospheric channel has reduced the received average signal power due to the extinction loss and inserted the power variance. When quantum-limited performance is approached ($P_b = 0$) the SNR becomes

$$\text{SNR} = \frac{1}{\left(\dfrac{\alpha \bar{P}_r}{F2B_c}\right)^{-1} + \Delta P_r} \tag{9.3.16}$$

where ΔP_r is the parameter in Eq. (9.3.14). Thus, the atmospheric scatter channel always reduces the quantum-limited SNR, but its value improves as more aperture averaging is used. In the limit, it is only the extinction loss of the channel that effects the quantum-limited SNR. It must be remembered, however, that opening the aperture in shot-noise-limited operation may increase the background noise (P_b term), effectively diluting the SNR gain from power averaging.

9.4 HETERODYNING OVER THE ATMOSPHERIC CHANNEL

When using heterodyning (coherent) detection over the atmospheric channel, the effect of beamfront coherence must be considered. Recall from Section 5.6 that heterodyning with random signal fields requires a modified calculation of the heterodyned signal power. This was given in Eq. (5.6.7) as

$$P_s = R_t(0)A_r \tag{9.4.1}$$

where $R_t(0)$ is the averaged received signal field intensity and A_r is the effective aperture area over which the received heterodyned power is collected. From Eq. (5.6.8),

$$A_r = \frac{1}{A} \int_A \int_A R_s(\mathbf{r}_1, \mathbf{r}_2)\, d\mathbf{r}_1 d\mathbf{r}_2 \tag{9.4.2}$$

where $R_s(\mathbf{r}_1, \mathbf{r}_2)$ is the spatial coherence function of the received signal field. In Figure 5.8, some limiting forms for A_r were given based on general forms for the spatial function $R_s(\rho)$. For the atmospheric scatter channel, we can use Eq. (9.2.7) to write

$$R_s(\mathbf{r}_1, \mathbf{r}_2) = e^{-(\rho/r_0)^2} \tag{9.4.3}$$

with again $\rho = |\mathbf{r}_1 - \mathbf{r}_2|$. The effective area A_r becomes

$$A_r = \frac{1}{A} \int_A \int_A e^{-(\rho/r_0)^2}\, d\mathbf{r}_1 d\mathbf{r}_2 \tag{9.4.4}$$

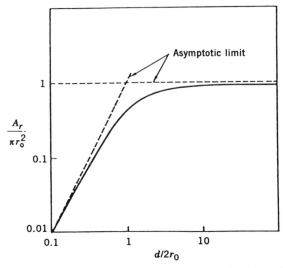

Figure 9.10. Effective signal collecting area A_r vs. normalized receiver diameter d. r_o is the coherence radius.

The integral is similar to the intensity integration in Eq. (9.3.11). Following the steps in Eq. (9.3.14) for the circular aperture, the integral can again be computed to generate the result in Figure 9.10. The results shows the dependence of the equivalent power collection area of the receiver in terms of the normalized ratio $d/2r_o$. For $d < 2r_o$, the area increases with d, and more power is collected. When $d > 2r_o$, the coherence diameter of the received field, the curves saturate, and the effective receiver area is limited to the coherence area. The atmospheric scattering breaks up the optical wavefront to such an extent that increasing the aperture diameter beyond $2r_o$ results in the collection of negligible additional coherent power. Thus, useful receiver heterodyning in space links is limited to the coherence area of the signal field and using an optical aperture larger than this area only adds additional background noise during heterodyning.

9.5   ATMOSPHERIC PULSE SPREADING

Besides the power loss and beam breakup, the atmosphere may also distort the optical waveshape during propagation. This is particularly true for high bandwidth signals such as narrow optical pulses, in which the atmospheric scattering can cause a multipath effect similar to the dispersion in fibers. Scattered pulse fields may be reflected toward the receiver and combine to produce a distorted optical pulse shape. This can be envisioned by the simplified diagram in Figure 9.11, showing a direct plane wave ray line and several scattered paths. The scattered paths are redirected by the particle

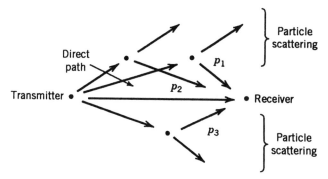

Figure 9.11. Direct and scattered paths with multiple scattering media.

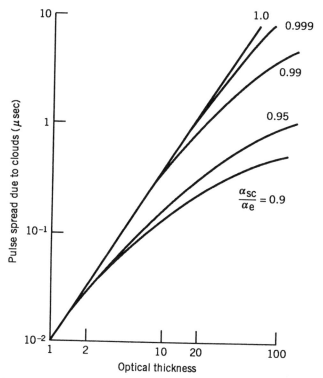

Figure 9.12. Pulse spreading caused by clouds. α_{sc} = scattering coefficient. α_e = extinction coemcient.

scattering, but some (paths p_1, p_2, p_3) are eventually scattered back to the collecting receiver. If an optical pulse is transmitted from the source, the pulse signals along the scattered paths arrive with delays relative to the direct path and combine to yield a wider, broadened optical field pulse from that transmitted. This channel dispersion effect is similar to that in fibers.

Experimental studies [10,11] have measured the pulse spreading after turbulent or cloud transmission. The amount of spreading has been shown to depend primarily on the optical thickness of the transmission path and on the amount of absorption in the medium. The more the scattering, the larger the expected path delays, and the more the pulse broadening. However, the higher the absorption coefficients in the medium, the more the path attenuation of the scattered propagation, making them less significant in the pulse combining. This absorption limits pulse delays to only those paths at relatively narrow scattering angles from the direct path.

These results are shown in Figure 9.12, based on measured cloud propagation, which plots pulse width response to delta function pulse propagation, after transmission over a given path thickness, with various absorption coefficients. It is evident that pulse spreading increases directly with optical thickness but is reduced with the higher absorption levels.

PROBLEMS

9.1 A channel has an absorption coefficient of $0.1\,\mathrm{km}^{-1}$ and a scattering coefficient of $0.05\,\mathrm{km}^{-1}$.

(a) What is the loss in decibels over a 10-km path?

(b) What is the extinction length of the channel?

(c) How many extinction lengths is the channel in (a)?

9.2 An optical rain channel at $\lambda = 1\,\mu\mathrm{m}$ has an root-mean-square scatter angle of $20°$. Plot the value of r_o in Eq. (9.2.8) as a function of channel scatter extinction length.

9.3 The size of scattering homogeneities in the atmosphere is often modeled as

$$x = (10^{-9}\,h)^{1/3} \quad 10\,\mathrm{m} < h < 1\,\mathrm{km}$$

$$= 1\,\mathrm{cm} \quad 1\,\mathrm{km} < h < 15\,\mathrm{km}$$

$$= 0 \quad \text{elsewhere}$$

where h is the height above the Earth.

(a) If an optical beamwidth of $10^{-5}\,\mathrm{rad}$ is transmitted from the Earth, determine the height at which the beam cross section is approximately equal to the size of the inhomogeneities.

(b) Determine the ratio of inhomogeneity size to beam width as a function of h for a beam with beam angle $10^{-5}\,\mathrm{rad}$ transmitted to the Earth from 400 miles above the Earth.

9.4 Beam scattering is often described by a billiard-ball model. The beam, considered as a propagating ray, undergoes a series of collisions with amplitude being multiplied by the gain constant $\exp(-a_i)$ at the ith collision, where a_i is a random variable.

    (a) Show, by the central limit theorem, that if the number of collisions is large, the resulting beam amplitude will have a log-normal probability density. [If $y = e^x$, then y is log-normal if x is Gaussian.]

    (b) Write the probability density of the log-normal amplitude in terms of the mean and variance of the sum of the a_i.

9.5 A 10-μm optical field is transmitted through a C-1 cloud channel (Figure 9.6).

    (a) What is the extinction length of the channel?

    (b) Repeat for a C-6 cloud.

9.6 **(a)** Use Table 9.1 to estimate the transmission loss in dB in propagating through a Cumulus cloud channel.

    (b) Repeat for a Cirrus cloud.

9.7 A horizontal optical atmospheric link is at an altitude of 6 miles. Determine the coherence distance r_o for both daytime and nightime operation at $\lambda = 1\ \mu$m.

9.8 A narrow optical pulse is transmitted over a channel of 20 scattering lengths. What is the estimated pulse width at the channel output if the absorption is 10 percent of the extinction?

REFERENCES

1. S. Karp, R. Gagliardi, S. Moran, and L. Stotts, *Optical Channels*, Plenum Press, New York, 1988.

2. A. Ishimaru, *Wave Propagation and Scattering in Random Media*, Vols. 1 and 2, Academic Press, New York, 1978.

3. V. Tartarski, *Wave Propagation in Turbulent Media*, McGraw Hill, New York, 1961.

4. H. van de Hulst, *Light Scattering by Small Particles*, Wiley, New York, 1957.

5. M. Kerker, *Scattering of Light and Electromagnetic Radiation*, Academic Press, New York, 1969.

6. D. Fried, "Optical heterodyne detection of distorted wavefronts", *Proc. IEEE*, 55, April 1967.

7. R. Fante, *Propagation of electromagnetic waves through turbulent plasmas"*, *IEEE Trans. Antennas Propagation*, May 1973.

8. R. Lutomirski and H. Yuma, "Propagation of finite optical beams in inhomogenious media", *Appl. Opt.*, 10, July 1971.

9. D. Fried "Aperture averaging of scintillation" *J. Opt. Soc. Am.*, 57, November 1967.

10. E. Butcher and R. Lerner, "Experiments in light pulses through clouds", *Appl. Opt.*, 12, Decemer 1973.

11. A. Ishimaru, "Diffusion of pulses in densely distributed scatters", *J. Opt. Soc. Am.*, 68, June 1978.

10

POINTING, ACQUISITION, AND TRACKING IN SPACE OPTICS

Before any data transmission can occur in a space communication system, it is necessary that the transmitter field power actually reach the receiver detector. This means that the transmitted field, in addition to having to overcome the effects of the propagation path, must also be properly aimed toward the receiver. Likewise, the receiver detector must be aligned with the angle of arrival of the transmitted field. The operation of aiming a transmitter in the proper direction is referred to as *pointing*. The receiver operation of determining the direction of arrival of an impinging beam is called *spatial acquisition*. The subsequent operation of maintaining the pointing and acquisition throughout the communication time period is called *spatial tracking*.

The problems of pointing, acquisition, and tracking become particularly acute when dealing with fields having narrow beamwidths and long propagation distances. Because both these properties characterize long-range optical space systems, such as intersatellite links and Earth–space links, these operations become an important aspect of the overall communication design problem. In this chapter, we discuss pointing, acquisition, and tracking in optical space links and present several procedures for system implementation. The effect of these subsystems on communication performance will also be considered.

10.1 THE OPTICAL POINTING PROBLEM

Recall that in our discussion of optical apertures and beamwidths in Chapter 1, it was shown that a typical optical beam in a space link could be confined to an angular beamwidth of less than 1 arcsecond. If this beam is to be detected at a receiver, then this beam must be pointed to within a fraction of this beamwidth. Alternatively, if the beam can be aimed toward a desired receiver (considered as a point) with an accuracy of only, say $\pm \psi_e$ radians, then the beamwidth must be at least $2\psi_e$ to ensure receiver reception, as shown in

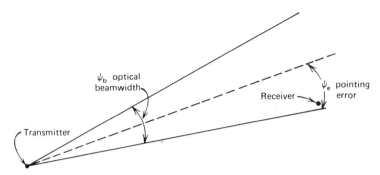

Figure 10.1. Beamwidths and pointing error in transmitter receiver systems.

Figure 10.1. To emphasize this result numerically, suppose this beam is from a satellite aimed at the Earth from a 22,000-mile altitude. A 50-μrad optical beam from this altitude will illuminate a distance on the Earth of $(50 \times 10^{-6})(22,000) \approx 1$ mile. That means the center of the satellite beam must be pointed to within a half mile of the Earth receiver. Contrast this to an RF satellite antenna having a beamwidth of approximately $10°$ and an Earth coverage of approximately 4000 miles. The RF pointing need be only within 2000 miles, a sizeable reduction in required accuracy.

The loss due to mispointing depends on the actual shape of the optical beamfront. If we assume a Gaussian-shaped transmitted beam, having a $1/e$ beamwidth of ψ_b radians, then the receiver power loss from a transmitter pointing error ψ_e is

$$P_r = \frac{C}{\psi_b^2} e^{-(2\psi_e/\psi_b)^2} \tag{10.1.1}$$

where C is a coefficient depending on the transmitter power, receiver area, and propagation distance (recall Eqs. 1.2.6 and 1.4.6). Note a significant power loss can occur when the pointing error exceeds the transmitter beamwidth. Increasing the beamwidth ψ_b to compensate for pointing errors reduces the exponential loss term in Eq. (10.1.1), but causes a reduction of the received power ($1/\psi_b^2$ term) from the reduced antenna gain caused by spreading the source power over a wider beamwidth.

Inaccuracies in pointing an optical beam in a specified direction over long distances are produced by several basic causes. The first major cause is the inability to exactly determine the desired direction. Uncertainty in line-of-sight direction is caused by reference frame errors, and pointing can only be established to within the accuracy that a fundamental coordinate system can be established. Usually, coordinate systems are oriented relative to a known star or celestial body, in which case it is important that compensation be made for reference motion. Generally this motion is not known precisely. In addition to real reference axis movement, often there is an apparent motion (e.g., the parallax motion of a star, due to displacement of the Earth from one side of its orbit to the other). Errors in frame reference translate directly to line-of-sight errors in pointing.

A second major cause of pointing errors is the error in the actual pointing apparatus. Often the telescopes or lenses are pointed by means of electronic or mechanical interconnections operated from a remote sensor. Errors in this mechanism from stress, noise, structure fabrication, and so on, will cause the beam to be pointed inaccurately. Such errors are called *boresight* errors.

A third error source is the inability to compensate exactly for transmitter and/or receiver motion. This may occur if either is in motion, or may be due simply to the rotation of the Earth, or hovering motion of a stationary satellite. When we attempt to predict this motion with system dynamical equations, coefficient errors lead directly to pointing errors. In addition, actual measurements of position and velocity, made through phase and Doppler frequency tracking, inherently contain errors due to system noise.

The Earth's atmosphere is a fourth cause of pointing errors when dealing with space links involving atmospheric propagation. Besides the attenuation (absorption) and scattering, pointing is drastically affected by the beam wander and beam spreading of turbulence, clouds, and thermal gradients. Beam wander is the bending of the light beam from its intended path during propagation and obviously directly affects pointing. Beam spreading causes increased beam divergence, leading to dilution of available power, and, therefore, has the effect of a pointing error. Because the atmosphere extends upward only a few hundred miles, it appears as a near field effect to an Earth-based transmitter, while appearing as a far field effect to a deep-space satellite transmitting to Earth. Therefore beam wander and divergence are more severe for Earth transmission than for Earth reception. Even slight angular shifts induced by a clear atmosphere on Earth-transmitted optical beams project to large error distances after propagating over a long path to a deep-space satellite. Up-link beam wander angles may typically vary from ± 1 to $\pm 15 \, \mu$rad, but may be as large as $50 \, \mu$rad for severe temperature gradients. Down-link beams passing through the atmosphere exhibit relatively little beam angular variations, and the predominate atmospheric effect is power absorption. The absence of atmosphere (e.g., a satellite-to-satellite link) greatly reduces the pointing inaccuracies.

Transmitter pointing may be further hindered if the transmission distances are such that the propagation transit time is significant and relative motion is involved. In this case, the transmitter must actually point the optical beam ahead of the receiver to allow reception. That is, the transmitter must allow for the additional motion that occurs during the beam transit time and point to the projected point. This pointing procedure is called *point ahead*. Point ahead is particularly important when Earth-based transmitters are communicating with space vehicles in orbit. The point-ahead operation is accomplished by directing the transmitted beam toward the point in space where the receiver will be when the transmitted beam arrives. This requires transmitting at a lead

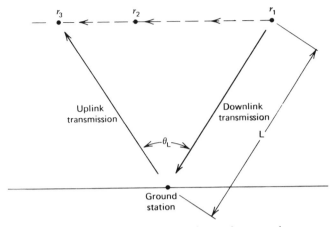

Figure 10.2. Point-ahead angles for moving transmitter.

angle with respect to the present line-of-sight position vector. Consider the geometry in Figure 10.2, showing relative motion between an Earth-based and satellite link. The satellite transmits at point $\mathbf{r}_1$ and, when received on Earth, defines a vector to the satellite position when it was transmitted. However, at time of reception, the satellite has moved to point $\mathbf{r}_2$. The Earth station in retransmitting, must compensate for the motion $\mathbf{r}_1$ to $\mathbf{r}_2$ and for the additonal motion to point $\mathbf{r}_3$, where the satellite will be when it receives. The angle from the Earth reception vector to the transmission vector defines the lead angle. Atmospheric effects, if present, cause the transmitter and satellite beams to distort and bend. Unfortunately these effects are not mutually compensating, since one represents a near-field and one a far-field effect. Both must be properly compensated for if the point ahead angle is to be exactly determined. Furthermore since the bending phenomena are generally time dependent, their effects may have to be continually updated.

A simplified expression for the point-ahead angle can be derived if it is assumed that the angle is small. This is generally true as long as the satellite motion is slow with respect to the speed of light. Let Z be the distance from the transmitter to the satellite and let τ be the round-trip transit time. Then $\tau = 2Z/c$, where c is the speed of light. The distance the satellite moves along its orbit is τv, where v is the satellite velocity. For the small angle assumption, the necessary lead angle θ_L in radians is approximately

$$\theta_L \cong \frac{2v}{c} \tag{10.1.2}$$

This angle must be adjusted for up- and down-link beam bending, if possible. Uncertainty in pointing effects due to other causes must be overcome by increasing beamwidth.

10.2   SPATIAL ACQUISITION

Spatial acquisition requires aiming the receiving lens in the direction of the arriving optical field. That is, it must align the normal vector to the aperture area with the arrival angle of the beam. Often alignment is acceptable to within some degree of accuracy. That is, the arrival angle can be within a specified solid angle from the normal vector. This acceptable angle is called the resolution angle (or resolution beamwidth) of the acquisition procedure, and is denoted Ω_r in subsequent discussion. The minimal resolution angle is obviously the diffraction-limited field of view, but, in practical design, desired resolution is generally larger. This allows for the possibility of the source blurring into many modes, and for compensating for pointing errors and ambiguities. Although resolution angle must be considered a design specification, its value plays an important role in subsequent analysis.

Acquisition can be divided into one-way and two-way procedures. One-way acquisition is shown in Figure 10.3a. A single transmitter, located at one point, is to transmit to a single receiver located at another point. If satisfactory pointing has been achieved (or equivalently, if the transmitter beamwidth covers the pointing errors), the optical beam will illuminate the receiver point. The receiver knows the transmitter direction to within some uncertainty solid angle Ω_u, defined from the receiver location. The receiver would like to aim its antenna normal to the direction of the arriving field to within some preassigned resolution solid angle Ω_r; that is, it wants its antenna normal vector pointed to within Ω_r steradians of the transmitter line-of-sight vector. In general, $\Omega_r \ll \Omega_u$, so that the receiver must perform an acquisition search over the uncertainty angle to acquire the transmitter with the desired resolution.

In two-way acquisition, both communicating stations contain both a transmitter and a receiver, as shown in Figure 10.3b. Both must spatially acquire a two-way communication link. In typical situations, one of the stations has somewhat accurate knowledge of the location of the other and can therefore transmit a beam wide enough to cover its pointing errors. It uses a receiving antenna with a similar field of view aimed along the line of sight of the transmitted beam. The second station may not have the a priori knowledge for pointing and must therefore search its uncertainty field of view Ω_{u_2} to acquire. After successful spatial acquisition with resolution Ω_{r_2} the second station transmits with beamwidth Ω_{r_2} to the first station, using the arrival direction obtained from the acquisition. The second station has now acquired and is pointed properly. The first station can now acquire the return beam with its desired resolution Ω_{r_1}. The link is now complete with the desired resolutions, and communication can begin. The operation can be repeated with narrower beams for further refinement, if desired. The first station would now narrow its transmit-and-receive beam, and the second station would reduce its resolution requirement Ω_{r_2}, reacquire, and retransmit with a narrower beam.

It is evident that the key operation in either one- or two-way spatial acquisition is the search over the uncertainty angle Ω_u to determine arrival

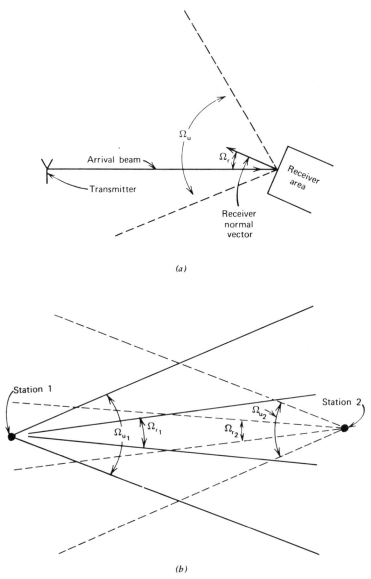

Figure 10.3. (*a*) One-way acquisition geometry. (*b*) Two-way acquisition geometry.

direction. Spatial acquisition is desired in as short a time period as possible, while maintaining a suitable fidelity, that is, probability of successful acquisition. In the following, we examine four common acquisition search procedures:

(1) Antenna scanning, in which the receiving system (antenna lens plus photodetector) is slewed over the uncertainty field of view to find the transmitted beam.

(2) Focal-plane scanning, in which the antenna lens and receiver are rigid, with a wide field of view, and the focal plane is scanned to locate the beam.

(3) Focal-plane arrays, in which an array of fixed detectors is used to cover the focal plane.

(4) Sequential searching, in which a fixed detector array is used, and the field of view is readjusted in sequential steps to "zoom in" on the transmitter.

10.2.1 Antenna Scanning

Consider a system using a receiver lens and photodetector with a fixed field of view of Ω_r. The receiving system is swept over the uncertainty region Ω_u, as shown in Figure 10.4. For simplicity we consider only an azimuth (horizontal) search, but the results can be easily extended to elevation searches as well. As the scanning is made the photodetector output is continuously monitored until the beam is believed to have been observed. This decisioning is accomplished by a threshold test on the detector output signal. The operation can be modeled as follows. Consider the transmitted field to be a monochromatic point source beacon, transmitted continuously and producing an average count rate at the receiver of $n_s = \alpha P_r$, where P_r is the receiver optical signal power. We assume a high gain counting model with the background adding a noise count rate of n_b. If the transmitted field is in the receiver field of view for T seconds, an average signal count of $K_s = n_s T$ is generated. The probability of correctly acquiring the beam (PAC) is then

$$\mathrm{PAC} = \sum_{k=k_T}^{\infty} \mathrm{Pos}(k, K_s + K_b) \tag{10.2.1}$$

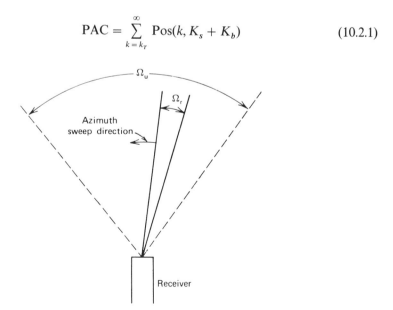

Figure 10.4. Antenna scanning: Ω_u = uncertainty field of view, Ω_r = resolution field of view.

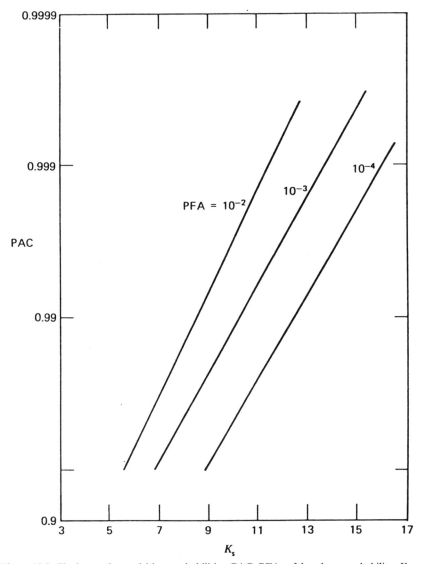

Figure 10.5. Single search acquisition probabilities, PAC. PFA = false alarm probability, $K_b = 1$.

where $K_b = n_b T$ is the noise count, k_T is the threshold count, and $Pos(k, m)$ is the Poisson count probability with mean count m. The threshold k_T is selected so that the probability of false acquisition (PFA) is a desired value, where

$$PFA = \sum_{k=k_T}^{\infty} Pos(k, K_b) \qquad (10.2.2)$$

A set of curves generated in this way are shown in Figure 10.5. The probabilities depend explicitly on the signal and noise counts, which, in turn, are directly related to T, the time the transmitter is in the receiver field of view. We can further relate T to the azimuth slewing rate S_L radians per second, by the approximation

$$T \approx \sqrt{\Omega_r/S_L} \qquad (10.2.3)$$

which is given in seconds. Equations (10.2.1) to (10.2.3) define the relationships among the key system parameters. Performance probabilities determine design values of K_s and K_b, from which detector field of view and slewing rate can be determined. Note that both noise and signal counts increase with T but only noise count increases with Ω_r. This implies that Ω_r should be as small as possible. Theoretically, Ω_r need be only wide enough to cover a single spatial mode (diffraction-limited operation), but propagation turbulence and diffraction blur the source, and larger resolution angles are generally needed to cover the transmitter uncertainty. We also see that reducing Ω_r requires a slower scan rate to generate the desired K_s, which lengthens the acquisition time.

The average acquisition time can be estimated by using a discrete version of the search model in Figure 10.4. We divide the uncertainty region Ω_u into Q disjoint subregions of angle Ω_r, where

$$Q = Q_u/\Omega_r \qquad (10.2.4)$$

We assume the receiver spends T seconds in each subregion position and searches over Ω_u until the field is detected. The probabilities at each position are given by Eq. (10.2.1) or Eq. (10.2.2), depending on whether or not the transmitted beam is present in that position.

The acquisition test continues until a threshold is crossed, with the performance governed by these probabilities. The acquisition probabilities are reminiscent of OOK error probabilities, and the system suffers from the same basic disadvantage—the noise level must be known to set the test threshold properly. Another serious disadvantage of a threshold acquisition search is that the acquisition time may become quite lengthy. Because PAC in Eq. (10.2.1) is the probability of successful acquisition in the correct subregion, the probability of acquiring in a single period search, PAC_1, is then obtained by averaging over all possible subregions. The probability of acquiring when the beam is in the jth subregion is $(PAC)(1-PFA)^{j-1}$, and the average probability over all subregions is then

$$\begin{aligned} PAC_1 &= \frac{1}{Q} \sum_{j=1}^{Q} (PAC)(1 - PFA)^{j-1} \\ &= \frac{PAC}{Q} \left[\frac{1 - (PFA)^Q}{PFA} \right] \end{aligned} \qquad (10.2.5)$$

where Q is the total number of subregions that may contain the target. If the acquisition is not successful in a single period, it must be repeated in the next period. The probability it will take i periods to acquire is then $(\text{PAC}_1)(1-\text{PAC}_1)^{i-1}$, and the average number of periods that will be searched is then

$$N_s = \sum_{i=1}^{\infty} i(\text{PAC}_1)(1 - \text{PAC}_1)^{i-1}$$

$$\cong \frac{1}{\text{PAC}_1} \tag{10.2.6}$$

Thus the average acquisition time is $TN_s \cong T/\text{PAC}_1$. If PFA in Eq. (10.2.2) is close to unity, the average test length may become quite long.

10.2.2 Focal-Plane Scanning

In focal-plane scanning, the search mechanism is different, but the overall effect is identical to antenna scanning. A fixed optical lens transfers the uncertainty field of view to the focal plane, and the searching is accomplished by a single detector scanned over the focal plane. Such an operation can be readily performed by an image dissector system. Focal-plane scanning is generally preferable over antenna scanning, because the receiving structure can be made rigid and no mechanically movable parts are needed. System analysis is identical to that of an antenna scanner, and, therefore, its performance is again described by Eqs. (10.2.1) to (10.2.5).

Both focal-plane and antenna scanning allow for raster storage in which comparison can be made after the complete focal field has been scanned. This would reduce the acquisition search procedure to a count comparison rather than a threshold test. That is, the scanner would scan the complete field of view while storing the collected signal. Decisioning is postponed until the end of the scan, at which time the most likely position is selected as the beam arrival angle [1]. If the discretized spatial model is used, the test appears as a comparison test over Q spatial cells to determine the maximum count. The acquisition probability is given by a Q-ary orthogonal count test identical to the PPM test in Eq. (6.6.2). The parameter Q in Eq. (10.2.4) is directly related to the desired resolution angle used for scanning. The probability of a successful acquisition in a single focal plane scan is therefore†

$$\text{PAC}_1 = \sum_{k_1=0}^{\infty} \text{Pos}(k_1, K_s + K_b) \left[\sum_{k_2=0}^{k_1-1} \text{Pos}(k_2, K_b) \right]^{Q-1} \tag{10.2.7}$$

where K_s and K_b are given in Eq. (10.2.1). It is convenient to define n_{bu} as the

---

†We omit cell equalities here in our probabilities, contending that an acquisition operation would rescan rather than guess among possible directions. Thus equalities among maximum counts are accepted as an incorrect acquisition.

noise count rate over the total uncertainty solid angle Ω_u. (For focal plane scanning, n_{bu} is the total noise over the receiver field of view.) Thus we write

$$K_s = n_s T$$
$$K_b = \frac{n_{bu} T}{Q}$$

(10.2.8)

in Eq. (10.2.7). In addition, we denote the total focal plane (uncertainty region) search time by T_t, which is related to T, the time spent observing each resolution position, by

$$T_t = QT$$

(10.2.9)

For a specified Q (i.e., a specified uncertainty region and desired resolution) and fixed power levels of the transmitter and background, the key system parameters, acquisition probability PAC_1, and total search time T_t are directly related to the observation time T. Figure 10.6 is a plot of PAC_1 in Eq. (10.2.7) for several values of $(n_s T, n_{bu} T)$ as a function of Q. At a particular value of Q,

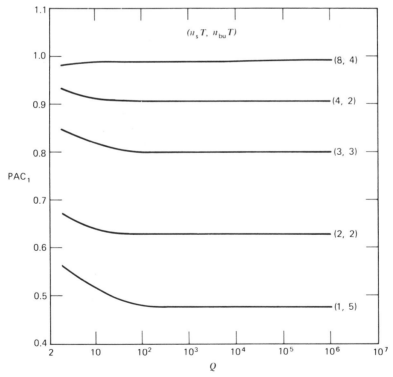

Figure 10.6. Scanning acquisition probabilities: Q = number of uncertainty cells. n_s = signal count rate, n_{bu} = background noise count rate.

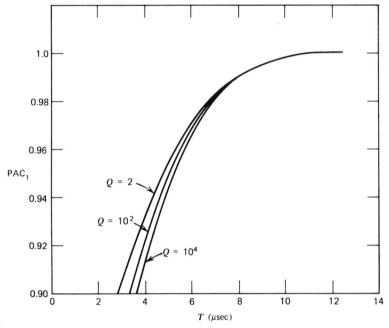

Figure 10.7. Scanning acquisition probabilities: $T =$ time per search position, $Q =$ number of search positions, $n_s = 2 \times 10^6$, $n_{bu} = 2 \times 10^6$.

we see that PAC_1 depends quite strongly on the signal and noise energies. For fixed power levels, this means that acquisition probability is critically related to the observation time T. This dependence can be further exhibited as shown in Figure 10.7, where PAC_1 is plotted versus T for several values of Q at a fixed power level. The results indicate a sudden drop in acquisition probability if T is not long enough. This result again stresses the importance of collected signal energy in optical detection problems. Surprisingly, the result is somewhat independent of Q when Q is large (typically $Q \geqslant 10^2$). Thus, large Q systems require abnormally long search times to maintain a desired acquisition probability. This is illustrated in Figure 10.8, in which Figure 10.7 is replotted, using Eq. (10.2.9). The curves manifest the time needed to perform a complete uncertainty region search for several values of Q. We immediately see a direct tradeoff in search time (T_t) and system resolution (Q) in attaining a prescribed PAC_1.

10.2.3 Focal-Plane Arrays

The use of detector arrays in the focal plane allows for parallel processing to be achieved, and therefore reduces the acquisition times. Each detector of the array would examine a certain portion of the uncertainty region, as shown in Figure 10.9. Collection of the individual detector outputs after a fixed obser-

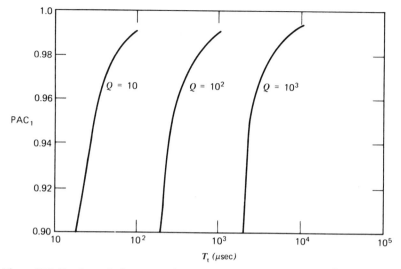

Figure 10.8. Total search time to acquire Q search positions, $n_{bu} = 2 \times 10^6$, $n_s = 2 \times 10^6$.

vation time permits a count comparison test for transmitter location. The detector with the largest count is considered to be viewing the received beam. Each detector must operate independently and its count must be properly cataloged, thereby increasing the receiver complexity. However, the acquisition time is reduced to a single observation time T, as opposed to Eq. (10.2.9), because the counting is done in parallel. We see, therefore, that detector arrays can be a powerful tool in fast acquisition applications.

Consider the case of an array having S detectors, so that the uncertainty angle can be subdivided into Ω_u/S resolution areas. It is obvious that S must be extremely large to obtain resolution cells on the order of those in the scanning methods. The focal plane is therefore divided into S areas, and the acquisition probability is identical to Eq. (10.2.7) with K_s and K_b defined in Eq. (10.2.8). However, we now have

$$\Omega_r = \Omega_u/S$$
$$T_t = T \tag{10.2.10}$$

The resulting single-scan acquisition probability, PAC_1, can therefore be derived directly from curves similar to Figure 10.7, with Q replaced by S. Because search time is now essentially independent of array size S, it is desirable to use as large an array as possible for the best resolution. The tradeoff is directly in terms of receiver complexity, because the array elements must be processed in parallel. Thus, although acquisition by scanning trades off acquisition resolution for search time, acquisition with focal plane arrays trades off resolution for receiver complexity.

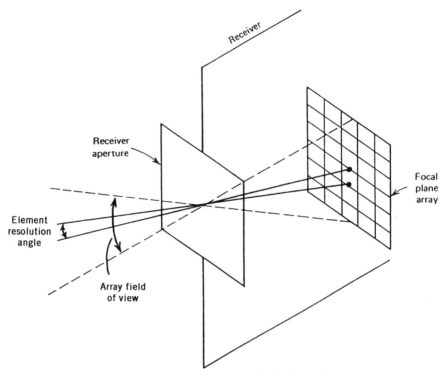

Figure 10.9. Focal-plane array for focal scanning.

10.2.4 Sequential Search with Arrays (Parallel Processing)

The difficulty with a single scan array search is that the achieved resolution (shown in Eq. 10.2.10) is generally larger than that desired, unless S is quite large. This can be improved, however, by repeating the array search until the desired resolution is achieved. Carrying out a sequence of searches with the same array is called *sequential acquisition*. A sequential spatial search is obtained by using a fixed detector array and continually adjusting the field of view so as to home in on the beam with the desired resolution. At each step of the search, the field of view is divided into sectors by the array, and a parallel decision is made as to which sector is observing the beam. The field of view is then reduced (magnification increased) to the decided sector, and the parallel array decisioning is repeated. The final achieved resolution will therefore depend on the number of times the test is repeated, but the overall operation is restricted by the allowable range of magnification that the receiving lens system will allow, that is, the "zoom" range of the receiver.

Consider a sequential search in which S detectors are used in the array, and the initial uncertainty is Ω_u. At the first step, the field of view is divided into S cells of resoluton angle Ω_u/S and a S-ary decision is made after T seconds with signal count energy K_s and noise count energy $K_b = n_{bu} T/S$. After the decision is made, the receiver field of view is readjusted to encompass the decided sector only. The test is repeated, with the field of view again divided into S sectors, this time with resolution angle Ω_u/S^2. At the ith step, the resolution angle is Ω_u/S^i, and the number of steps, r, needed to reduce the original uncertainty Ω_u to the desired resolution Ω_r is that for whch $\Omega_r = \Omega_u/S^r$ or,

$$r = \frac{\log(\Omega_u/\Omega_r)}{\log S} \tag{10.2.11}$$

The probability of detecting the correct sector at the ith step is then

$$\text{PAC}_i = \sum_{k_1=0}^{\infty} \text{Pos}\left(k_1, n_s T_i + \frac{n_{bu} T_i}{S^i}\right)\left[\sum_{k_2=0}^{k_1-1} \text{Pos}\left(k_2, \frac{n_{bu} T_i}{S^i}\right)\right]^{S-1} \tag{10.2.12}$$

where T_i is the observation time of the ith decision. The acquisition probability is then the probability of correct detection at each step, so that

$$\text{PAC} = \prod_{i=1}^{r} \text{PAC}_i \tag{10.2.13}$$

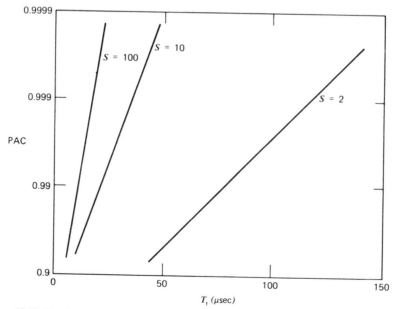

Figure 10.10. Total search time to acquire an S-detector array and sequential acquisition (fixed time per test, $n_s = 2 \times 10^6$, $n_{bu} = 2 \times 10^6$).

The total acquisition time T_t is the time to perform all r tests. Thus, neglecting the time to adjust the magnification, the parallel processing requires a total time

$$T_t = \sum_{i=1}^{r} T_i \qquad (10.2.14)$$

Equations (10.2.11) to (10.2.14) describe the performance parameters of sequential acquisition with parallel processing. For convenience, the operation is generally carried out as either a constant-detection probability test, in which PAC_i is adjusted to be equal at each i, or as a constant observation time test, in which T_i is the same at each i. Because the field of view, and therefore, the noise count, is reduced at each step, a constant probability test uses less time at each i. Conversely, a constant observation time will operate with improved acquisition probability at each step.

Figure 10.10 summarizes the results for a fixed time sequential test. For a fixed S, $T_i = T$ is adjusted for a desired observation time, and the resulting overall PAC in Eq. (10.2.13) is computed for a sequential test. Figure 10.10 shows a typical plot of PAC versus the required time to perform all r tests for several array sizes, S. The result is cross-plotted in Figure 10.11, relating array size to search time for a fixed PAC. Note that search time is uniformly decreased as the array size is increased, while achieving the given PAC.

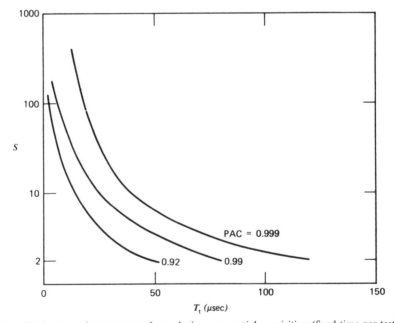

Figure 10.11. Array size versus total search time, sequential acquisition (fixed time per test).

However, the improvement attained may not be significant for the larger array sizes. That is, in Figure 10.9, it may not be necessary to consider arrays larger than, say, 100 elements. Note also the significant decrease in total search time T_t for a sequential test, when contrasted with the results of focal-plane scanning in Figure 10.4. This can be directly related to the reduction in the number of tests that must be performed. Scanning requires a search time directly proportional to the number of resolution cells that must be examined (Eq. 10.2.9), whereas sequential searching is proportional to the log of this number (Eq. 10.2.11).

In a fixed probability sequential search, the acquisition probability at each step is adjusted to a desired value. Thus, if PAC is the desired overall acquisition probability, and if r steps will be performed in the test, then we select

$$PAC_i = (PAC)^{1/r} \qquad \text{for each } i \qquad (10.2.15)$$

and adjust each T_i to achieve this in Eq. (10.2.12). The total search time is then found from Eq. (10.2.14).

10.2.5 Sequential Search with Fixed Arrays (Serial Processing)

In the foregoing fixed array test, parallel processing was assumed. That is, all the detectors of the array are examined simultaneously for their outputs, from which a maximum is selected. The field of view corresponding to this maximum is then selected as containing the transmitter. The test is then repeated over this latter field of view. However, parallel processing over large numbers of detectors may become physically difficult. An alternative is to serially scan the array for the maximal output at each iteration rather than parallel examination. The receiver complexity is now eased at the expense of a slightly longer search time. For a fixed observation time of T seconds, an array with S detectors now requires ST seconds to perform a single iteration, with r separate iterations needed to complete the search. The total search time is now, from Eq. (10.2.11),

$$T_t = rST = \left(\frac{S}{\log S}\right) T \log\left(\frac{\Omega_u}{\Omega_r}\right) \qquad (10.2.16)$$

For a fixed T (specified PAC) and fixed field of view Ω_u and resolution Ω_r, it is interesting to note that T_t is minimized for the value S such that $dT_t/dS = 0$ which occurs for $\log_e S = 1$, or $S = e$. For square arrays, S must be a squared integer. Hence search time is minimized for $S = 4$ (the nearest square to e) corresponding to a 2×2 array. The sequential test is then one that divides the field of view in quadrature, serially examines each quadrant for that most likely to contain the transmitter, and then repeats the test for the decided quadrant.

Note that with serial processing the design objective has completely reversed to the use of small detector arrays.

In discussing sequential acquisition, we have implicitly required a change of magnification (field of view) at each step. It is important to recognize that time is required to change lens power. In addition, there is a maximum magnification range, $\mathcal{M}$, over which we can expect a telescope system to operate. At the minimum power (first step of the acquisition search), we would cover a field of $D_s = \Omega_u/\Omega_{dL}$ modes, or a solid angle of Ω_u steradians. At the maximum power (last step) we cover a field of S modes. Therefore the power change must occur in $\log_S(D_s) - 1$ steps, and we require that at each step the power change by the amount S. We then see that the magnification $\mathcal{M}$ is given by

$$\mathcal{M} = [D_s/S]^{1/2} \tag{10.2.17}$$

For example, if $D_s = 10^5$, then a magnification of $\mathcal{M} = 31.6$ is required with an array of 100 elements. However, decreasing the array to 10 elements will require an increased $\mathcal{M}$ of approximately 100. If we try to change the magnification mechanically, we can almost guarantee that the required time for adjustment will be much greater than the scan times involved. Thus devices or techniques that can significantly reduce magnification adjustment time would be extremely beneficial to sequential acquisition operations in optical search systems. A possible technique might be the use of electronic magnification using larger arrays. In such systems, we use a large array to cover the desired field of view, but we electronically regroup the detectors to effectively change the field of view. In this way we avoid the difficulty associated with the vast amount of components needed to parallel process a large detector array. For the sequential application here, we would only use S processing components at each step, but the processors could be sequentially reconnected to different array elements with proper switching networks, to accomplish the effective change in field of view. It would, of course, still be necessary for each of the array elements to have sufficient gain if quantum-limited performance were desired. In general, several orders of magnitude in search time would be lost if the individual array elements had insufficient internal gain.

In conclusion, then, we see that the operation of spatial acquisition involves a tradeoff of system design objectives. Once we specify a receiver aperture (diffraction limited field of view Ω_{dL}) and an uncertainty angle (Ω_u), the total spatial dimension, $D_s = \Omega_u/\Omega_{dL}$, of the search is specified. From our knowledge of orthogonal mode detection, we know that the theoretically best performance, in terms of shortest acquisition time and minimum resolution for a fixed acquisition probability, is obtained when each of the D_s modes is searched in parallel. Any effort to decrease the total number of modes decreases the system resolution. Any effort to decrease the number of modes observed at one time increases acquisition time. From an engineering point of view it may in fact be much more convenient to implement lower resolution or longer search time systems. It must be remembered that the definition of a long acquisition time depends on the system mission. For example, short acquisition time for a communication system that will operate for several hours is not as critical as rapid acquisition of a reentering vehicle or missile. We would of course always desire D_s to be as small as possible by reducing a priori uncertainty and building better components. Reduction of background noise energy during the observation time of any mode requires directly a reduction of optical bandwidth. Since the modulation frequencies will generally be low, we are confined to the achievable minimal optical bandwidth in determining noise per mode. To observe D_s modes simultaneously requires a very large array. The alternative sequential procedures would use as large an array as economically possible, and implement the sequential search algorithms. The resulting performance, while suboptimal, is nevertheless quite good and would probably satisfy most applications.

10.3 SPATIAL TRACKING

After pointing and spatial acquisition have been achieved, there remains the task of maintaining the transmitted beam on the detector area in spite of beam wander or relative transmitter–receiver motion. This operation of keeping the receiver aperture properly oriented relative to the arriving optical field requires spatial tracking. This tracking is achieved by generating instantaneous pointing error voltages that are used to continually realign the optical hardware.

After successful acquisition of the incoming light beam, the beacon field should be focused at the center of the acquisition array, which is coaxially aligned to the position error sensor of the tracking subsystem. The acquisition threshold removes the array processing and enables the tracking operation using the centered, focused beam. The tracking subsystem then generates the error signals as the focused beam moves off-center, due to either line-of-sight beam motion or receiver platform jitter.

The tracking subsystem usually uses two separate (azimuth and elevation) closed loops, as shown in Figure 10.12. The tracking error is determined instantaneously for both azimuth and elevation coordinates by means of the position error sensor, which generates the error signals. The error sgnals are then used to control the alignment axis of the receiver lens. This is accomplished by some type of control loop dynamics, generally with separate servo loops for individual control of the azimuth and elevation pointing. The loop control functions are typically of the form of some type of low-pass integration filtering that smooths the error signals for position control. The filter bandwidths must be wide enough to allow the tracking loop to follow the expected beam motion, yet allowing minimal noise effects within the loop.

Let (θ_z, θ_l) be the azimuth and elevation angle of the line of sight vector to the transmitter point, with respect to a selected receiver coordinate system. Let (ϕ_z, ϕ_l) be the corresponding angles of the normal vector to the receiver aperture area. Then the instantaneous angular errors in pointing the receiver

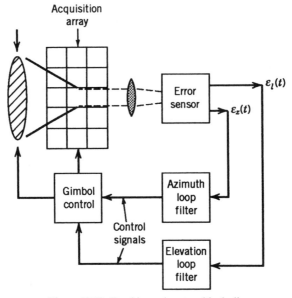

Figure 10.12. Tracking subsystem block diagram.

to the transmitter are given by

$$\psi_z(t) = \theta_z - \phi_z$$
$$\psi_l(t) = \theta_l - \phi_l \tag{10.3.1}$$

where the dependence on t emphasizes the changing errors in time. Let $\varepsilon_z(t)$ and $\varepsilon_l(t)$ be the error voltage generated from the optical sensor for control of the azimuth and elevation angles, respectively. These error signals are used to correct the pointing of the (ϕ_z, ϕ_l) angles. Hence,

$$\phi_z(t) = \overline{\varepsilon_z(t)}$$
$$\phi_l(t) = \overline{\varepsilon_l(t)} \tag{10.3.2}$$

where the overbars denote the average effect of the loop filtering. Combining Eqs. (10.3.1) and (10.3.2) yields the pair of system equations

$$\psi_z(t) = \theta_z(t) - \overline{\varepsilon_z(t)}$$
$$\psi_l(t) = \theta_l(t) - \overline{\varepsilon_l(t)} \tag{10.3.3}$$

The equations represent a pair of coupled differential equations for the angular pointing errors. The terms $[\theta_z(t), \theta_l(t)]$ represent the movement of the line-of-sight vector to the transmitter, and therefore appear as forcing functions in Eq.

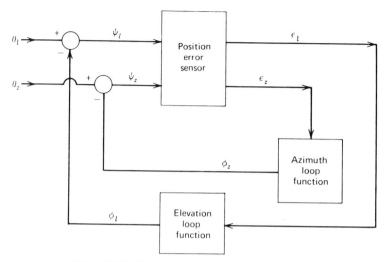

Figure 10.13. Equivalent azimuth-elevation tracking loops.

(10.3.3). The specific form of the equations depends on the relationship of the error voltage to the position error (which is specified by the properties of the optical sensor) and the type of filtering within the loop. Note that the form of the equations suggests the equivalent system in Figure 10.13. Here, the position angles are the variables of the equivalent loop, which involves the coupled feedback loops driven by the movement of the transmitter vector. The loop filtering in the equivalent system is identical to that in the actual tracking loop. Note that the optical sensor is an integral part of the equivalent loop as well as the actual loop.

The optical sensor is used to generate the error-control signal from the received optical beam. The most common type of sensor is the quadrant detector, in which four separate photodetectors are used to determine instantaneous position error. The received optical field is focused onto the center of the detectors placed in a quadrant arrangement in the focal plane, as shown in Figure 10.14a. Error voltages are generated by properly comparing the detector outputs. When the received plane wavefield arrives normal to the lens, it is focused exactly to the center of the quadrant, and theoretically all detectors receive equal energy. Offsets in arrival angle causes imbalance in detector inputs, which can then be used to generate correcting voltages. An error signal in azimuth and elevation is derived by combining two detector outputs and comparing. Let us represent the individual detector current outputs as $y_i(t)$, so that the error signals are

$$\varepsilon_z(t) = [y_1(t) + y_2(t)] - [y_3(t) + y_4(t)] \qquad (10.3.4a)$$

$$\varepsilon_l(t) = [y_1(t) + y_4(t)] - [y_2(t) + y_3(t)] \qquad (10.3.4b)$$

The detector outputs are shot noise processes whose mean values are related

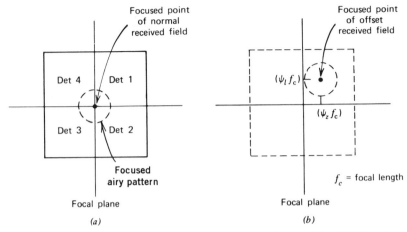

Figure 10.14. The quadrant error sensor. (a) Normal-arriving focused field; (b) Off-angle arriving focused field.

to the average field intensities collected over their surfaces. Let the received transmitter field be a plane wavefield arriving normal to the receiver aperture, and let $I_s(t, \mathbf{r})$ be its diffracted field in the focal plane. Let the background add a constant intensity I_b over the focal plane. The mean errors in Eq. (10.3.4a) due to the reception of the transmitter and background fields is then

$$
\overline{\varepsilon_z(t)} = \overline{[y_1(t) + y_2(t)]} - \overline{[y_3(t) + y_4(t)]}
$$

$$
= (ge)\left[\alpha \int_{A_1 + A_2} [I_s(t, r) + I_b]\, dr - \alpha \int_{A_3 + A_4} [I(t, \mathbf{r}) + I_b]\, dr\right]
$$

(10.3.5)

where A_i is the area of the ith detector and g is the detector gain, assumed equal for each detector. Note that, if the detectors are perfectly balanced (equal gain and areas), the background term cancels out, and the mean error functions depends only on the differences in the integrated signal intensities.

For a circular lens of diameter d, we can approximate the Airy disc in the focal plane by the intensity function

$$
I_s(t, \mathbf{r}) = I_s \qquad |\mathbf{r}| \leqslant \frac{1.22\lambda f_c}{d}
$$

$$
= 0
$$

(10.3.6)

where λ is the wavelength and f_c is the focal length. If the field arrives exactly normal, the diffraction pattern is centered on all four detectors, and Eq. (10.3.5) produces a zero error. If the signal field arrives at an offset angle (ψ_z, ψ_l) in azimuth and elevation, the intensity pattern in Eq. (10.3.6) is centered at the points $(f_c\psi_z, f_c\psi_l)$ in the azimuth and elevation coordinates of the focal plane, as shown in Figure 10.14b. The integrals in Eq. (10.3.5) therefore correspond to areas under sectors of an offset circle in the focal plane. For small displacements such that the Airy pattern is still encompassed by the quadrant area, Eq. (10.3.5) integrates to

$$\overline{\varepsilon_z(t)} = uP_r\left\{1 - \frac{2}{\pi}\cos^{-1}\left(\frac{\psi_z d}{1.2\lambda}\right) + (0.53)\left(\frac{\psi_z d}{\lambda}\right)\left[1 - \left(\frac{\psi_z d}{1.2\lambda}\right)^2\right]^{1/2}\right\} \qquad (10.3.7)$$

where $u = ge\alpha$ is the detector responsivity, and P_r is the total received transmitter power on all four detectors (refer to Problem 10.5). Note that Eq. (10.3.7) does not depend on ψ_l, and the mean azimuth error signal is generated only from the azimuth angle error. Similarly, the mean elevation signal depends only on elevation angle errors, as in Eq. (10.3.7), with ψ_z replaced by ψ_l. Thus, the quadrant detector uncouples the tracking operation, and each tracking loop in Figure 10.13 can be considered a separate loop in its tracking behavior. Other types of optical error sensors, involving lobing prisms and rotating reticles, have been suggested as alternatives to the quadrant array [2].

A plot of the bracketed expression in Eq. (10.3.7) is shown in Figure 10.15 and shows how the mean error signal in azimuth (or elevation) is generated from the tracking error. As the Airy spot moves off center, an increasing error voltage is produced, which saturates and then falls to zero as the spot slides off the detector array. In tracking terminology, the normalized function $S(\psi)$ in Figure 10.15 is called the loop *S-curve* and indicates the conversion from angle errors to error voltages.

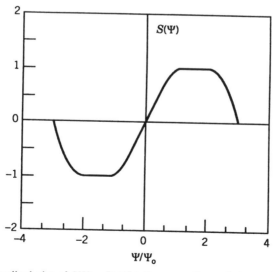

Figure 10.15. Normalized plot of $S(\Psi) = [\bar{\varepsilon}_q(t)]/uP_r$ versus offset pointing angle Ψ (azimuth or elevation). $\Psi_0 = 1.2\lambda/d$.

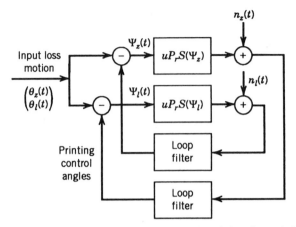

Figure 10.16. Equivalent block diagram for azimuth and elevation pointing control.

The model of the complete joint tracking operation using the loop S-curves is shown in Figure 10.16. The difference angles between the transmitter vector and the receiver normal vector defines the instantaneous pointing errors $[\psi_z(t), \psi_l(t)]$, which are converted to error voltages by the S-curve for both azimuth and elevation. Detector noise (shot noise, dark current, and thermal noise) from each detector of the quadrant array is added to the mean error signals. The combined noisy error signals are then filtered and fed back for gimbol adjustment and control of the pointing angle. The overall tracking operation in azimuth and elevation is therefore governed by the pair of tracking equations

$$\psi_z(t) + F[uP_rS(\psi_z(t)) + n_z(t)] = \theta_z(t) \qquad (10.3.8a)$$

$$\psi_l(t) + F[uP_rS(\psi_l(t)) + n_l(t)] = \theta_l(t) \qquad (10.3.8b)$$

where $F[\cdot]$ represents the filtering operation, and the noise signals $[n_z(t), n_l(t)]$ represent the combined detector noise in each loop. The above tracking equations represent a pair of nonlinear, stochastic equations that describe the joint tracking operation in azimuth and elevation. The additive noise terms in each loop are the result of combining the detector noises from the four quadrant detectors, and the resulting spectral level of the loop input noise is

$$N_{OL} = (ge)^2\alpha(P_r + 4P_b) + 4N_{0c} \qquad (10.3.9)$$

where P_b and N_{0c} are the background power and thermal noise level per detector. When the tracking angles are small, the loop S-curve in Figure 10.15 can be considered linear, and we approximate

$$S(\psi) \cong (0.53d/\lambda)\psi \qquad (10.3.10)$$

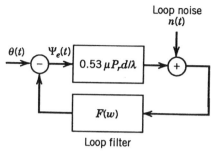

Figure 10.17. Linearized tracking loop for azimuth or elevation pointing control.

where the coefficient is the approximate slope of the S-curve at the origin. With the loop linearized, the tracking operation for either Eq. (10.3.8a) or Eq. (10.3.8b) can be modeled by the linear tracking loop in Figure 10.17. Such a linear loop can now be described by the closed-loop transfer function

$$H_L(\omega) = \frac{G_L F(\omega)}{1 + G_L F(\omega)} \qquad (10.3.11)$$

which depends on the loop gain factor $G_L = 0.53 u d P_r / \lambda$ and on the loop filtering $F(\omega)$. It can be seen that the error process in the closed loop is obtained by effectively filtering the detector noise with the linear transfer function $-H_L(\omega)/G_L$ and by filtering the input angle motion with the transfer function $[1 - H_L(\omega)]$. The total mean-square tracking error variance in either azimuth or elevation is then

$$\sigma_e^2 = \frac{1}{2\pi} \int_{-\infty}^{\infty} |1 - H_L(\omega)|^2 S_\theta(\omega) \, d\omega + \frac{1}{2\pi} \int_{-\infty}^{\infty} \left(\frac{N_{0L}}{G_L^2} \right) |H_L(\omega)|^2 \, d\omega \qquad (10.3.12)$$

where $S_\theta(\omega)$ is the frequency spectrum of the input angle motion $\theta(t)$ (transmitter dynamics and platform jitter), and N_{0L} is the noise spectral level in Eq. (10.3.9). The first term is the uncompensated error variance due to line of sight motion, whereas the second term is the noise-induced error. From this error variance equation, we see that input angle variations within the loop bandwidth [range of ω where $H_L(\omega) \cong 1$] will be tracked out by the loop, whereas those outside the loop bandwidth will contribute directly to tracking errors. This means that tracking errors due to transmitter and receiver platform motion can be reduced by increasing the tracking loop bandwidth $[H_L(\omega)]$ so as to exceed the significant frequencies of the expected motion spectrum $S_\theta(\omega)$. Hence, loop filter design is a direct tradeoff of improved tracking capability versus increased loop noise [3, 4].

The tracking error variance due to loop noise can be written by substituting from Eq. (10.3.9) as

$$\sigma^2 = \left(\frac{2N_{0L}}{G_L^2}\right) B_L \qquad (10.3.13)$$

where B_L is the loop noise bandwidth defined by

$$B_L = \frac{1}{2\pi} \int_0^\infty |H_L(\omega)|^2 \, d\omega \qquad (10.3.14)$$

If we combine the results from Eqs. (10.3.13) and (10.3.14), the overall angle error variance caused by the noise can be rewritten as

$$\sigma^2 = \frac{(1.8\lambda/d)^2}{\text{SNR}_L} \qquad (10.3.15)$$

where

$$\text{SNR}_L = \frac{(eg\alpha P_r)^2}{2N_{0L} B_L}$$
$$= \frac{(eg\alpha P_r)^2}{[e^2\alpha(P_r + 4P_b) + 4N_{0c}]2B_L} \qquad (10.3.16)$$

This is the effective photodetected output SNR_L in the loop noise bandwidth B_L. Thus the tracking error root-mean-square value due to noise appears as a fraction of the receiver parameter $(1.8\lambda/d)$ and is inversely related to the loop SNR_L. Recall that λ/d is the approximate diffraction-limited beamwidth associated with a receiver having an aperture of diameter d. Hence, with reasonable SNR_L, the linearized loop will track with root-mean-square pointing error angle that is within the diffraction-limited field of view.

The power P_r in Eq. (10.3.16) is the average power collected at the receiver aperture from the received transmitter beam. If the optical beam where pulsed, with received pulse power P_p, pulse width τ and pulse repetition frequency PRF, then P_r in Eq. (10.3.16) whould be replaced by the average pulse power

$$P_r = (P_p \tau)\text{PRF} \qquad (10.3.17)$$

Hence, Eq. (10.3.16) can be applied to either continuous or pulsed optical fields.

When the photodetector noise processes are taken as zero-mean Gaussian noise and the azimuth and elevation loops are linearized and uncoupled, the tracking error components in azimuth and elevation evolve as independent joint Gaussian error processes, each with the variance in Eq. (10.3.13). If the loops track the line-of-sight motion perfectly, the magnitude error

$$|\psi_e| = [\psi_z^2 + \psi_l^2]^{1/2} \tag{10.3.18}$$

is therefore due entirely to the noise and becomes a Rayleigh distributed random variable at each t. The instantaneous pointing error magnitude $|\psi_e(t)|$ then has the probability density

$$p_{\psi_e}(x) = \frac{x}{\sigma^2} \exp[-(x^2/2\sigma^2)] \tag{10.3.19}$$

at any t. Thus the pointing-error magnitude of the beam-tracking receiver varies randomly and has the probability density in Eq. (10.3.19) from the photodetection noise in the tracking loop sensors. These statistical models have been verified by reported experimental studies as well [5–9].

10.4 DOUBLE-ENDED OPTICAL BEAM TRACKING

In Section 10.3, we analyzed a single-ended beam-tracking system, in which a receiver established a line of sight by tracking the arrival optical beam from a transmitter that was pointed perfectly toward the receiver. In this section, we extend to a double-ended tracking operation, where two terminals simultaneously track the optical beam from the other terminal. Pointing-error vectors can therefore develop at both ends of the system, and the pointing accuracy at one end effects the errors at the other end. The pointing-error vectors at the two ends therefore evolve as joint variables that are both temporally and statistically related. Such tracking scenarios will occur, for example, in establishing two-way optical communications between two moving spacecraft or orbiting space vehicles [10, 11].

The double-ended beam-tracking geometry is shown in Figure 10.18a. The optical terminals T_1 and T_2 are to simultaneously track optical beams transmitted toward each other and point their own beams back in the same direction We assume relative line-of-sight motion $\theta_1(t)$ observed from T_1 relative to its own coordinate system, and $\theta_2(t)$ observed from T_2. Let $[\psi_1[(t), \psi_2(t)]$ be the corresponding pointing error vectors of each terminal. The above angles are each vector processes with both azimuth and elevation components.

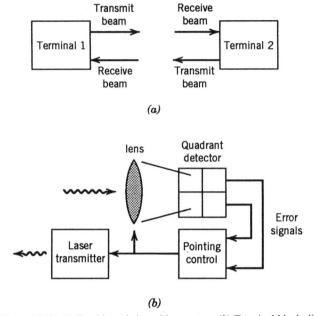

Figure 10.18. (*a*) Double-ended tracking system. (*b*) Terminal block diagram.

We again assume that each receiver uses a direct detection tracking system consisting of a focusing lens with a quadrant detector array, generating simultaneous azimuth and elevation pointing-error voltages for pointing control, as shown in Figure 10.18*b*. Error voltages from the quadrant array are produced to keep the incoming beam aligned on the quadrant crosshairs at each terminal. Each terminal then retransmits its laser beam back along the received beam direction. Hence, each terminal tracks and points at the same time. We assume that no point ahead is needed and, for simplicity, assume that identical tracking loops, S-curves, and beam shapes are used at each end.

The measured power at time t, over the receiver aperture at T_2, due to the field arriving from T_1, is then

$$\text{received power at } T_2 = P_r G(|\psi_1(t - t_d)|) \tag{10.4.1}$$

where $G(|\psi_1|)$ is the normalized transmit beam pattern loss from T_1 (assumed to be symmetrical around boresight), t_d is the propagation delay from T_1 to T_2, and P_r is the received power for perfectly pointed transmission from T_1. Thus the pointing error at T_1 occurring t_d seconds earlier affects the received power at terminal T_2 and time t through the transmitter beam pattern loss function. The tracking system at T_2 generates the pointing error signals that are filtered, as was shown in Figure 10.16. Extending the earlier system equations from Eqs.

(10.3.8), we generate the tracking equation (azimuth or elevation) at T_2

$$\psi_2(t) + F[uP_r G(|\psi_1(t - t_d)|)S(\psi_2(t)) + n_2(t)] = \theta_2(t) \qquad (10.4.2)$$

where $n_2(t)$ is again the detector noise processes, $F[\cdot]$ is the loop filtering, and $S(\psi_2)$ is the tracking loop S-curve. Likewise, for terminal T_1 at the same time t, we have

$$\psi_1(t) + F[uP_r G(|\psi_2(t - t_d)|)S(\psi_1(t)) + n_1(t)] = \theta_1(t) \qquad (10.4.3)$$

Equations (10.4.2) and (10.4.3) represent pairs of interconnected, stochastic beam-tracking equations that relate the joint pointing errors during double-ended tracking. The equations are driven by the line-of-sight dynamics and the receiver noise processes, and depend on the beam patterns, error sensors, and loop filters at each end.

These system equations can be analyzed from a pure stability point of view to determine ranges and conditions for simultaneous solutions to exist (for example, see Problem 10.9). Such approaches depend on the type of loop filtering, and the mathematical models for the beam shapes and S-curves. A detailed study in this direction is beyond our scope here, but stability studies of this type have been discussed [12–14].

The coupled system equations show that the instantaneous tracking errors at one end depend on the pointing errors at the other end. The noise causes the errors to evolve as random processes in time. If we condition on the tracking error at T_1 at t_d seconds earlier, [letting $x = |\psi_1(t - t_d)|$], then the statistics of the tracking error at T_2 can be estimated. The conditional (on x) variance of the component (azimuth or elevation) tracking error due to noise at T_2 and time t becomes

$$\sigma_e^2 \,|\, x = \frac{(1.8\lambda/d)^2}{\text{SNR}_L} \qquad (10.4.4)$$

where now SNR_L is the conditional SNR,

$$\text{SNR}_L = \frac{[\alpha P_r G(x)]^2}{\{\alpha[P_r G(x) + 4P_b] + 4n_c\}2B_L} \qquad (10.4.5)$$

Here $n_c = N_{0c}/(ge)^2$ is the detector thermal noise count rate, and P_r is again the received power at T_2 with perfect pointing from T_1. Depending on the relative values of signal power and receiver noise power, the SNR may be rewritten. If $\alpha P_r G(x) \gg 4(\alpha P_b + n_c)$, the receiver is quantum limited, and

$$\text{SNR}_L = \left(\frac{\alpha P_r}{2B_L}\right) G(x) \qquad (10.4.6)$$

When $\alpha P_r G(x) \ll 4(\alpha P_b + n_c)$, the receiver is background limited, and

$$\text{SNR}_L = \left[\frac{(\alpha P_r)^2}{4(\alpha P_b + n_c)2B_L} \right] G^2(x) \tag{10.4.7}$$

For intermediate cases, these expressions can be generalized, with the exponent of the beam function $G(x)$ varying between 1 and 2. That is, we can always rewrite Eq. (10.4.5) in the general form

$$\text{SNR}_L = (\text{SNR}_o)G^q(x) \tag{10.4.8}$$

where now

$$\text{SNR}_o = \frac{(\alpha P_r)^2}{[\alpha(P_r + 4P_b) + 4n_c]2B_L} \tag{10.4.9}$$

and q falls between 1 and 2. (see Problem 10.10). The SNR_o is the receiver detected SNR_L with perfect pointing from T_1. This simplification allows us to write Eq. (10.4.4) as

$$\begin{aligned} \sigma_e^2 \,|\, x &= \left[\frac{(1.8\lambda/d)^2}{\text{SNR}_o} \right] G^{-q}(x) \\ &= \sigma_o^2 G^{-q}(x) \end{aligned} \tag{10.4.10}$$

with σ_o^2 the receiver pointing variance at T_2 with perfect pointing from T_1. Thus the conditional pointing variance at T_2 is always increased because of pointing errors from T_1 as the received beam pattern falls off according to the beam pattern function $G(x)$. A similar statement can be made for the tracking at T_1 due to pointing errors from T_2.

The overall joint probability density of the pointing errors can now be estimated by reusing the results in Eq. (10.3.19), again assuming Gaussian noise and linearized, uncoupled tracking loops for azimuth and elevation. The pointing error magnitude at T_1 [denoted by $y = |\psi_2(t)|$] is then conditionally Rayleigh distributed with density

$$p_2(y \,|\, x) = \frac{y}{\sigma_o^2 G^{-q}(x)} \exp[-(y^2/2\sigma_o^2)G^{-q}(x)] \tag{10.4.11}$$

with q obtained from Eq. (10.4.8) and σ_o^2 given in Eq. (10.4.10). The joint density of the error magnitude variables x and y can then be obtained as

$$p(x, y) = p_2(y \,|\, x)p_1(x) \tag{10.4.12}$$

where $p_1(x)$ is the probability density of the variable $x = |\psi_1(t - t_d)|$ occurring at T_1. The latter can be approximated by assuming that the pointing error at T_1 has reached a steady state t_d seconds earlier. However, a steady-state tracking condition is achieved only under certain conditions. For example, if terminal T_1 has excess noise in its tracking loops, the power received at T_2 will be reduced, increasing T_2's tracking variance. This in turn will reduce the power from T_2 back to T_1, further reducing T_1's variance, which in turn further reduces T_2's variance, and so on.

We can estimate a steady-state condition by examining these variance changes. If a steady state is reached, the variance change during a transmission cycle will no longer increase. If a steady-state variance is to occur, it can be determined by using a recursive computation on successive variance calculations at each end, as proposed by Peters and Sasaki [13]. By assuming the tracking systems at each end are identical, we can consider iterations on the same loop every t_d seconds. Let ψ_i denote the error at time it_d after initiation. The variance at step $i + 1$ is obtained as the average value of the variance at step i, where the average is taken over the probability of the pointing error at step i. Thus, assuming that we start at the initial variance σ_o^2 in Eq. (10.4.9),

$$\sigma_{i+1}^2 = \sigma_o^2 \int_{-\infty}^{\infty} G^{-q}(x) p_i(x)\, dx \qquad (10.4.13)$$

where $p_i(x)$ is the probability density of $x = |\psi_i|$. When $G(x)$ is a normalized transmitted Gaussian beam pattern with a beamwidth λ/d and $p_i(x)$ is Rayleigh distributed with parameter σ_i^2, Eq. (10.4.13) integrates to

$$\sigma_{i+1}^2 = \frac{\sigma_o^2}{1 - 2q(d/\lambda)^2 \sigma_i^2} \qquad (10.4.14)$$

In this case, if a steady state exists, $\sigma_{i+1}^2 = \sigma_i^2 \triangleq \sigma_{ss}^2$, which occurs only if

$$\sigma_{ss}^2 = \frac{1 - \sqrt{1 - 8q(d\sigma_o/\lambda)^2}}{4q(d/\lambda)^2} \qquad (10.4.15)$$

A real steady-state solution requires

$$\sigma_o^2 \leqslant \frac{(\lambda/d)^2}{8q} \qquad (10.4.16)$$

This is the maximum tracking variance at each end that can be tolerated in a double-ended tracking operation. Equation (10.4.16) is equivalent to a

TABLE 10.1 Values of Maximum σ_o and σ_{ss} in Double-Ended Tracking with Gaussian and Circular Beam Shapes

	Gaussian Beam[a]		$[2J_1(x)/x]^2$ Beam[a]	
	$q=1$	$q=2$	$q=1$	$q=2$
$\dfrac{\sigma_o}{(\lambda/d)}$	0.113	0.08	0.171	0.131
$\dfrac{\sigma_{ss}}{(\lambda/d)}$	0.159	0.113	0.191	0.151

[a] $q=1$ is quantum limited and $q=2$ is noise limited.

condition on SNR_o of

$$\text{SNR}_o \geqslant (1.8)^2 8q \tag{10.4.17}$$

for which

$$\sigma_{ss}^2 \leqslant \frac{(\lambda/d)^2}{4q} \tag{10.4.18}$$

Hence in quantum-limited tracking ($q = 1$), a SNR of approximately 14 dB is needed to prevent the tracking variances to increase without bound. For background-limited operation ($q = 2$), this increases to approximately 17 dB. As long as SNR_o is above these critical values, the loop tracking theoretically will converge with a steady-state variance given by Eq. (10.4.18).

For the general case of a non-Gaussian beam shape, Eq. (10.4.12) must be solved by numerical methods. Table 10.1 lists some results of these computations, showing the maximum values of σ_o and σ_{ss} for both a Gaussian beam and a beam described by a circular $[2J_1(x)/x]^2$ pattern, with both $q = 1$ and $q = 2$, and a beam diameter of λ/d. Note again that the variances are always a small fraction of the diffraction-limited beamwidths. Higher power levels (higher SNR_o and smaller σ_o) are required with Gaussian beams due to the faster fall-off at beam edges with pointing errors.

The steady-state joint probability density for the pair of tracking error magnitudes occurring during double-ended tracking can now be obtained by using Eq. (10.4.12), assuming a steady state had been reached t_d seconds earlier. Again letting $x = |\psi_1(t - t_d)|$ and $y = |\psi_z(t)|$, we finally have

$$p(y, x) = \left(\frac{xyG^q(x)}{\sigma_o^2 \sigma_{ss}^2}\right) \exp\left[-\left(\frac{y^2}{2\sigma_o^2}\right)G^{-q}(x) - \left(\frac{x}{2\sigma_{ss}^2}\right) \right] \tag{10.4.19}$$

where σ_o^2 is the error variance in Eq. (10.4.10) and σ_{ss}^2 is the steady-state variance in Eq. (10.4.18). The coefficient q depends on the power levels and has the value $q = 1$ for quantum-limited receivers and $q = 2$ for the noise-limited case. For operation in between these cases, $1 < q < 2$, and must be obtained by numerically solving Eq. (10.4.8). The above probability density will be used in the next section for the evaluation of digital decoding in the presence of double-ended beam tracking.

10.5   EFFECT OF BEAM TRACKING ON DATA TRANSMISSION

While beam tracking is taking place at each receiver in a two-way space link, as in Figure 10.18*a*, data can be transmitted by directly modulating the tracking beam. After acquisition is completed, the tracking operation holds the focused received field close to the quadrant array center, while generating the tracking error signals. Data can then be transmitted by field modulation, creating a modulated optical carrier, or a pulsed optical carrier, depending on the type of encoding. By summing the outputs of all detectors of the quadrant array at the receiver, a separate data demodulation channel can be generated, as shown in Figure 10.19. In this sum channel, the total power on the array is available for decoding.

The tracking subsystem at each receiver therefore generates the azimuth and elevation correction signals, while the sum channel signal is simultaneously used for data processing. Hence the double-ended tracking can continue to take place while data is transmitted. Pointing errors from transmitter to receiver cause the received beam power to be reduced because of beam pattern loss when operated off boresight. Pointing errors of the receiver shift the diffraction pattern off-center on the quadrant array, causing power losses from photodetection integration in the focal plane. Hence, the tracking operation at each end directly effects data transmission because of these combined power losses. In this section, we examine this relationship.

Assume the pointing-error magnitudes (say x at the transmit end and y at

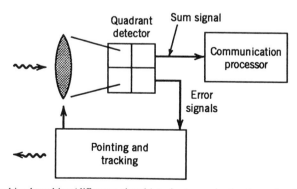

Figure 10.19. Combined tracking (difference signals) and communication (sum signals) subsystems.

the receive end) remain constant over the pulse integration time T_p of a digital link. The collected signal count in the sum channel can then be written as

$$K_s(x, y) = K_{s0}[G_1(x)G_2(y)] \tag{10.5.1}$$

where K_{s0} is the signal count under perfect pointing, $G_1(x)$ is the transmit beam pattern loss, and $G_2(y)$ accounts for the power loss from receiver mispointing causing the diffraction pattern shift. The signal count in Eq. (10.5.1) is collected along with the background and noise count, K_b and K_n. The counts then determine the bit error probability depending on the encoding format, as discussed in Chapter 6. Let PE(x, y) denote the resulting PE with the K_s in Eq. (10.5.1) inserted. Because the pointing errors during any pulse or bit time are random variables, the decoding PE must be obtained from

$$PE = \int_{-\infty}^{\infty} \int_{-\infty}^{\infty} PE(x, y)p(x, y) \, dx \, dy \tag{10.5.2}$$

where $p(x, y)$ is the joint probabilities of the pointing errors. That is, the decoding PE must be averaged over the pointing errors being generated by the tracking subsystems.

As an example, the PE for a direct detection M-ary PPM is given by the approximation in Eq. (6.6.2) as

$$PE \cong \frac{M}{2} Q\left(\frac{\sqrt{SNR}}{2}\right) \tag{10.5.3}$$

where

$$SNR = \frac{K_s^2}{K_s + 2K_b + 2K_n} \tag{10.5.4}$$

As in Eq. (10.4.8) we use the substitution

$$SNR = SNR_o[G_1(x)G_2(y)]^q \tag{10.5.5}$$

where SNR_o is the perfectly pointed SNR, and q is between 1 and 2. The PE in Eq. (10.5.3) is then

$$PE(x, y) = \frac{M}{2} Q\left\{\frac{\sqrt{SNR}}{2}[G_1(x)G_2(y)]^{q/2}\right\} \tag{10.5.6}$$

Using the joint density in Eq. (10.4.18), Eq. (10.5.6) becomes

$$
PE = \int_{-\infty}^{\infty} \int_{-\infty}^{\infty} \frac{M}{2} Q \left\{ \frac{\sqrt{SNR_o}}{2} [G_1(x)G_2(y)]^{q/2} \right\} \left[\left(\frac{xyG^q(x)}{\sigma_o^2 \sigma_{ss}^2} \right) \right.
$$
$$
\left. \cdot \exp\left[-\left(\frac{y^2}{2\sigma_o^2} \right) G_1^{-q}(x) - \left(\frac{x^2}{2\sigma_{ss}^2} \right) \right] \right] dx \, dy
$$

(10.5.7)

For specific functions $G_1(x)$ and $G_2(y)$, PE can usually be integrated numerically. An important example is again the Gaussian-shaped loss functions,

$$
G_1(x) = \exp[-(2xd_t/\lambda)^2]
$$
$$
G_2(y) = \exp[-(2yd_r/\lambda)^2]
$$

(10.5.8)

where d_t and d_r are the transmit and receive optical diameters. Equation (10.5.7) can first be simplified by inserting the change of variables $u = x/\sigma_{ss}$ and $v = y/\sigma_o$ so that

$$
2xd_t/\lambda = u\gamma_t
$$
(10.5.9a)

$$
2yd_r/\lambda = v\gamma_r
$$
(10.5.9b)

with

$$
\gamma_t = 2\sigma_{ss}d_t/\lambda
$$
$$
\gamma_r = 2\sigma_o d_r/\lambda
$$

(10.5.10)

Thus, Eq. (10.5.7) becomes

$$
PE = \frac{M}{2} \int_{-\infty}^{\infty} \int_{-\infty}^{\infty} Q \left[\frac{\sqrt{SNR_o}}{2} \exp\left\{ -[u\gamma_t)^2 + (v\gamma_r)^2] \frac{q}{2} \right\} \right]
$$
$$
\cdot \left\{ uv \exp\left[\frac{-(u\gamma_t)^2}{2} \right] \exp[-(v^2/2)e^{-(u\gamma_t)^2}] \right\} du \, dv
$$

(10.5.11)

We immediately note that PE is a function only of the normalized aperture diameters in Eq. (10.5.10) and the parameters q, M, and SNR. Figure 10.20 shows the result of a numerically integrated PE for the case $M = 4$, with $\gamma_t = \gamma_r = \gamma$, $q = 2$, as a function of SNR_o. As γ increases (larger apertures), PE improves at fixed SNR_o, but higher SNR_o is needed to maintain a fixed PE as the beamwidths narrow. Clearly, there are various combinations of power

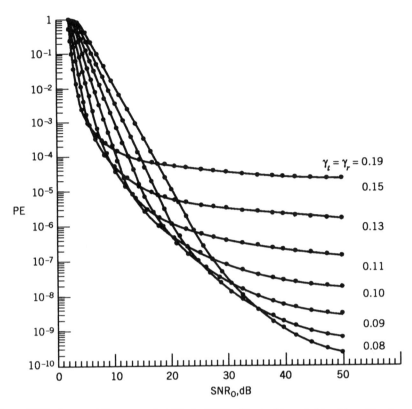

Figure 10.20. Average bit error probability in PPM with double-ended tracking. $M = 4$. $q = 2$. SNR_o is the tracking loop SNR_L with perfect pointing. Gaussian shaped loss functions. $\gamma_t = \gamma_r$ = normalized aperture gain.

(SNR_o) and optic size (γ) that produce a specific PE. This can be seen from Figure 10.21, which shows the values of SNR_o and γ needed to produce a $PE = 10^{-9}$ with various operating conditions.

These results imply that a tradeoff optimization among these parameters is possible in specific applications. Because minimization of required power is generally of most importance, one generally attempts to find values of γ that produce a given PE with minimal SNR_o. That is, we design the optics to relax the laser requirements. Reported results using this optimization procedure can be found in references 12 and 15.

In summary, we see that optical beam space communications, in conjunction with continuous beam tracking and pointing, produces a somewhat interconnected performance measure (bit error probability or SNR). Evaluation of these measures requires combining the link communication parameters with the statistical operations of the tracking errors. In this chapter, we have attempted to show procedures for carrying out this analysis.

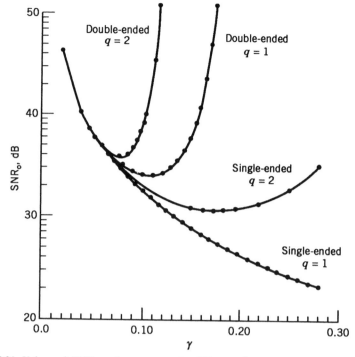

Figure 10.21. Values of SNR$_o$ and $\gamma = \gamma_r = \gamma_t$, for PE $= 10^{-9}$ in 4-PPM with double-ended tracking. Single-ended means no tracking error at transmitter.

PROBLEMS

10.1   The received power in a perfectly pointed space link is given by Eq. (10.1.1) with $\psi_e = 0$ and a bandwidth of $\psi_b = 10^{-4}$ rad.

    (a) Determine the power loss that will occur with a pointing error of $\psi_e = 2 \times 10^4$ rad.

    (b) Recompute this loss if the beamwidth is increased to 4×10^{-4} rad, with the same ψ_e as in (a).

    (c) Sketch a curve to show how the loss varies as ψ_b is varied from 10^{-5} to 10^{-3} rad.

10.2   **(a)** An orbiting satellite circles the Earth every 2 h at an altitude of approximately 1000 miles. The satellite–Earth link has a pointing error totaling 50 μrad, and the uncertainty in the altitude is ± 10 miles. Determine a suitable look-ahead angle and beamwidth to maintain communications from Earth.

    (b) Derive an equation for up and down look-ahead angles when both stations are in motion. Assume the stations move parallel to each other and each has separate pointing and velocity errors.

10.3   An optical receiver operates with a received count rate of 2×10^6 and a noise count per mode of 0.2. We assume an uncertainty field of view of 10^7 spatial modes and a desired resolution of $0.1 \, \mu$rad. The acquisition probability is given by 0.98. Use Figure 10.7 and determine the maximum slewing rate for an azimuth search.

10.4   A point source optical transmitter operating at 10^{14} Hz is located somewhere within a $1° \times 1°$ field of view. The receiver has a $3 \, \text{cm} \times 3 \, \text{cm}$ area.

   (a) Find the number of diffraction-limited spatial modes that must be searched.

   (b) If a resolution of 50 arcseconds $\times$ 50 arcseconds is desired, find the value of Q in Eq. (10.2.4).

10.5   Derive (10.3.7) by determining the areas under an offset circle of radius r_0 that fall within symmetrically located quadrants.

10.6   **(a)** A quadrant error sensor uses four detectors with unequal gains G_i, $i = 1, 2, 3, 4$. Determine the bias error produced when receiving a constant intensity field of I_0 over a focal plane area r_0.

   (b) If $G_1 = G_2 \neq G_3 = G_4$, is there a best arrangement of the detectors for minimizing the bias error? Explain.

10.7   In Figure 10.18, assume the loop filter $F(\omega) = 1/j\omega$, a pure integrator.

   (a) Determine the loop filter function $H(\omega)$ in Eq. 10.3.11).

   (b) Determine the corresponding loop noise bandwidth B_L.

   (c) Show that the loop equations (10.3.8) convert to first-order differential equations in $\psi(t)$.

10.8   Using Eq. (10.3.19), determine the probability that at any t the receiver normal will be pointed to within a distance of m miles (in any direction) of a transmitter located L miles away.

10.9   Consider Eqs. (10.4.2) and (10.4.3). Neglect noise and let $F(\omega) = 1/j\omega$, writing each as a first-order differential equation in $\psi_1(t)$ and $\psi_2(t)$, involving the G and S functions.

   (a) For the condition $d\theta_1(t)/dt = \Lambda$, $d\theta_2(t)/dt = -\Lambda$, find the condition on G and S for a steady-state solution to exist.

   (b) Show how to determine this value.

10.10   Write Eq. (10.4.5) in the form

$$\frac{a^2 G^2}{aG + b}$$

   (a) For a given value of a, G, and b, show that this is equal to the expression

$$\left(\frac{a^2}{a + b}\right) G^q$$

where

$$q = \log_G \left[\frac{G^2(a + b)}{aG + b}\right]$$

(b) Let $G = 0.9$ and compute q for each of the following conditions:

$$a = 100 \quad b = 1$$
$$a = 100 \quad b = 100$$
$$a = 100 \quad b = 10^4$$

10.11 Let the Airy spot on a quadrant array have a Gaussian shape, $I_s e^{-(\rho/r_0)^2}$

  (a) Assume the array detector area has width W. Determine the total power on the array when the spot is centered.

  (b) Compute the difference in Eq. (10.3.5) when the spot is offset by a horizontal amount r.

  (c) For a focal length f_c, determine the S-curve in azimuth (or elevation) in terms of the offset angle ψ.

REFERENCES

1. G. Picchi, D. Santerini, and G. Prati, Algorithms for spatial laser beacon acquisition, *IEEE Trans. Aerospace Electron Syst.*, 22, 106–114, March 1986.

2. R. M. Gagliardi and S. Karp, *Optical Communications*, Wiley, New York, (1976).

3. Y. T. Chan, A. G. C. Hu, and J. B. Plant, A Kalman filter based tracking scheme with input estimation, *IEEE Trans. Aerospace Electron Syst.*, 15, 237–244, March 1979.

4. I. M. Teplyakov, Acquisition and tracking of laser beams in space communications, *Acta Astronaut*, 7, 341–355 (1980).

5. T. M. Duncan and T. H. Ebben, Measurement of pointing error distributions in tracking loops of optical intersatellite links, *Proc. SPIE*, 756, 54–61, January 1987.

6. R. D. Nelson, T. H. Ebben, and R. G. Marshalek, Experimental verification of the pointing error distribution of an optical intersatellite link, *Proc. SPIE*, 885, 132–142, January 1988.

7. E. A. Swanson and J. K. Roberge, Design considerations and experimental results for direct-detection spatial tracking systems, *Opt. Eng.*, 28, 659–666, June 1989.

8. A. F. Popescu, P. Huber, and W. Reiland, Experimental investigation of the influence of tracking errors on the performance of free-space links, 'Proc. SPIE, 885, 93–93, January 1988.

9. R. M. Gagliardi and M. Sheikh, Pointing error statistics in optical beam tracking, *IEEE Trans. Aerospace Electron Syst.,* 16, 674–682, September 1980.

10. M. Katzman, *Laser Satellite Communications,* Prentice-Hall, Englewood Cliffs, NJ, 1988.

11. R. M. Gagliardi, V. A. Vilnrotter, and S. J. Dolinar, Optical deep space communication via relay satellite, Jet Propulsion Laboratory Publication, Pasadena, Calif., August 15, 1981.

12. T. S. Wei and R. M. Gagliardi, Cooperative optical beam tracking performance analysis, *Proc. SPIE,* 887, 176–183, January 1988.

13. R. A. Peters and M. Sasaki, An iterative approach to calculating the performance of two coupled optical intersatellite link tracking subsystem, *Proc. SPIE,* vol. 756, 78–85, January 1987.

14. H. Golstein, *Classical Mechanics,* 2nd ed., Addison-Wesley, Reading, MA, 1980.

15. C. C. Chen and C. Gardner, Impact of random pointing and tracking errors on the design of coherent and incoherent optical intersatellite communication links, *IEEE Trans. Commun.,* COM-37, 252–260, March 1989.

INDEX